GUANIDINES
Historical, Biological, Biochemical,
and Clinical Aspects of the Naturally
Occurring Guanidino Compounds

GUANIDINES

Historical, Biological, Biochemical, and Clinical Aspects of the Naturally Occurring Guanidino Compounds

Edited by

Akitane Mori

Okayama University Medical School
Okayama, Japan

B. D. Cohen

The Bronx Lebanon Hospital Center
Bronx, New York

and

A. Lowenthal

Born-Bunge Foundation
Universitaire Instelling Antwerpen
Wilrijk, Belgium

PLENUM PRESS • NEW YORK AND LONDON

Library of Congress Cataloging in Publication Data

International Symposium on Guanidino Compounds (1983: Tokyo, Japan)
    Guanidines: historical, biological, biochemical, and clinical aspects of the naturally occurring guanidino compounds.

    "Proceedings of the International Symposium on Guanidino Compounds, held at the Sixth Annual Meeting of the Japan Guanidino Compounds Research Association, September 5–7, 1983, in Tokyo, Japan"—T.p. verso.
    Includes bibliographies and index.
    1. Guanidino compounds—Congresses. I. Mori, Akitane, 1930–   . II. Cohen, Barry D. III. Lowenthal, A. (Armand) IV. Japan Guanidino Compounds Research Association. Meeting (6th: 1983: Tokyo, Japan) V. Title. [DNLM: 1. Guanidines—congresses. QU 60 I6074g 1983]
QP801.G83I58   1983   599′.019′24   85-3466
ISBN 0-306-41920-3

QP
801
.G83
I58
1983

Proceedings of the International Symposium on Guanidino Compounds,
held at the Sixth Annual Meeting of the Japan Guanidino Compounds Research Association,
September 5–7, 1983, in Tokyo, Japan

©1985 Plenum Press, New York
A Division of Plenum Publishing Corporation
233 Spring Street, New York, N.Y. 10013

Printed in the United States of America

PREFACE

In 1978, we had the first research meeting of guanidino compound analysis in Okayama, Japan. The purpose of the meeting was to standardize the methods of analyzing guanidino compounds, because the analytic methods, even, for example, that of normal plasma, were quite different from laboratory to laboratory at that time. More than ten laboratories joined in this research project. Thereafter, the purpose of the annual meetings was extended to cover general biochemistry of guanidino compounds, and every meeting has served to promote our mutual growth and progress. The last meeting of the Japan Guanidino Compounds Research Association in 1982 brought together more than 80 members, and about 30 papers were presented there. This, the 6th annual meeting, brings the International Symposium on Guanidino Compounds here to Tokyo.

Recently, many scientists are engaged in the research of guanidino compounds, especially in the field of clinical biochemistry. I am sure that the observation of guanidinosuccinic acid in the urine of uremic patients by Dr. Cohen's research group in 1963 ignited the recent exploration of the role of guanidino compounds in renal diseases.

The study of guanidino compounds does, however, have a long history. The first knowledge of guanidine was obtained by Strecker in 1861 (Annalen der Chemie und Pharmacie, 118, 151-177, 1981). He decomposed guanine with potassium chlorate and hydrogenchloride, and after isolating parabanic acid, oxaluric acid and xanthine from the reaction mixture, he isolated the sulfonic acid salt of an unknown base which he called "guanidine". We can find many reports on guanidine synthesis published during the following years.

The natural occurrence of guanidine was first reported by Schulze in 1892, who isolated it from the germ of a kind of pea (Ber. Dtsch. Chem. Ges. 65, 658-661, 1982). Today, guanidino compounds are known to be ubiquitous substances in plants as well as animals. These guanidino compounds are classified and reviewed by Dr. Robin in this symposium, including new guanidino

v

compounds she found, such as audouine, hirudonine, phascoline, phascolosomine and many phosphagenes.

We are very happy that such well-known scientists in guanidino compound research as Dr. Ronin and Dr. Cohen attended the meeting here in Tokyo and presented special lectures about their specialities.

In addition to their lectures, this symposium covers the following four sections:

The first section concerns analytical methods for guanidino compounds. Recently, high performance liquid chromatography for the determination of guanidino compounds has been developed and is now utilized widely in many laboratories. An analyzing system of the newest kind, a new idea for a fluorogenic reagent, and a new enzymatic method for methylguanidine estimation are reported in this section.

In the second section the metabolism of guanidino compounds in normal and abnormal organs, and in metabolic diseases, are discussed. We have many problems to be solved concerning the metabolism of guanidino compounds, including such basic problems as the metabolism of arginine to guanidino compounds, and the biosynthetic and regulatory mechanisms of guanidinosuccinic acid, guanidinoacetic acid or methylguanidine, especially regarding renal and hepatic failure. Current papers on argininemia also are considered in this section.

The physiological, pharmacological and toxicological aspects of guanidino compounds are discussed in the third section. γ-Guanidinobutyric acid induces convulsions in rabbits as reported by Mori and his colleagues in 1966. Subsequently it was observed that many guanidino compounds such as taurocyamine, glycocyamine, N-acetylarginie, methylguanidine, α-guanidino-glutaric acid, homoarginine and α-keto-δ-guanidinovaleric acid could induce seizure activity in experimental animals. Several papers about methylguanidine and taurocyamine, i.e. guanidino-ethane sulfonate, are presented in the session. The relationship of taurine to taurocyamine is one of the most interesting themes to be discussed.

In the last section, the involvement of guanidino compounds in acute and chronic renal failure are discussed. This is an urgent theme because of the vast number of patients with renal failure. Presently, guanidinosuccinic acid, guanidinoacetic acid and methylguanidine are used as indicators of renal function in clinical laboratories. Implications of these guanidino compounds in the pathogenesis of acute and chronic renal failure, including uremia, are discussed. Some prophylactic and thera-

peutic treatments of guanidino compound anomalies are also reported in this section.

Finally, we thank the Ministry of Health and Welfare of
Japan, the Pharmaceutical Association in Tokyo and Osaka, and
the many contributors for their encouragement and financial
support of our symposium.

A. Mori

DEDICATION

It's been over twenty years since I first began meddling in uremic plasma looking for a hypoglycemic factor and stumbled upon a guanidino compound. I've carried on a love affair with the guanidines ever since but never realized there was anyone else out there interested until two years ago when I met Akitane Mori. He introduced me to a great mob of people dabbling in guanidines. I felt like a poor orphan raised by strangers who suddenly discovers his true family.

The orphan metaphor is all too appropriate since the guanidines are classic examples of the "medical orphan", an etiology in desperate search of a disease. The ensuing pages represent an effort to document that search. This text results from the proceedings of the International Symposium on Guanidino Compounds held in Tokyo, Japan on September 5-7, 1983 and is a product of the hard work and imagination of Professor Mori and his Organizing Committee. I would, therefore, dedicate this volume to the Guanidine Family and to Professor Mori who fathered this strange brood.

B.D. Cohen

ACKNOWLEDGEMENTS

The editorial work was performed by a team of the Department of Neurochemistry, Institute for Neurobiology, Okayama University Medical School under the direction of Dr. I. Yokoi. To all these collaborators we wish to express our gratitude.

CONTENTS

III. PHYSIOLOGICAL, PHARMACOLOGICAL AND TOXICOLOGICAL
ASPECTS OF GUANIDINO COMPOUND

IV. INVOLVEMENT OF GUANIDINO COMPOUND IN ACUTE AND
CHRONIC RENAL FAIURE

I: ANALYTICAL METHOD FOR GUANIDINO COMPOUNDS

NEW HIGH-SPEED FULLY AUTOMATED GUANIDINO COMPOUND ANALYZER

Sakae Higashidate, Tetsuya Maekubo,
Muneo Saito, Masaaki Senda and Tadao Hoshino*

JASCO, Japan Spectroscopic Co., Ltd.
No. 2967-5 Ishikawa-Cho, Hachioji City, Tokyo 192, Japan

*Pharmaceutical Institute, School of Medicine
Keio University, No. 35 Shinano-Machi, Shinjuku-ku
Tokyo 160, Japan

INTRODUCTION

The introduction of dialysis has made rapid progress in the field of research and clinical treatment of renal failure, and has contributed very much to physiological elucidation of uremia. In uremia, concentrations of serum creatinine, BUN (Blood urea nitrogen), uric acid and guanidino compounds increase, and abnormalities of water, electrolytes, and acid-base equilibrium are seen. In addition, there are still many unknown substances and unknown factors, which play an important role in this disease. These substances are generally called uremic toxins. It is essential to investigate uremic toxins for the elucidation of uremia and the establishment of proper dialysis. Many compounds from low molecular weight to middle molecular weight have been investigated as suspicious candidates for the uremic toxin. Among these compounds, guanidines such as methylguanidine (MG), guanidinosuccinic acid (GSA) and guanidine (G) have been suggested. These guanidino compounds are also considered as causes for abnormalities in brain metabolism.

The method for the detection of guanidino compounds was reported by Sakaguchi in 1925, for the individual determination of arginine. This method was modified by many researches for more stable and more sensitive analysis. Analysis for guanidino compounds has been carried out using the Sakaguchi reaction after separation with paper or thin-layer chromatography. However, this

method required a long analysis time, and had limitations of sensitivity. Guanidino compounds have also been determined by gas
chromatography. However, this method involves a complex sample
preparation which includes derivatizing the sample to produce vol-
atile compounds before injection. The automatic quantitative
analysis of guanidino compounds in plasma and brain tissue[1, 2, 3] has
been reported using a modified automated amino acid analyzer with
the Sakaguchi reaction. Instead of the Sakaguchi reaction, a fluorescent reaction of guanidino compounds with 9,10-phenanthrene-
quinone (PQ) was reported, in 1972, by Yamada and Itano[4] for the
manual determination of arginine. In 1977, JASCO established a
highperformance liquid chromatographic (HPLC) system with post-
column derivatization utilizing PQ fluorogenic reagent, and introduced the JASCO model G–520 guanidino compound analyzer. The model
G–520 required two hours for complete separation of nine guanidino
compounds and 30 minutes for regeneration of the column. In 1979,
Yamamoto et al.[5] and Mori et al.[6, 7] reported the analysis of guanidino compounds using the post-column derivatization with PQ. In
1981, we succeeded in shortening the analysis time to one hour including regeneration by using a new column and modifying the chromatographic conditions.

     Fluorogenic reagents other than PQ have also been investigated.
Kinoshita et al. reported[8], in 1981, a new fluorescent detection
method for guanidino compound analysis utilizing the ninhydrin
reagent in HPLC. Ohkura et al. (The 4th Annual Meeting of Japan
Guanidino Compounds Research Association, 1981, Tokyo) also reported
a new high sensitive fluorescent reagent, benzoin-DMF, for pre-
column and post-column drivatization in HPLC.

     We describe, in this paper, the new high-speed fully automated
guanidino compound analyzer, using a shorter ion exchange column
(6 mm I. D. x 35 mm L.) packed with 5 µm of strongly acidic cation
exchange resin. We have succeeded in reducing the analysis time
drastically down to only 15 minutes including regeneration.

EXPERIMENTAL

Chemicals

     A mixture of guanidino compounds was obtained from Wako Pure
Chemicals (Osaka, Japan). This mixture contains nine guanidino
compounds; guanidine (G) (200 µM), methylguanidine (MG) (50 µM),
guanidinoacetic acid (GAA) (50 µM), guanidinosuccininc acid (GSA)
(100 µM), and creatinine (CRN) (500 µM). It was diluted 10 times
with 0.2 N sodium citrate, pH 2.2 before use as a standard mixture.
Sodium citrate, sodium hydroxide, hydrochloric acid and dimethylformamide (DMF) were all obtained from Wako Pure Chemicals
(Osaka, Japan). All the reagents were analytical grade. The 9,10-

Table 1. Composition of Eluents

(min)	Eluents				
	First (4.5)	Second (2.8)	Third (2.4)	Fourth (2.3)	Fifth (3.0)
pH_+	3.00	3.50	5.25	10.00	–
Na concentration (N)	0.4	0.4	0.4	0.4	1.0
Na-citrate (g/l)	39.2	39.2	39.2	39.2	–
Hydrochlorate (36%)(ml/l)	26.6	22.7	8.0	–	–
Borate (g)	–	–	–	2.0	–
NaOH (g/l)	–	–	–	–	40.0
2N NaOH (ml)	–	–	–	10.8	–

phenanthrenequinone (PQ) was obtained from Tokyo Kasei Chemicals (Tokyo, Japan) and used after recrystallization.

Eluent Buffers and Reagent Solutions

The eluent buffers are listed in Table 1. The pH of each eluent buffer is adjusted with concentrated HCl or 2.0 N NaOH. All the eluent buffers, 1 N and 2 N sodium hydroxide solutions were prepared using distilled water and then filtered with a 0.45 m membrane filter (Toyo Rhoshi Co., Ltd, Tokyo, Japan) prior to use. Since 9,10-phenanthrenequinone is insoluble in water, the PQ solution was prepared by dissolving 500 mg of PQ in 1000 ml of dimethyl formamide.

Chromatographic Conditions

Chromatographic conditions are listed in Table 2. The separation column is the μ-Guanidinopak (JASCO, Tokyo, Japan) packed with strongly acidic cation exchange resin with a mean particle size of 5 μm. The column dimensions are 6 mm in inner diameter by 35 mm in length. The column temperature is kept at 70°C. The flow rate of the eluent is 1.1 ml/min and those of PQ and sodium hydroxide solutions are both 0.5 ml/min. The temperature of the heating bath is 70 °C. The reaction coil is of 0.5 mm inner diameter by 5 m length. The sample injection volume was 100 μl. Other conditions are as shown in Table 2.

Instrumental

Fig. 1 shows a flow diagram of the new high-speed fully automated guanidino compound analyzer. Each eluent is selected stepwise by the eluent exchanger according to the eluting condition.

Table 2.   Chromatographic Conditions

Column   : μ-Guanidinopak (6 mm I.D. x 35 mm L.)
Column temperature   : 70 °C
Eluent flow rate   : 1.1 ml/min
PQ reagent flow rate   : 0.5 ml/min
NaOH reagent flow rate   : 0.5 ml/min
Heating bath temperature : 70 °C
Reaction coil   : 0.5 mm I.D. x 5 m L.
Fluorometric detector   : FP-110C (Ex 365 nm, Em 495 nm)
Sample volume   : 100 μl
Buffers   : 1. pH 3.00   0.4 N Na^+ (4.5 min)
                            2. pH 3.50   0.4 N Na^+ (2.8 min)
                            3. pH 5.25   0.4 N Na^+ (2.4 min)
                            4. pH 10.0   0.4 N Na^+ (2.3 min)
                            5. 1 N NaOH   (3.0 min)
Reagents   : 1. 2 N NaOH
                            2. 9,10-Phenanthrenequinone/DMF
                               (500 mg/1)

The selected eluent flows into the eluent pump P_1 (JASCO Model
TWINCLE) and is delivered via the autosampler (JASCO Model AS-L350)
to the separation column. The column effluent, containing separated
guanidino compounds, is mixed with the dimethylformamide solution
of PQ that is delivered by the pump P_2 (JASCO Model SP-024).   Then,
the effluent is mixed with 2 N sodium hydroxide solution that is
delivered by the pump P_3 (same as P_2) and introduced into the
reaction coil which is kept at 70 °C, where a fluorescent reaction
takes place.   In the reaction, a strongly fluorescent product
2-amino-1-hydro-phenanthro[9,10-d]imidazole (API) is yielded.   API
is then continuously monitored by the fluorometric detector (JASCO
Model FP-110C) with the excitation wavelength at 365 nm and the
emission at 495 nm.   Finally, the amount of each guanidino compound
is determined by the data processor (JASCO Model DP-L230).   The
whole operation process is performed automatically by the computer-
ized sequential programmer (JASCO Model UP-200).

Sample Preparation

Samples, serum, dialysate and urine were deproteinized by
ultrafiltration using a membrane filter (CF-25, Amicon Corp.), soon
after they were obtained because improper handling of the sample
causes inaccurate results[9].   Each filtrate was then mixed with an
equal volume of the sample dilution buffer, 0.2 N sodium citrate,
pH 2.2.   An aliquot of 100 μl of the mixture was injected[10].

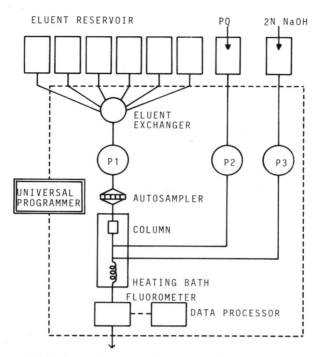

Fig. 1. Flow diagram of the analyzer

RESULT AND DISCUSSION

Standard Chromatogram

A typical chromatogram of the standard mixture of nine guanidino compounds, prepared by the method described above, is shown
in Fig. 2. The amount of each peak is 1 nmole for G-Tau, GSA and
GBA, 0.5 nmole for GAA, GPA and MG, 2 nmole for Arg and G, and 5
nmole for CRN, respectively. The separation conditions such as pH,
ionic strength of eluent buffer and column size were as shown in
Tables 1 and 2. All nine guanidino compounds are eluted with excellent separation, the resolutions Rs<1.1, even though the total
elution time is only thirteen minutes.

Reproducibility

Table 3 shows the reproducibility for each retention time and
Table 4 indicates that for each peak area. The average and coefficient of variation for each retention time were obtained from ten
successive analyses of the standard mixture. The coefficients of
variation (CV %) are 0.08 % for MG and G, 0.48 % for GSA and 0.53 %
for GAA, respectively. The values for the others are between

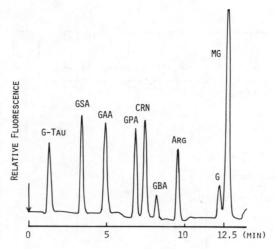

Fig. 2.   Chromatogram of standard mixture

Table 3.   Reproducibility of Retention Time

	$\bar{X}$ (min)	σ_{n-1} (min)	CV (%)
G–Tau	1.37	0.008	0.60
GSA	3.44	0.016	0.48
GAA	5.03	0.026	0.53
GPA	6.92	0.012	0.17
CRN	7.54	0.016	0.21
GBA	8.29	0.007	0.08
Arg	9.70	0.014	0.15
G	12.54	0.007	0.06
MG	13.08	0.007	0.05

(n=10)

0.15 % and 0.60 %. The coefficient of variation for each peak area
was obtained from the chromatograms that were used in the calcula-
tion of reproducibilities of retention times. The CV values for
peak areas are 1.62 % for GSA, 0.50 % for GAA and 0.26 % for MG,
respectively. Those values for the others are between 0.58 % and
2.46 %. It is concluded that the quantitative precision of the new
system is good enough to meet general requirements in clinical
diagnostic analysis.

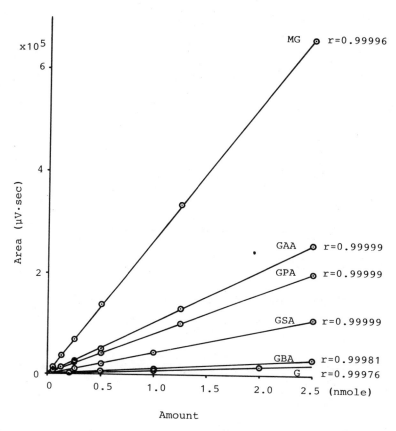

Fig. 3.   Calibration curves for guanidino compounds

Table 4.   Reproducibility of Peak Area

	$\bar{\chi}$	σ_{n-1}	CV (%)
G-Tau	35110	417	1.19
GSA	45242	734	1.62
GAA	47284	235	0.50
GPA	38349	420	1.10
CRN	50162	749	1.49
GBA	11705	288	2.46
Arg	35958	207	0.58
G	18924	186	0.98
MG	135690	347	0.26

(n=10)

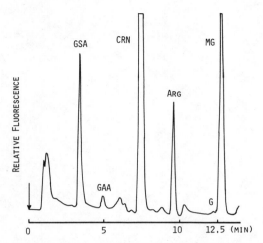

Fig. 4.   Chromatogram of serum from uremic patient

Table 5.   Detection Limit (pmole, S/N=2)

GSA	4	GAA	2
GPA	2	CRN	20
GBA	20	ARG	8
G	30	MG	1

Calibration Curves and Detection Limits

Fig. 3 shows the calibration curves for each guanidino compound. The calibration curves are confirmed to be linear up to 2.5 nmole for GAA, GPA and MG, up to 5 nmole for GSA, GBA and CRN, and up to 10 nmole for Arg and G, respectively. Table 5 shows the detection limit of each guanidino compound defined by the signal-to-noise ratio, S/N=2. The detection limits are 1 pmole for MG, 2 pmole for GAA and GPA, and 4 pmole for GSA, respectively. The detection limits for the others are between 8 pmole and 30 pmole. These results will cover the necessary dynamic range in physiological samples, except for CRN, which often occurs in quantities of more than 5 nmole.

Applications

Typical examples of serum, dialysate and urine from uremic patients are shown in Figs. 4, 5 and 6. Fig. 4 shows a chromatogram of guanidino compounds in serum of a uremic patient before peritoneal dialysis. The serum was deproteinized by ultrafiltra-

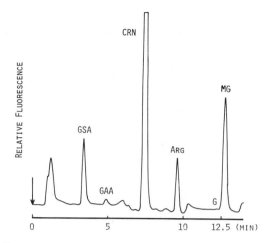

Fig. 5. Chromatogram of dialysate from uremic patient

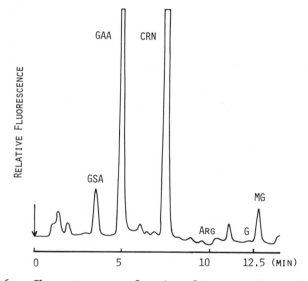

Fig. 6. Chromatogram of urine from uremic patient

tion using Amicon CF-25. Then, filtrate was mixed with an equal volume of the sample dilution buffer, 0.2 N sodium citrate, pH 2.2. One hundred microliters of the mixture was injected into the system. As shown in Fig. 4, GSA, GAA, Arg, G and MG are separated without interference by other substances. Fig. 5 shows a chromatogram of the dialysate from the same patient as in Fig. 4 at the beginning

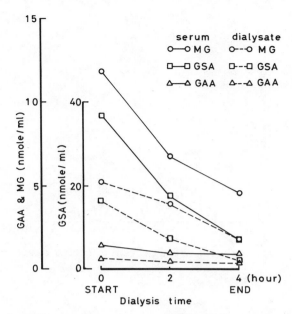

Fig. 7.   Changes of guanidino compounds concentrations
          during peritoneal dialysis

of dialysis.   The sample was prepared by the same method as that
used in serum.   GSA, GAA, Arg, G and MG are also eluted with good
separation.   Fig. 6 shows a chromatogram of urine from a patient
with chronic glomerulonephritis.   The sample was prepared by the
same method.   As shown in Figs. 4, 5 and 6, all nine components
of guanidino compounds are analyzed in fifteen minutes without
sacrificing separation and sensitivity.

     By using this analyzer, the fifteen-minute analysis time
makes possible sixteen analyses during a four-hour dialysis,
providing the analyst with a detailed composition profile as a
function of time.   Fig. 7 shows changes in the concentrations of
MG, GSA, GAA in plasma and dialysate from a patient during a
four-hour peritoneal dialysis.   The concentrations of MG and
GSA in plasma were observed to be rapidly decreasing as dialysis
proceeded.   In this way, the new high-speed fully automated
guanidino compound analyzer can be used for the bedside determination of these componds, and is expected to accelerate the
progress of research work.

REFERENCES

1. M. Matsumoto, H. Kishikawa and A. Mori, Guanidino compounds
   in the sera of uremic patients and in the sera and brain of
   experimental uremic rabbits, Biochemical Medicine,
   16: 1 (1976).
2. R. Shainkin, Y. Berkenstadt, Y. Giatt and G. M. Beryne,
   An automated technique for the analysis of plasma guanidino
   acids, and some findings in chronic renal disease,
   Clin. Chem. Acta, 60: 45 (1975).
3. A. Mori, M. Hiramatsu, K. Takahashi and M. Kohsaka, Guanidino
   compounds in rat organs, Comp. Biochem. Physiol.,
   51B: 143 (1975).
4. E. A. Itano and S. Yamada, 2-Aminophenanthroimidazole, fluorescent product of the reaction of phenanthrenequinone with
   arginine, Anal. Biochem., 43: 483 (1972).
5. Y. Yamamoto, T. Manji, A. Saito, K. Maeda and K. Ohta,
   Ion exchange chromatographic separation and fluorometric
   detection of guanidino compounds in physiologic fluids,
   J. Chromatogr., 162: 327 (1979).
6. A. Mori, Y. Katayama, S. Higashidate and S. Kimura, Fluorometrical analysis of guanidino compounds in mouse brain,
   J. Neurochem., 32: 643 (1979).
7. A. Mori, M. Akagi, Y. Katayama and Y. Watanabe, α-Guanidinoglutaric acid in cobalt-induced epileptogenic cerebral
   cortex of cats, J. Neurochem., 35(3): 603 (1980).
8. Y. Hiraga and T. Kinoshita, Post-column derivatization of
   guanidino compounds in high-performance liquid chromatography using ninhydrin, J. Chromatogr., 226: 43 (1981).
9. T. Hoshino, M. Sakuma, E. Ohosawa, W. Yamayoshi, K. Sakakibara,
   T. Satoho, H. Sakurai and S. Toyoshima, Effects of hydrogen
   ion concentration on determination of free guanidino
   compounds in plasma, Jap. J. Clin. Chem., 9: 198 (1980)
   (in Japanese)
10. T. Hoshino, Recommended deproteinizing methods for plasma
    guanidino compound analysis by liquid chromatography, in:
    "Urea Cycle Diseases," A. Lowenthal, A. Mori and B. Marescau,
    eds., Plenum Publishing Co., New York (1983).

AUTOMATIC GUANIDINE ANALYZER USING BENZOIN

AS A FLUOROGENIC REAGENT

Yiau-Lin Huang, Masaaki Kai and Yosuke Ohkura

Faculty of Pharmaceutical Sciences, Kyushu University
62, Maidashi, Higashi-ku, Fukuoka 812, Japan

INTRODUCTION

In the current analytical methods for guanidino compounds, ion-exchange chromatographic methods coupled with automatic colorimetric or fluorimetric detection, including high-performance liquid chromatography (HPLC), is the most popular because of its simplicity in operation, though other chromatographic methods based on paper[1,2], thin-layer[3,5] and gas chromatography[5,8] have been reported. The colorimetric detection of these compounds by means of the Sakaguchi or Voges-Proskauer reaction in ion-exchange chromatography has a limited sensitivity and thus necessitates a large amount of sample[9,11]. On the other hand, fluorimetric detection based on the post-column derivatization in HPLC using ninhydrin or 9, 10-phenanthraquinone as a fluorogenic reagent can offer a method sensitive enough to measure the compounds at the picomole level[12,14].

Recently we have developed a manual fluorimetric method for the selective determination of guanidino compounds using benzoin[15,16]. This reagent is a non-fluorescent material, but reacts with the guanidino moiety of the compounds in an alkaline medium and gives highly fluorescent derivatives, 2-substituted amino-4,5-diphenylimidazoles[17]. Therefore this reaction is applicable for fluorimetric detection as both the pre- and post-column derivative in HPLC of guanidino compounds. The application of the reaction to the pre-column derivatization of the compounds in HPLC was described in a separate paper[18]; this method is most sensitive and useful for determination of guanidino compounds at the femto mole level, however the limited separation of the fluorescent derivatives did not permit the assay of biogenic guanidino compounds

15

without a clean-up procedure for removal of oligo-peptides in biological sample.

In this paper, we describe the HPLC conditions for a rapid simultaneous separation of guanidino compounds on a cation-exchange column and the application of the benzoin reaction to the post-column fluorescent derivatization of the compounds in order to assemble an automatic analyzer for routine assay of biogenic guanidino compounds in human urines and sera from normal subjects or uremic patients. Fourteen guanidino compounds, which were employed as representative compounds for the investigations, are expressed as the following abbreviations, taurocyamine, TC; guanidinosuccinic acid, GSA; creatine, CR; guanidinoacetic acid, GAA; N-α-acetyl-arginine, AcARG; argininic acid, ARA; guanidinopropionic acid, GPA; creatinine, CRN; guanidinobutyric acid, GBA; arginine, ARG; phenylguanidine, PG; guanidine, G; methylguanidine, MG; agmatine, AGM.

EXPERIMENTAL

Chemicals and solution

All chemicals were of analytical reagent grade, unless otherwise noted. Deionized and distilled water was used. Benzoin (Wako, Osaka, Japan) was recrystallized from absolute methanol. Trishydroxymethylaminomethane (Tris) (Wako, Osaka, Japan) was recrystallized from 60 % aqueous methanol to remove fluorescent impurities. Standard solutions of the guanidino compounds were prepared in 0.05 M hydrochloric acid.

Three aqueous eluents are required for separation of the guanidino compounds. Buffer A (pH 3.5): Dissolve 5.25 g of trisodium citrate dihydrate, 2.7 g of sodium chloride, 8.1 g of citric acid and 7.2 mg of sodium pentachlorophenol as a preservative in ca. 400 ml of water and dilute with water to 500 ml (the final concentration of each component, 36, 92, 77 and 0.05 mM, respectively). Buffer B (pH 5.0): Dissolve 13.35 g of trisodium citrate dihydrate, 20 g of sodium chloride, 3.05 g of citric acid and 7.2 mg of sodium pentachlorophenol in ca. 400 ml of water, and dilute with water to 500 ml (the final concentration of each component, 91, 684, 29 and 0.05 mM, respectively). Potassium hydroxide in ca. 400 ml of water and dilute with water to 500 ml. The solutions were throughly de-gassed in the usual manner before use.

Chromatographic system and the operation

Fig. 1 is a schematic diagram of the HPLC analyzer constructed for analysis of the guanidino compounds.

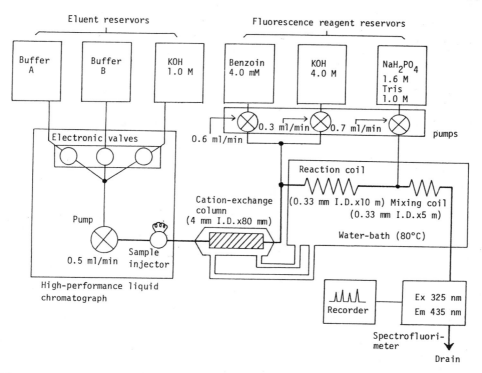

Fig. 1.   Schematic diagram of HPLC analyzer of guanidino compounds.

     A cation-exchange column (4 mm I.D. x 80 mm; packing material,
Hitachi 2619 resin; particle size, 5 μm) was used in the HPLC with
a stepwise gradient elution of pH and/or ion strength by using the
buffers A and B, and 1.0 M potassium hydroxide.   The temperature
of the water-jacketed column was kept at 80 °C by a Hitachi con-
stant-temperature circulator.   The mobile phase (buffers A and B,
and 1.0 M potassium hydroxide) was pumped at a flow rate of 0.5
ml/min by a Hitachi 638-30 high-performance liquid chromatograph
which had a programming controller of the electronic valves placed
prior to the pump-inlet for various gradient elutions.   The buffer
A was first run into the column for 2 min, the mixture of the
buffers A and B (1:1, v/v) for 2 min, the buffer B for 11 min and
1.0 M potassium hydroxide for 10 min to separate the guanidino
compounds; and then successively the column was equilibrated with
the buffer A for 20 min before the start of the next sample
(Fig. 2).   The above change of the eluents was automatically
controlled with the electronic programmer of the chromatograph.

     The eluate from the column was conducted to a fluorescence

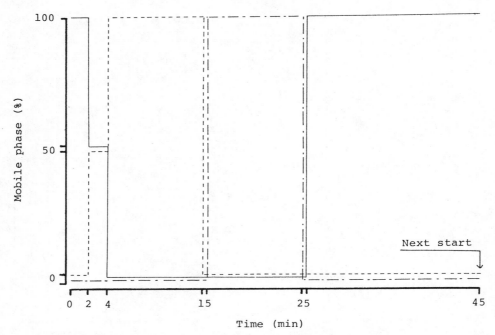

Fig. 2. Eluation mode of the stepwise gradient.

reactor system through a teflon tube (0.33 mm I.D. x 20 cm). All
the coils in the reaction system are made of teflon. Benzoin
(4.0 mM) which was dissolved in 60 % aqueous methylcellosolve and
potassium hydroxide (4.0 M) were first added to the eluate stream
at a tee connector by a Hitachi reagent-delivery pump for amino
acid analyzer and a Hitachi 663-C chemical pump at flow rates of
0.6 and 0.3 ml/min, respectively, and then the mixture was passed
through a reaction coil (0.33 mm I.D. x 10 m) immersed in a 80 °C
water-bath. After the fluorescence reaction, a mixture of sodium
dihydrogen phosphate (1.6 M) and Tris (1.0 M) was added to the
reaction mixture at a flow rate of 0.7 ml/min by a Hitachi
reagent-delivery pump for amino acid analyzer and the mixture was
passed through a mixing coil (0.33 mm I.D. x 5 m). The fluorescence intensity from each guanidino compound in the last eluate
was monitored at 435 nm emission against 325 nm excitation with
both the slit-widths of 5 nm by a Hitachi 650-10 LC spectrofluorimeter equipped with a flow cell (18 μl) and a xenon lamp.

Preparation of physiological fluids

    Urine and serum specimens were obrained from healthy volun-
teers in our laboratory and patients with chronic renal failure
being hemodialyzed at Japan Red Cross Fukuoka Hospital (Fukuoka,
Japan).

Urine samples: To 100 μl of urine centrifuged at 1000 g for 5 min, 800 μl of 7.5 nmol/ml phenylguanidine was added. An aliquot (100 μl) of the mixture was applied to the HPLC. Serum samples: A 100 μl of serum was mixed with 50 μl of 1.8 M perchloric acid and 25 μl of 25 nmol/ml phenylguanidine. The mixture was centrifuged at 1500 g for 10 min. A 100 μl of supernatant was neutralized by adding 50 μl of 0.6 M potassium carbonate and then the potassium perchlorate produced was removed to avoid precipitation of the salt in the lines of the HPLC analyzer. The pH of the resultant supernatant was adjusted to ca. 1.7 with ca. 20 μl of 0.7 M hydrochloric acid. An aliquot (100 μl) of the final mixture was applied to the HPLC.

RESULTS AND DISCUSSION

Separation of guanidino compounds

Conventional separation of guanidino compounds has been performed by chromatography using strong cation-exchange resins, based on the separation technique used for basic amino acids[19]. A resin of this type was used in our investigations. The resin particles were small and arranged to be of the same diameter (5 μm) since well-regulated small particles of resin serve an increasing theoretical plate number[20]. Thus the resin packed in a short column (4 mm I.D. x 80 mm) gave a high resolution of the guanidino compounds in the HPLC. Fig. 3 shows the chromatogram of a standard mixture of fourteen guanidino compounds attained by the HPLC analyzer. Complete separation of the compounds on the column is achieved within 35 min in a single run with a stepwise gradient elution according to the procedure described in Experimental.

The guanidino compounds tested (except for the strongly basic compounds such as PG, G, MG and AGM) were retained on the column and resolved by using a sodium citrate buffer (0.1 M, pH 4.0) with isocratic elution, and their retention times were affected by the pH and ion strength in the buffer. The ion strength was controlled by the addition of sodium chloride to the buffer. The rise of pH and/or ion strength in the buffer resulted in an early elution of the guanidino compounds. On the other hand, PG, G, MG, and AGM were strongly retained on the column and not eluted with the acidic buffer. However, with an alkaline solution such as 1.0 M potassium hydroxide, they were resolved in reasonable retention times. From the above preliminary studies on the separation of the guanidino compounds, the combination of three eluents, buffer A (pH 3.5), buffer B (pH 5.0) and 1.0 M potassium hydroxide as mobile phase, was employed in the HPLC (the constituent of each eluent and the elution mode for the stepwise gradient are described in Experimantal). Buffer A is mainly used for the separation of TC and GSA, the buffer B for CR, GAA, AcARG, ARA, GPA, CRN, GBA

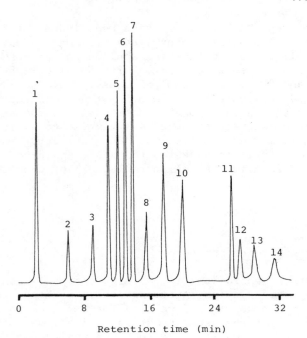

Retention time (min)

Fig. 3.   Chromatogram of a standard mixture of guanidino compounds.
          For the separation and detection conditions of the HPLC
          analyzer, see text.   Peaks and amount of guanidino
          compounds (per injection volume of 100 µl):
          1 =TC (800 pmol), 2 =GSA   (800 pmol), 3 =CR (40 nmol),
          4 =GAA (800 pmol), 5 =AcARG (800 pmol), 6 =ARA (800 pmol),
          7 =GPA (800 pmol), 8 =CRN   (800 nmol), 9 =GBA (800 pmol),
          10=ARG (800 pmol), 11=PG   (800 pmol), 12=G   (800 pmol),
          13=MG (800 pmol), 14=AGM   (800 pmol).

and ARG and 1.0 M potassium hydroxide for PG, G, MG and AGM.

     The retention times of the guanidino compounds are also in-
fluenced by the column temperature in HPLC. The rising temperature facilitates early elution of guanidino compounds, especially
the late-eluting compounds such as G, MG, and AGM, without deterioration of their resolution and peak shapes. For example, when
the column was operated at an ambient temperature (23 °C), MG was
eluted late at ca. 44 min, but at the 80 °C the compound was
eluted at 29.2 min under the conditions of the HPLC. Consequently
the column was operated at 80 °C not only to shorten the analytical time but also to obtain a reproducible retention time for each
guanidino compound. A good durability of the column was also
observed; the column can be used for more than 2000 analyses of
the samples.

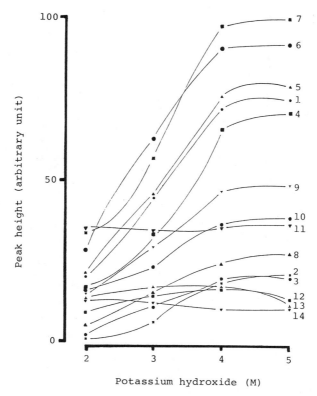

Fig. 4.   Effect of potassium hydroxide concentration on the fluo-
          rescence development of guanidino compounds. For the
          separation conditions, peak numbers and the amounts of
          the guanidino compounds in the HPLC, see Fig. 3.

Fluorescence reactor and detection

In the manual method[16], the fluorescence reaction of the
guanidino compounds with benzoin needs a strongly alkaline medium
and a heating condition to minimize the reaction time, and also
the produced fluorescent derivatives fluoresce most intensely in a
weakly alkaline solution (pH 8.5-10.5). Thus the eluate from the
column was first mixed with benzoin and potassium hydroxide, and
then mixture was heated at 80 °C through a reaction coil for ca.
45 sec and made weakly alkaline by adding a mixture of sodium dihydrogen phosphate and Tris. The excitation and emission maxima
of the fluorescence from all the guanidino compounds tested,
obtained by this detection system, were around 325 and 435 nm,
respectively. These data agreed with those in the manual method
previously described.

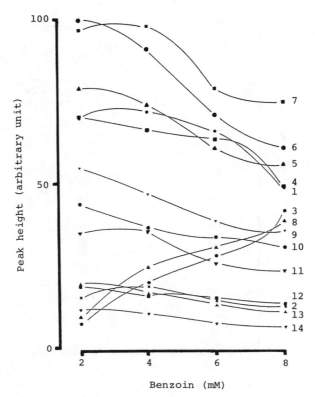

Fig. 5.   Effect of benzoin concentration on the fluorescence development of guanidino compounds. For the separation conditions, peak numbers and the amounts of the guanidino
compounds in the HPLC, see Fig. 3.

Reaction conditions of the post-column fluorescent derivatization were examined to construct the fluorescence reactor system.
A potassium hydroxide solution ranging in concentration from 4.0
to 5.0 M is required to obtain a maximum fluorescence intensity,
corresponding to the peak height of each guanidino compounds, as
shown in Fig. 4; 4.0 M potassium hydroxide was used in the reactor
system. Benzoin concentration also influences development of the
fluorescence from each guanidino compound. With increasing concentration of benzoin in a range of 2.0 to 8.0 mM, the fluorescence intensities from the compounds other than CR and CRN
decrease slightly, but both the intensities from CR and CRN
increase (Fig. 5). In the system, 4.0 mM benzoin was selected to
obtain fairly large intensities of fluorescence from the guanidino
compounds except for CR and CRN, because concentrations of CR and
CRN in human urine and serum are usually higher than those of the
other guanidino compounds. An elevated reaction temperature is

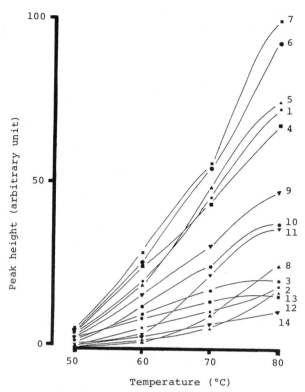

Fig. 6. Effect of reaction temperature on the fluorescence development of guanidino compounds. For the separation conditions, peak numbers and the amounts of the guanidino
compounds in the HPLC, see Fig. 3.

required for development of the fluorescences from all the
guanidino compounds tested as shown in Fig. 6. When the reaction
temperature was higher than 80 °C, an irregularity of the base
line occurred frequently on the chromatogram, which may be caused
by air bubbles generated in the reaction coil. Thus a compromised
temperature of 80 °C was used for the reaction in the reactor
system, whereas this temperature was kept by use of the same
water-bath as that used for the column. In order to attain the
intense fluorescence from the benzoin derivatives, it was
necessary to adjust the pH of the mixture after the benzoin
reaction to a weakly alkaline pH by adding a pertinent acidic
salt. When a mixture of sodium dihydrogen phosphate (1.6 M) and
Tris (1.0 M) was added to the reaction mixture, the pH of the
final eluate was sufficiently lowered and ranged from 8.5 to 10.7
during the operation of the HPLC analyzer, even with the step-wise
gradient elution of pH in the mobile phase. At these hydrogen

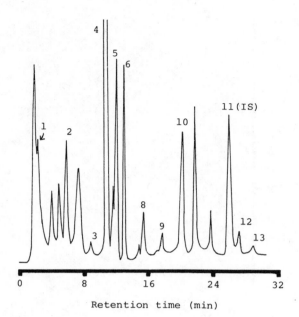

Fig. 7.   Chromatogram of guanidino compounds in urine from a
          healthy man. Phenylguanidine (6 nmol) was spiked to
          100 μl of urine as an internal standard and the sample
          was treated as described in Experimental. For the
          separation conditions, peak numbers and the amounts of
          the guanidino compounds in the HPLC, .see Fig. 3.

ion concentrations the derivatives of all the guanidino compounds
tested fluoresce at least five times as strongly as those without
the adjustment of pH.

     A linear relationship was observed between the peak height
and the amount of each guanidino compound in an injection volume
(100 μl) up to at least 8 nmol for monosubstituted guanidino
compounds, 100 nmol for CR and 8 μmol for CRN under the established conditions of the fluorescence reactor system. The lower
limits of detection are 5 pmol for ARA and GPA, 8pmol for TC, GAA
and AcARG, 12 pmol for GBA, ARG and PG, 30 pmol for GSA, G and MG,
50 pmol for AGM, 1 nmol for CR, and 20 nmol for CRN. The limit
is defined as the amount in 100 μl of the injection volume giving
a peak height at twice the noise level.

Analysis of biogenic guanidino compounds in human urine and serum

     The typical chromatograms obtained by the HPLC analyzer are
shown in Fig. 7 for a normal urine and Fig. 8 for sera of a normal
subject and a patient with chronic renal failure. The guanidino

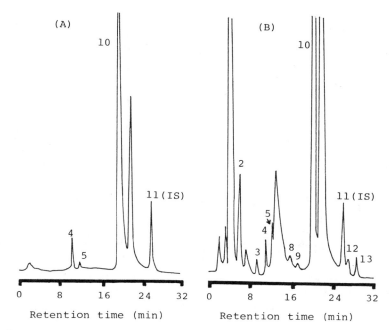

Fig. 8. Chromatograms of guanidino compounds in sera from (A)
         healthy man and (B) remic patient. Phenylguanidine
         (625 pmol) was spiked to 100 μl of each serum as an
         internal standard and the samples were treated as
         described in Experimental. For the separation condi-
         tions, peak numbers and the amounts of the guanidino
         compounds in the HPLC, see Fig. 3.

compounds in their samples were identified on the basis of the
retention times by comparison with the standard compounds and also
by co-chromatography of the standards and sample with different
elution of the molbile phase, i.e. using a lower pH and/or ion
strength than those of the eluents used for the recommended procedure. This elution served more sufficient separation of the
guanidino compounds but their elutions were delayed in the HPLC.
In addition, their retention times were not affected by the biological matrix of the samples. Unidentified peaks were also
observed in their chromatograms, though the benzoin reaction works
on only the compounds with a guanidino moiety in the biological
samples for fluorescence detection [15,16]. The peaks are probably
from native fluorescent substances (including drugs administered to
the patient during therapy), unknown guanidino compounds and/or
peptides with arginyl residues which could not be removed by deproteinization. The oligopeptides with one or two arginyl
residues such as tuftsin, angiotensins 1, 11 and 111, kallidin,
bradykinin, LH-releasing hormone, substance P and neurotensin were

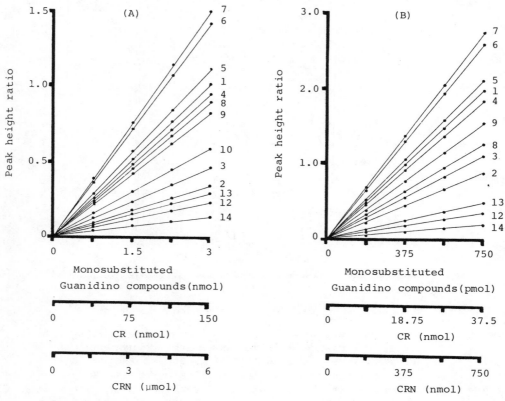

Fig. 9. Calibration curves for guanidino compounds in (A) urine
         and (B) serum. (A), To portions (100 μl) of a pooled
         urine, 0-3 nmol of the monosubstituted guanidino compounds,
         0-150 nmol of CR, 0-6 μmol of CRN and 6 nmol of phenyl-
         guanidine (IS) were added. (B), To portions (100 μl) of
         a pooled serum, 0-750 pmol of the monosubstituted gua-
         nidino compounds, 0-37.5 nmol of CR, 0-750 nmol of CRN
         and 625 pmol of phenylguanidine (IS) were added. The
         curve for ARG was not made because the peak of endogeneous
         ARG in the serum was too high as compared with that of the
         internal standard. For the HPLC conditions and curve
         numbers in the graphs, see Fig. 3.

examined for their retention times in the HPLC. These peptides
were eluted at a retention time of ca. 22 min which is near the
elution of ARG. Therefore, in the chromatogram of the patient
serum, one of the unknown big peaks around 22 min may be ascribed
to peptides since several peptides have been suspected in the toxic
manifestations of uremia[21,22]. Other unidentified peaks in their
chromatograms were not studied.

Table 1. Concentrations of guanidino compounds in urines from healthy men

Age	Guanidino compounds* (μmol/day)										
	TC	GSA	CR	GAA	AcARG	ARA	CRN	GBA	ARG	G	MG
26	11.34	24.94	ND	323.00	17.51	8.50	11333	7.36	13.04	11.34	ND
25	6.66	28.34	250.00	516.25	15.00	15.84	15834	17.50	30.84	18.34	10.00
23	ND	29.41	ND	340.00	25.50	7.94	17000	14.74	17.00	18.14	6.80
23	16.15	20.80	ND	544.67	10.45	6.65	13300	8.55	17.10	15.20	5.70
24	ND	22.40	210.00	583.00	26.60	12.60	13066	7.94	27.02	15.82	6.54
28	ND	16.44	316.60	437.00	11.40	5.39	8866	4.75	16.15	8.24	3.80
34	ND	42.54	513.34	366.63	21.26	23.46	24934	5.17	28.60	16.14	8.80
28	ND	38.00	570.00	709.30	25.33	11.40	17732	6.33	30.40	17.73	7.60
29	ND	9.34	ND	405.30	3.87	8.70	11600	5.80	7.73	17.40	9.67
Mean	11.38	26.06	371.98	477.48	17.42	11.16	14851	8.67	20.87	15.37	7.36
SD	4.75	10.10	160.76	135.00	7.90	5.61	4742	4.45	8.46	3.42	2.09

* Other guanidino compounds could not be successfully determined because of their minute amounts.
ND, not detected.

Table 2. Concentrations of guanidino compounds
in sera from healthy men

Age	Guanidino compounds* (nmol/ml)				
	CR	GAA	AcARG	G	MG
26	11.22	1.61	0.92	ND	ND
29	23.21	2.18	1.02	ND	ND
34	18.03	2.45	0.80	ND	ND
22	32.86	2.50	0.92	ND	0.46
24	16.43	3.24	1.14	ND	ND
25	8.19	3.17	0.72	0.69	0.46
24	19.18	2.73	0.98	ND	0.38
26	9.63	2.51	0.79	0.31	0.77
Mean	17.34	2.55	0.19	0.50	0.52
SD	8.11	0.52	0.14	0.27	0.17

*Other guanidino compounds, excluding ARG,
could not be successfully determined because
of their minute amounts. ND, not detected.

For a precise and facile quantification of the guanidino
compounds, phenylguanidine was used as an internal standard.
Both the calibration curves on urine and serum, which were made by
plotting the ratios of net peak heights of the spiked guanidino
compounds against the peak height of the internal standard, were
linear in the relationship between the ratios and amounts of the
guanidino compounds added to urine or serum. The calibration
curves are shown in Fig. 9. The correlation coefficients (r) of
all the curves were greater than 0.997 and no change of the slopes
in the graphs was observed depending on the urine or serum used.
The recovery of each guanidino compound added to 100 µl of urine
in the amounts of 3-6 nmol for monosubstituted guanidino compounds, 300 nmol for CR and 6 µmol for CRN was in a range of 95 to
105 % (CV, 6.2 %), and to 100 µl of serum in the amounts of 0.75
nmol for monosubstituted guanidino compounds, 37.5 nmol for CR and
750 nmol for CRN was in a range of 97 to 104 % (CV, 6.3 %). The
values are the average of ten independent analyses. The precision
of the method for the determination of the biogenic guanidino
compounds was also examined by performing five analyses separately
on a pooled urine and serum. The coefficients of variation on the
assay of the compounds (the compound and concentration (nmol/ml),
in parenthesis) in the urine were 5.3 (GSA, 15.6), 6.0 (CR, 451.0),
8.1 (GAA, 312.0), 3.1 (AcARG, 10.9), 4.6 (ARA, 6.6), 9.1 (CRN,
6393.2), 1.1 (GBA, 5.4), 3.1 (ARG, 18.5), 3.1 (G, 10.2) and 3.5 %

Table 3. Concentrations of guanidino compounds in sera from
patients with chronic renal failure*

	GSA	CR	GAA	AcARG	CRN	GBA	G	MG
			Guanidino	compounds*	(nmol/ml)			
1	4.64	40.50	3.05	2.56	325.87	0.19	2.25	2.82
2	13.90	31.25	2.07	1.71	506.75	ND	2.25	5.19
3	22.21	53.47	1.92	1.65	451.34	ND	2.46	6.66
4	6.95	121.42	1.97	1.57	377.45	ND	1.97	3.03
5	15.53	80.21	1.60	1.75	599.00	ND	2.66	7.21
6	4.13	104.25	1.06	2.27	585.03	ND	2.41	2.89
7	25.20	182.03	2.]2	ND	677.22	ND	2.62	3.06
8	3.39	64.21	2.18	0.57	200.00	0.40	2.07	2.18
9	25.51	64.67	2.39	1.80	646.67	0.30	3.03	5.17
10	20.05	86.02	2.58	2.34	860.21	0.41	2.21	3.69
Mean	14.15	82.80	2.09	1.80	522.95	0.33	2.39	4.19
SD	8.89	44.46	0.54	0.58	192.12	0.10	0.31	1.75

*Other guanidino compounds, excluding ARG, could not be success-
fully determined because of their minute amounts.
ND, not detected.

(MG, 3.1) and on the assay of the compounds in the serum were 3.9
(GSA, 7.0), 5.8 (CR, 121.4), 8.4 (GAA, 2.0), 9.6 (AcARG, 1.6), 3.2
(CRN, 377.5), 3.6 (G, 2.0) and 3.8 % (MG, 3.0).

The concentrations of the guanidino compounds in urines and
sera from healthy men, and sera from the patients with chronic
renal failure on maintenance hemodialysis were determined by this
method (Tables 1-3). Relatively many guanidino compounds were
identified in the urine as compared with those in sera from healthy
men. Higher concentrations of GSA, CR, CRN, G and MG than those
in normal sera were observed in the patient sera. GAA and AcARG
were present in the normal sera but not significantly different
from those in the patient sera. The mean values of the individual
guanidino compounds are in good agreement with the published data.

The present method for the automatic determination of the
guanidino compounds is rapid and gives satisfactory sensitivity in
the analysis of physiological fluids; the sensitivity permits use
of less than 100 μl of urine and serum from normal subject or
uremic patient, and this HPLC analyzer is another tool for clinical studies of guanidino compounds in the uremic syndrome.

SUMMARY

An automatic analyzer on the basis of high-performance liquid chromatography has been developed for quantification of biogenic guanidino compounds in human physiological fluids. Fourteen guanidino compounds in the fluids are mutually separated within 35 min on a cation-exchange column with a stepwise gradient elution of pH and/or ion strength in the mobile phase and then converted automatically to their fluorescent derivatives with benzoin. The method in this system is simple, rapid and sensitive; the lower limits of detection are 5-50 pmol for monosubstituted guanidino compounds, 1 nmol for creatine and 20 nmol for creatinine in a 100-μl injection volume.

ACKNOWLEDGMENTS

We wish to thank Drs. S. Ganno, K. Tsukada and M. Ito, Naka-works Hitachi Co. Ltd., for fundamental construction of the analyzer, and Dr. M. Yamamoto, Japan Red Cross Fukuoka Hospital, for supply of the patient sera.

REFERENCES

1. A. S. Jones and T. W. Thompson, Detection of guanidine compounds on paper chromatograms, J. Chromatogr., 10:248 (1963).
2. S. Giovannetti, M. Biagini and L. Cioni, Evidence that methylguanidine is retained in chronic renal failure, Experientia, 24:341 (1968).
3. M. Rink and D. Krebber, Dunnschichtchromatographische bestimmung von Kreatinin und Kreatin im harn, J. Chromatogr., 25:80 (1966).
4. F. Di Jeso, Qualitative and quantitative thin-layer chromatography of guanidine derivatives and differentiation of phosphagens from other phosphorus compounds, J. Chromatogr., 32:269 (1968).
5. A. Volkl and H. H. Berlet, Thin-layer chromatography of guanidino compounds, J. Clin. Chem. Clin. Biochem., 15:267 (1977).
6. H. Patel and B. D. Cohen, Gas-liquid chromatographic measurement of guanidino acids, Clin. Chem., 21:838 (1975).
7. T. Kawabata, H. Ohshima, T. Ishibuchi, M. Matsui and T. Kitsuwa, Gas chromatographic dtermination of methylguanidine, guanidine and agmatine as their hexafluoroacetylacetonates, J. Chromatogr., 140:47 (1977).
8. A. Mori, T. Ichimura and H. Matsumoto, Gas chromatography-mass spectrometry of guanidino compounds in brain, Anal. Biochem., 89:393 (1978).

9. P. P. Kamoun, J. M. Pleau and N. K. Man, Semiautomated method for measurement of guanidinosuccinic acid in serum, Clin. Chem., 18:355 (1972).

10. R. Shainkin, Y. Berkenstadt, Y. Giatt and G. M. Berlyne, An automated technique for the analysis of plasma guanidino acids, and some findings in chronic renal disease, Clin. Chem. Acta, 60:45 (1975).

11. M. Matsumoto, H. Kishikawa and A. Mori, Guanidino compounds in the sera of uremic patients and in the sera and brain of experimental uremic rabbits, Biochem. Med., 16:1 (1976).

12. Y. Yamamoto, T. Manji, A. Saito, K. Maeda and K. Ohta, Ion exchange chromatographic separation and fluorimetric detection of guanidino compounds in physiologic fluids, J. Chromatogr., 162:327 (1979).

13. Y. Yamamoto, A. Saito, T. Manji, K. Maeda and K. Ohta, Quantitative analysis of methylguanidine and guanidine in physiologic fluid by high-performance liquid chromatography-fluorescence detection method, J. Chromatogr., 162:23 (1979).

14. Y. Hiraga and T. Kinoshita, Post-column derivatization of guanidino compounds in high-performance liquid chromato-graphy using ninhydrin, J. Chromatogr., 226:43 (1981).

15. Y. Ohkura and M. Kai, Fluorimetric determination of mono-substituted guanidino compounds with benzoin-dimethyl-formamide reagent, Anal. Chim. Acta, 106:89 (1979).

16. M. Kai, T. Miura, K. Kohashi and Y. Ohkura, New method for the fluorimetric determination of guanidino compounds with benzoin, Chem. Pharm. Bull., 29:1115 (1981).

17. M. Kai, M. Yamaguchi and Y. Ohkura, Fluorescent products of the reaction of monosubstituted guanidino compounds with benzoin-dimethylformamide, Anal. Chim. Acta, 120:411 (1980).

18. M. Kai, T. Miyazaki, M. Yamaguchi and Y. Ohkura, High-performance liquid chromatography of guanidino compounds using benzoin as a pre-column fluorescent derivatization reagent, J. Chromatogr., 268: 417 (1983).

19. D. J. Durzan, Automated chromatographic analysis of free monosubstituted guanidines in physiological fluids, Can. J. Biochem., 47:657 (1969).

20. R. E. Majors, Effect of particle size on column efficiency in liquid-solid chromatography, J. Chromatogr. Sci., 11:88 (1973).

21. J. Bergström, P. Fürst and L. Zimmerman, Uremic middle molecules exist and are biologically active, Clin. Nephrol., 11:229 (1979).

22. J. Menyhart and J. Grof, Many hitherto unknown peptides are principal constituents of uremic middle molecules, Clin. Chem., 27:1712 (1981).

SIMPLE METHODS FOR THE DETERMINATION OF METHYLGUANIDINE AND GUANIDINOSUCCINIC ACID IN BIOLOGICAL FLUIDS

Toshiyuki Nakao*, Seiji Fujiwara and Tadashi Miyahara

Second Department of Internal Medicine
Jikei University School of Medicine Tokyo, Japan
*Saiseikai Central Hospital, Tokyo, Japan

It is recognized that guanidino derivatives significantly accumulate in uremic patients and contribute to the toxic manifestations of uremia. Among those compounds, both methylguanidine (MG) and guanidinosuccinic acid (GSA) have been especially interesting for their role as uremic toxins[1]. As the implications of MG and GSA as uremic toxins has become more obvious, routine measurement of those substances in biological fluids has assumed increasing importance in managing uremic patients to observe the effects of treatment.

Various methods[2-13] have been developed to determine these compounds in biological fluids and high performance liquid chromatography is recently proposed. However these methods are all cumbersome for routine practice, so measurement of these substances is carried out only in a few laboratories.

In order to make the measurement available as a routine clinical examination, we simplified the previous column chromatographic method and designed a new "mini-column" with ion-exchange column chromatography and fluorimetry.

METHODS

Resin

MG. Dowex 50W x 12, 100 - 200 mesh (Dow Chemical Company) was used. The resin obtained in the hydrogen form was converted to the sodium form by washing with 3 M NaOH filtering on a funnel.

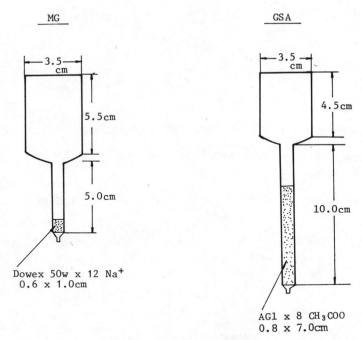

Fig. 1. "Mini-column" for the isolation of MG or GSA.

The resin was then transferred to a beaker, immersed in 0.1M NaOH and stored until using.

GSA. AG1- x 8, 200 - 400 mesh, acetate form (Bio-Rad Laboratories) was used. It was immersed in 0.1 M CH_3COOH in a breaker and stored until use.

Column

MG. We used a newly designed mini-column (0.6 x 5 cm) with a reservoir (3.5 x 5.5 cm) at the top of the column as shown in fig. 1. The column was packed to a height of 1.0 cm with the resin and was equilibrated with 0.1 M NaOH.

GSA. Mini-column (0.8 x 10 cm) with a reservoir (3.5 x 4.5 cm) at the top of the column was designed as shown fig. 1. The resin was packed in the column to a height of 7.0 cm and was equilibrate with 0.1 M CH_3COOH.

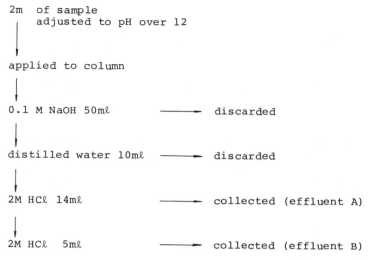

Fig. 2. Outline of the procedure for isolation of MG from samples.

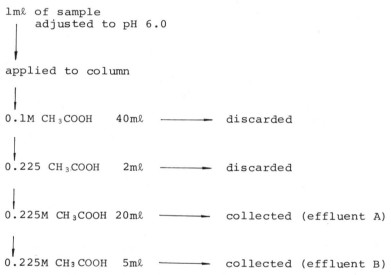

Fig. 3. Outline of the procedure for isolation of GSA from samples.

Procedure of isolation

Outline of the procedure is illustrated in fig. 2. and fig. 3.

MG. A two milliliter aliquot of serum or urine was adjusted to a pH over 12 with less than 10 µl of NaOH and then applied

directly to the column. After the specimen had entered the column,
elution was begun with 50 ml of 0.1 M NaOH, followed by 10 ml of
distilled water and those effluents were all discarded. 14 ml of
2 M HCl was then applied, and the effluent in which MG was isolated
was collected all together in a test tube (effluent A). Finally
5 ml of 2 M HCl was run through the column and the effluent was
also collected in a separate test tube, which was used as solvent
blank (effluent B).

GSA. One milliliter of serum or urine was adjusted to pH 6.0
with less than 10 μl of glacial acetic acid or NaOH, and then
applied directly to the column. After the specimen had entered
the resin, elution was begun with 40 ml of 0.1 M CH_3COOH, followed
by 3 ml of 0.225 M CH_3COOH and those effluents were all discarded.
Twenty ml of 0.225 M CH_3COOH was then applied, and the effluent
was collected in a test tube in which GSA was isolated (effluent A).
Finally, 5ml of 0.225 M CH_3COOH was run through the column and the
effluent was also collected in a separate test tube, which was used
for a solvent blank (effluent B).

Fluorimetric measurement

To each 2 ml of effluent A, B and standard, 0.5 ml of 6M KOH
and the same volume of freshly prepared 0.5 % w/v ninhydrin solution in water was added. After 5 minutes had elapsed, fluorescence was measured with Hitachi Spectrofluorimeter Model MPF 3.
Excitation was produced by a xenon arc lamp at a wave length of
395 nm and the emission was read at 495 nm. Concentrations of MG
and GSA were determined from calibration curves of 5, 10, 15 and
25 μg per 100 ml of MG (SIGMA) and 25, 50, 100 and 200 μg per
100 ml of GSA (SIGMA) run simultaneously.

RESULTS

The results are summarized in Table 1. The analytical recovery of added known amounts of MG was 95.4 ± 4.0 % and that of
GSA was 94.2 ± 8.0 %. The coefficient of variation for eight re-
plicate determinations of MG was 9.2 % and that of GSA was 8.5 %.
The lower limit of sensitivity of the method was 0.3 μg/dl for MG
and 8.0 μg/dl for GSA. The mean plasma concentration of MG in
five healthy subjects was 7.1 ± 3.1 μg/dl and that of GSA was
274.7 ± 36.0 μg/dl. In five maintenance hemodialized patients,
the mean plasma concentration of MG was 70.9 ± 40.9 μg/dl and that
of GSA was 1128.3 ± 373.2 μg/dl. By analysis of high performance
liquid chromatography of the effluent A from the mini-column for
MG, admixture of guanidine was identified in trace amounts and
also admixture of trace amounts of MG in the effluent A from the
mini-column for GSA was identified.

Table 1.   Summarized results of MG or GSA determination by present
           "mini-column" methods.

		MG	GSA
Recovery rate	(%)	95.4±4.0	94.2±8.0
Reproduciability C.V.	(%)	9.2	8.5
Lower limits of detection (μg/dl)		0.3	8.0
Concentration in plasma (μg/dl) Healthy subjects (n=5)		7.1 ±3.1	274.7 ±36.0
Hemodialized patients (n=5)		70.9±40.9	1128.3±373.2

DISCUSSION

The quantitative determination of MG and GSA using "mini-column" with ion-exchange chromatography and fluorimetry is found to be a very useful and sensetive method for the determination of MG and GSA in normal and uremic subjects. The principles of isolation of MG or GSA by ion-exchange chromatography were derived from those which Menichini[4] and Grof[7] had previously reported. The ability of guanidino compounds to combine with ninhydrin in strongly alkaline solutions to form a highly fluorescent addition-al product[14] forms the basis of a simple, sensitive means of quantitative measurement.

Judging from the result of our studies, the mini-column method should fulfill the requirement for an easy and reliable procedure. Using this mithod we could assay many samples simultaneously within a short time.

Contamination by other guanidino compounds was seen in the effluent from the column. However, it was in small amounts so that the assay was not affected.

In conclusion, the mini-column method with fluorimetry has yielded good results in the determination of MG and GSA. This method offers a simplified procedure suitable for routine use in clinical practice.

REFERENCES

1. J. Bergström and P. Fürst, Uremic toxins, in: "Replacement of
    renal function by dialysis," W. Drukker, F. M. Parson and
    J. F. Maher, ed., Martinus Nijhoff, Hague, (1978) p 334.
2. I. M. Stein, G. Perez, R. Johnson and N. B. Cummings, Serum
    levels and urinary excretion of methylguanidine in chromic
    renal failure, J. Lab. Clin. Med., 77:1020 (1971).
3. L. R. I. Baker and R. D. Marshall, A Reinvestigation of
    methylguanidine concentration in sera from normal and
    uremic subjects, Clin. Sci., 41: 563 (1971).
4. G. C. Menichini, M. Gonella, G. Barsotti and S. Giovannetti,
    Determination of methylguanidine in serum and urine from
    normal and uremic subjects, Experientia, 27:1157 (1971).
5. I. M. Stein, B. D. Cohen and R. S. Kornhauser, Guanidino-
    succinic acid in renal failure, experimental azotemia and
    inborn errors of the urea cycle, New Eng. J. Med.,
    280:926 (1969).
6. P. P. Kamoun, J. M. Pleau and N. K. Man, Semiautomated
    method for measurement of guanidinosuccinic acid in serum,
    Clin. Chem., 18:355 (1972).
7. J. Gróf, A. Tankó and J. Menyhárt, New method for measure-
    ment of guanidinosuccinic acid in serum and urine,
    Clin. Chem., 20:574 (1974).
8. R. Shainkin, Y. Berkenstadt, Y. Giatt and G. M. Berlyne, An
    automated technique for the analysis of plasma guanidino
    acids, and some findings in chronic renal disease,
    Clin. Chem. Acta, 60:45 (1975).
9. H. Patel and B. D. Cohen, Gas-liquid chromatographic measure-
    ment of guanidino acids, Clin. Chem., 21:838 (1975).
10. G. Perez, A. Rey, M. Micklus and I. Stein, Cation-exchange
    chromatography of guanidine derivatives in plasma of
    patients with chronic renal failure, Clin. Chem.,
    22:240 (1976).
11. S. Eksborg, B. Persson, L. Allgen, J. Bergström, L. Zimmerman
    and P. Fürst, A selective method for determination of
    methylguanidine in biological fluids. Its application in
    normal subjects and uremic patients, Clin. Chem. Acta,
    82:141 (1978).
12. Y. Yamamoto, A. Saito, T. Manji, H. Nishi, K. Ito, K. Maeda,
    K. Ohta and K. Kobayashi, A new automated analytical
    method for guanidino compounds and their cerebrospinal
    fluid levels in uremia, Trans. Am. Soc. Artif. Inter.
    Organs, 24:61 (1978).
13. M. D. Baker, H. Y. Mohammed and H. Veening, Reversed-phase
    ion-pairing liquid chromatographic separation and fluoro-
    metric detection of guanidino compounds, Anal. Chem.,
    53:1658 (1981).
14. R. B. Conn and R. B. Davis, Green fluorescence of guanidinium
    compounds with ninhydrin, Nature, 183:1053 (1959).

ENZYMIC DETERMINATION OF METHYLGUANIDINE IN URINE

Motoo Nakajima, Kazuo Nakamura and Yoshio Shirokane

Research & Development Division
Kikkoman Corporation
399 Noda, Noda-shi, 278
Japan

INTRODUCTION

Methylguanidine is known to accumulate in the body fluids
of uremic patients[1,2] and has proved to be a uremic toxin[3].
Determination of its concentration in body fluids would, therefore, be useful in clinical practice. Recently, the determination of methylguanidine has been carried out by means of automated
high-performance liquid chromatography with a fluorometric detection method for guanidino compounds[4]. The guanidino compounds
analyzer is well suited for the simultaneous determination of
several guanidino compounds. However, it is very expensive and
inappropriate for analyses of many samples. In clinical practice,
the enzymic determination of metabolite concentration has been
widely employed because of its simplicity and specificity.

To establish the enzymic determination of methylguanidine we
found two new enzymes, methylguanidine amidinohydrolase (MGAH)[5]
and methylamine oxidase (MAOD), and developed the assay system
with these enzymes[6]. Due to very high substrate specificity for
methylguanidine of MGAH, this system is excellent and reliable.
The methylguanidine in biological fluids is converted to formaldehyde by these two enzymes. The formaldehyde is measured by
a fluorometric method which increases sensitivity. The schemes
of the enzymic reactions and fluorescence reaction are as follows:

Table 1. Some properties of MGAH and MAOD

Properties	MGAH	MAOD
Substrate Specificity	Highly specific for Methylguanidine	Monomethylamine> Ethylamine> Propylamine
Km	1.3×10^{-3} M	6.7×10^{-4} M
Optimum pH	12.5	7.5-9
Optimum temp.	60 °C (pH 10)	45 °C (pH 7)
pH stability	6-12	6-7
Temp. stability	60 °C (pH 7)	50 °C (pH 7)

1)   MGAH
   Methylguanidine + water = methylamine + urea

2)   MAOD
   Methylamine + water + oxygen = formaldehyde + ammonia
                                             + hydrogen peroxide

3) Fluorometric determination of formaldehyde

We used a crude preparation of MGAH from *Alcaligenes* sp. N-42 and MAOD from *Bacillus* sp. N-104, respectively. Some properties of two enzymes are shown in Table 1.

We reported an enzymic method for the determination of methylguanidine in plasma and serum of patients with chronic renal failure at the 5th annual meeting of the Japan Guanidino Compounds Research Association . In the case of the determination of methylguanidine in blood, a deproteinization procedure was needed to remove the inhibitor in blood for MAOD and it was accomplished using centrifugal ultrafiltration with Amicon CF-25 after the enzymic reaction with MGAH. As the reaction reagent for the fluorometric determination of formaldehyde we employed acetoacetic acid methylester. The reaction was performed at 45 °C for 15 min, because the time over 40 min was required to complete the reaction at 37 °C. Good reproducibility (n=5) for the enzymic analyses of methylguanidine in a plasma pool from patients with chronic renal failure was shown and gave a 6.4 % coefficient of variance (CV) within each batch with a 37.9 µg/dl mean methylguanidine concentration. Analytical recovery by this procedure was found to be 92.2 % to 98.3 %. A comparision of the enzymic method with the high-performance liquid chromatography method for guanidino compounds gave the following linear regression equation.

```
Urine sample                                    0.5  ml
    |  ← Buffer (A)                              0.25 ml
    |  ← Water                                   0.25 ml
    ↓  ← MGAH (or Water)                         0.1  ml
First enzyme reaction    -----   37°C, 15 min.
    |  ← Buffer (B)                              1    ml
    ↓  ← MAOD                                    0.05 ml
Second enzyme reaction   -----   37°C, 5 min.
    |  ← Buffer (C)                              2    ml
    ↓  ← MAC reagent                             0.1  ml
Fluorescence reaction    -----   37°C, 15 min.
    ↓
Cooling in a water bath  -----   1 min.
    ↓
Fluorometric determination
           Excitation wavelength: 375 nm
           Emission wavelength: 465 nm

For each sample blank, use 0.1 ml of water instead of MGAH
```

Fig. 1.   Enzymic determination procedure for methylguanidine
           in urine.

$y = 0.647 x + 11.74$
$r = 0.92 (n = 34)$.
where y: the amount of methylguanidine by the HPLC method,
      x: the amount of methylguanidine by the enzymic method,
      r: correlation coefficient,
      n: sample number.

Here, we describe a modification of the enzymic determination
method for methylguanidine and the application to urinary samples.

PROCEDURE

The enzymic determination procedure for methylguanidine in
urine is shown in Figure 1.   In this case, deproteinization was
unnecessary.

We found that methyl 3-aminocrotonate (MAC) reacted with
formaldehyde and the reaction product had fluorescence.   This
fluorescence reaction with 3-aminocrotonate can be used for a
rapid reaction and sensitive determination of formaldehyde.
Therefore, acetoacetic acid methylester used in the old method was
replaced by methyl 3-aminocrotonate.   As the result of this replacement, the reactions could be performed at 37 °C.

REAGENTS AND METHYLGUANIDINE STANDARD

Buffer (A) : Sodium hydrogen carbonate-sodium carbonate
             buffer, 0.2 mol/1, pH 10.

Buffer (B) : Potassium dihydrogen phosphate-disodium hydrogen
             phosphate, 0.1 mol/1, pH 7.0.

Buffer (C) : Potassium dihydrogen phosphate-disodium hydrogen
             phosphate, 0.1 mol/1, pH 5.5.

MGAH   : 6 units/ml.

MAOD   : 5 units/ml.

MAC reagent: Methyl 3-aminocrotonate (Aldrich Chemical Co.,
             Milwaukee), 26 g/1 in ethanol.

Methylguanidine standard solution:
             1 mg/1 (1.6 mg methylguanidine sulfate prepared
             by ourselves[8] per liter of distilled water).

RESULTS AND DISCUSSION

Hydrolysis of methylguanidine in urine by MGAH

        We reported that the free and protein-bound methylguanidine
in the serum from patient with chronic renal failure was completely decomposed by MGAH[7]. As shown in HPLC chromatograms (Figure 2),
the peak of methylguanidine in 24-hour urine from a patient with
chronic renal failure completely disappeared by incubation with
MGAH at 37 °C for 5 min. Although this urine sample contained
thymol as antiseptic, the enzyme reaction was not influenced.

Linearity

        Figure 3 shows the calibration curve by the enzymic method.
Good linearity was obtained in a wide range of methylguanidine
concentrations.

Precision

        Reproducibility of the enzymic assays was determined by
multiple analyses of two different 24-hour urines from patients
with chronic renal failure (Table 2). Precision of this method
was appareciable good.

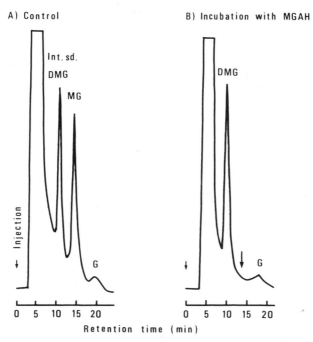

Fig. 2.   HPLC chromatograms showing hydrolysis of urinary methyl-
           guanidine by MGAH. Enzymic reaction : 5 min, 37 °C.
           HPLC conditions: column, Shodex HC-095, 120 x 3 mm ID.,
           60 °C; eluent, 0.3 N NaOH in 50 % methanol, 0.4 ml/min;
           reagent, 1 N NaOH, 0.6 ml/min and 0.6 % ninhydrine,
           0.8 ml/min; reaction temperature, 60 °C; fluoromonitor,
           FLD-1 (Shimazu). Int. Std.: internal standard. DMG:
           dimethylguanidine. MG: methylguanidine. G: guanidnine.

Analytical recoveries

We added 0.01 ml of methylguanidine solution at two different
concentrations to 0.45 ml of urine to give final concentrations of
148.9 μg/dl and 167.5 μg/dl. As shown in Table 3, analytical re-
coveries were good.

Accuracy

The methylguanidine values determined by the enzymic method
(x) were compared with those of the high-performance liquid chromatography (y). As shown in Figure 4, good correlation was found
between the two determination. It gives a linear regression equation, $y = 1.07 x - 20.5$ with $r = 0.97$.

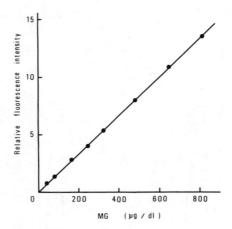

Fig. 3.   Calibration curve of methylguanidine by the enzymic
          method.

Table 2. Precision of the enzymic determination of
         methylguanidine in urine from patients with
         chronic renal failure

	Sample A	Sample B
Number of Repeated Assay	10	4
Mean value, µg/dl	264.3	129.1
SD, µg/dl	8.11	5.26
CV, %	3.07	4.08

SD: standard deviation
CV: coefficient of variance

Healthy urine analysis

Forty five spot urine specimens were obtained from apparently healthy volunteers aged 10 to 44 years. The methylguanidine
concentration in each urine sample was determined by the enzymic
method. The mean value and SD for the methylguanidine concentration were 56.2 µg/dl and 29.4 µg/dl, respectively. In the 4th
annual meeting of the Japan Guanidino Compounds Research
Association, the average methylguanidine concentration determined
by several guanidino compound analyzers in healthy 24-hour urine
specimens was reported as 32.4 µg/dl and with 35.4 µg/dl SD[9].
These values can be considered in good agreement with our results
in spite of the difference in analytical methods and methods for
urine collection.

An analytical example of urinary methylguanidine excretion per day
of an inpatient with chronic renal failure

Figure 5 shows an example of daily methylguanidine excretion

Table 3. Analytical recovery of methylguanidine from urine

Added (μg/dl)	Methylguanidine (μg/dl)		Recovery (%)
	in sample	recovered	
0	129.1		
20.2	148.9	19.8	98.0
39.8	167.5	38.4	96.5

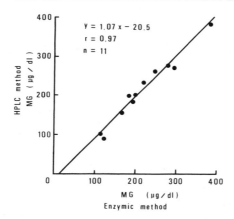

Fig. 4.   Correlation between methylguanidine concentrations in
24-hour urine of patients with chronic renal failure
measured by enzymic and HPLC methods. Where y: MG by the
HPLC method, x: MG by the enzymic method, r: correlation,
n: sample number.

in urine of an inpatient with chronic renal failure. In the case
of this patient, the urinary methylguanidine excretion per day
ranged from 3.2 mg/day to 5.2 mg/day for about 3 weeks. This value
is about 10 times higher than the reported average normal urinary
excretion value[9] of 0.39 mg/day plus or minus SD of 0.43 mg/day.

We believe that this enzymic determination method of methylguanidine in blood and urine will be very useful in clinical
practice.

AKNOWLEDGEMENTS

We thank Dr. M. Ishizaki and Mr. H. Kitamura (Sendai insurance
Hospital) for their kind presentation of urine specimens of patients
with chronic renal failure.

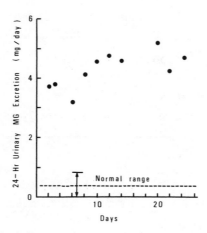

Fig. 5.   An analytical example of urinary methylguanidine excretion per day of an inpatient with chronic renal failure.

REFERENCES

1. I. M. Stein, G. Perz, R. Johnson and N. B. Cummings, Several levels and urinary excretion of methylguanidine in chronic renal failure, J. Lab. Clin. Med., 77: 1020 (1971).
2. G. C. Menichini and S. Giovannetti, A new method for measuring guanidine in uremia, Experientia, 29: 506 (1973).
3. S. Giovannetti and G. Barsotti, Dialysis of methylguanidine, Kidney Int., 6: 177 (1974).
4. Y. Yamamoto, A. Saito, T. Manji, H. Nishi, K. Ito, K. Maeda, K. Ohta and K. Kobayashi, A new automated analytical method for guanidino compounds and their cerebrospinal fluid levels in uremia, Trans. Am. Soc. Artif. Intern. Organs, 24: 61 (1978).
5. M. Nakajima, Y. Shirokane and K. Mizusawa, A new amidino-hydrolase, methylguanidine amidinohydrolase from Alcarigenes sp. N-42, FEBS Lett., 110: 43 (1980).
6. M. Nakajima, K. Nakamura, Y. Shirokane and K. Mizusawa, Mono-methylamine-oxidizing enzyme and a process for its manufacturing, Japan Kokai Tokkyo Koho, 58-71886 (1983).
7. M. Nakajima, Y. Shirokane and K. Mizusawa, Enzymic determination of methylguanidine (11) Determination of methylguanidine concentration in blood of patients with chronic renal failure and hemodialysis, Abstract of 5th Annual Meeting of the Japan Guanidino Compounds Research Association, Suita, 5: 25 (1982).
8. R. Phillips and H. T. Clarke, The preparation of alkyl-guanidines, J. Am. Chem. Soc., 45: 1755 (1923).
9. K. Isoda, Normal values of guanidinocompounds in serum, erythrocyte and urine (by fluorometric HPLC), Abstract of 4th Annual Meeting of Japan Guanidino Compounds Research Association, Tokyo, 4: 99 (1981).

II. METABOLISM OF GUANIDINO COMPOUND IN NORMAL AND ABNORMAL ORGANS,
AND IN METABOLIC DISEASES

DEVELOPMENTAL CHANGES IN GUANIDINO COMPOUNDS LEVELS

IN MOUSE ORGANS

Yoko Watanabe, Shoichiro Shindo, and Akitane Mori

Department of Neurochemistry
Institute for Neurobiology
Okayama University Medical School
2-5-1 Shikatacho
Okayama 700, Japan

INTRODUCTION

Many kinds of guanidino compounds are known to be existent in mammalian organs[1], and they are thought to be related to nitrogen metabolism. There have been many reports of guanidino compound levels in human body fluids and animal organs changing under various pathological conditions. Methylguanidine (MG) and guanidinosuccinic acid (GSA) in serum are elevated in uremic patients[2,3]. Increased excretion of α-keto-δ-guanidinovaleric acid in urine was observed in patients with hyperargininaemia[4]. High concentrations of tauro-cyamine in the cerebrospinal fluid of some epileptic patients have been observed[5]. Yokoi et al.[6,7] showed the homoarginine (HArg) level to be high in liver, kidney and serum of rats administered alcohol chronically. These observations suggest that guanidino compound levels in organs may change according to physiological conditions. It is possible that the levels of guanidino compounds may also change during maturation when physiological conditions are thought to be undergoing change. The data reported herein are the results of a study of developmental changes in the levels of guanidino compounds in mouse organs.

METHODS

Male ddY mice were used except for fetuses and one-day-old mice which were used without distinction of sex. Three to six-week-old mice were purchased from Shizuoka Agricultural Co-op. Assoc. for Laboratory Animals (Shizuoka, Japan) and maintained on

food (Oriental Yeast Co. Ltd., Tokyo, Japan) and water ad libitum
in a light controlled (lights on 07:00-19:00), temperature regulated (25±2 °C) room until sacrifice. Fetal mice were obtained from
mice in the last stage of pregnancy which were also purchased from
Shizuoka Agricultural Co-op. Assoc. for Laboratory Animals.
Suckling (from birth to 3 weeks) and 4-week-old mice were born in
our laboratory's breeding room, and brought up with their parents
until 3 weeks of age. All animals were sacrificed by decapitaion
between 13:00 and 13:30, and the liver, kidney and brain were
rapidly removed on ice. The pancreas and skeletal muscle from the
thigh were also obtained from mice of more than 3 weeks of age.
Since the organs of fetal and suckling mice were too small to
determine guanidino compounds, the organs obtained from 2 to 11
mice were pooled for one experiment. These organs were weighed
and stored frozen at -80 °C until the extraction procedure was
carried out.

Each weighed tissue sample was homogenized with 10 volumes of
1 % picric acid. After centrifugation at 1100 x g for 15 minutes
at 4 °C, the precipitate was washed with 0.01 N-HCl solution. The
washing was combined with the original supernatant and the solutions were passed through a column of Dowex 2x8 (Cl⁻form) to
remove excess picric acid. The column was washed with 5 resin
volumes of dilute HCl solution (pH 2.2). The colorless eluate was
evaporated to dryness at 40 °C, and the residue was dissolved in
dilute HCl solution (pH 2.2) for guanidino compound analysis. The
samples were stored at -20 °C until analysis.

The guanidino compound concentrations were determined with a
high performance liquid chromatograph, JASCO G-520 (Japan Spectroscopic Co., Hachioji, Tokyo, Japan)[8]. Authentic guanidino compounds were obtained as previously described[8].

RESULTS

As shown in Fig. 1, the guanidinoacetic acid (GAA) level in
kidney was low prior to birth and increased constantly up to 3
weeks after birth, and then decreased to the adult level. The
GAA contents in young adult mice (4 and 6 weeks) were higher than
those of old mice. The change in GAA level in liver closely re-
sembled that in kidney, though the level was much lower than in
kidney. The GAA level in brain reached a maximum level one week
after birth and thereafter rapidly leveled off to the adult level
within 6 weeks. Guanidino compounds in pancreas and muscle were
studied from 3 weeks of age. The GAA contents in pancreas were
higher in young adult mice (3 to 6 weeks) than in old mice (12 to
52 weeks). In muscle, the highest level of GAA was found at 3
weeks of age, and then its level consistently decreased.

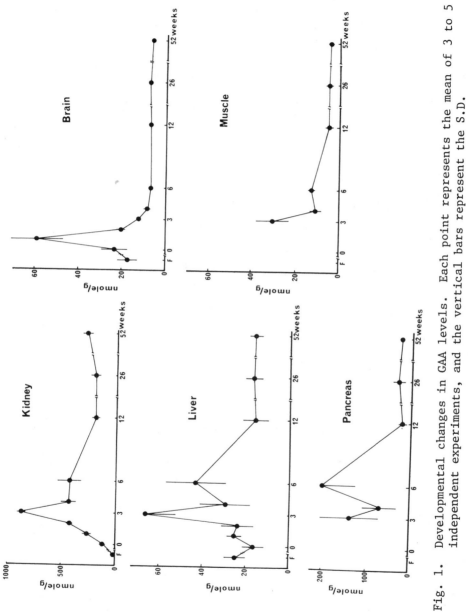

Fig. 1. Developmental changes in GAA levels. Each point represents the mean of 3 to 5 independent experiments, and the vertical bars represent the S.D.

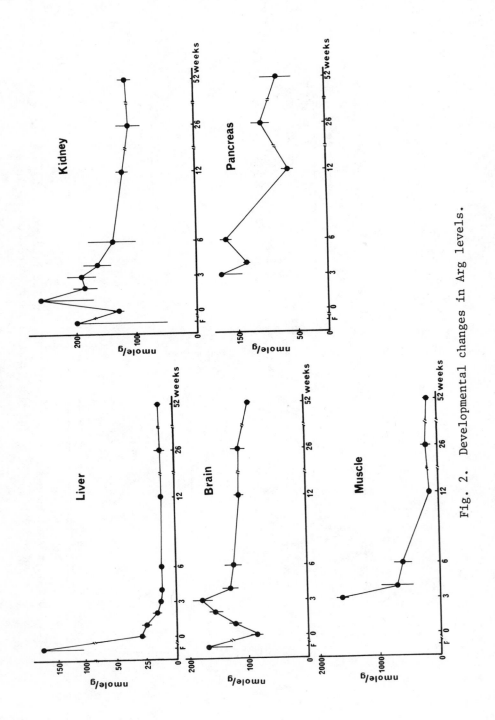

Fig. 2. Developmental changes in Arg levels.

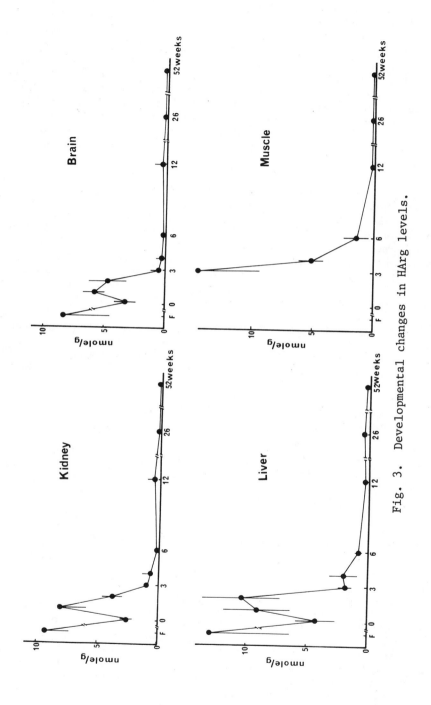

Fig. 3. Developmental changes in HArg levels.

The change in arginine (Arg) levels in shown in Fig. 2. The
Arg level in kidney was observed, the highest at one week of age
decreasing thereafter and reaching the adult level at 6 weeks. In
liver, the Arg levels before birth were markedly higher than after
birth, and the levels tended to decrease after birth to the adult
level within 3 to 4 weeks. The fetal Arg level in brain was also
higher than one day after birth, however, the Arg level in brain
increased up to 3 weeks after birth and then dropped to the adult
level. The Arg content in muscle was the highest at 3 weeks of
age, decreased up to 12 weeks and then remained steady up to 52
weeks. In pancreas, Arg levels in young adult mice were found to
be higher than in old mice.

HArg was detected in liver, kidney, brain and muscle (Fig. 3),
but it was not detected, or only in trace volumes if detected, in
pancreas. The HArg contents in adult mouse tissues were very low,
but the fetuses and sucklings had a high concentration of HArg.
The developmental changes of HArg were quite similer in liver,
kidney and brain, with its levels being high before birth, decreasing just after birth and then increasing rapidly. The maximum
levels of HArg after birth were observed at one or 2 weeks, after
which the level dropped to the low adult level within 6 weeks.

GSA was determined in liver. In other organ tissues, it was
either not detected or deteced only in trace volumes. GSA levels
in liver were very low until 2 weeks after birth, increased thereafter and maintained relatively high levels in adult mice (Fig. 4).

β-Guanidinopropionic acid (GPA) and γ-guanidinobutyric acid
(GBA) were detected only in liver. The GPA levels were low in
fetuses and one-day-old mice, then slowly increased, retaining a
high level in 3 to 6-week-old mice. The GPA levels at 12 and 52
weeks of age were lower than at the young adult stage (Fig. 4).
The GBA level in fetuses was also low and remained so for one week
after birth. It then rose to a maximum level at 4 weeks, decreased
somewhat thereafter (Fig. 4).

A small amount of MG were detected in all organs, however,
there was no change during maturation (data is not shown).

DISCUSSION

It was found from our experiments that the level of some
guanidino compounds changed dramatically during maturation.

The Arg level in liver was the highest before birth, but decreased rapidly after that. On the other hand, the highest level
of Arg was observed at one to three weeks after birth in the
brain, kidney, pancreas and muscle. The Arg level decreased in

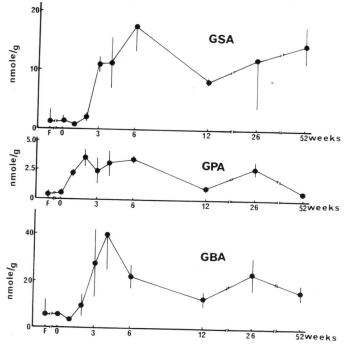

Fig. 4.   Developmental changes in GSA, GPA and GBA levels of liver.

these organs after that, but the rate of the decrease in kidney, pancreas and brain was not as rapid as in liver and muscle.

Argininosuccinate synthetase and argininosuccinase have been detected in brain, kidney, pacreas and muscle[9]. Moreover, Arg is considered to be an essential amino acid, especially at a young age[10]. In the suckling period, Arg could be an important amino acid for the brain, kidney, pancreas, muscle and liver. However, the liver has all the enzymes of the urea cycle, and the urea cycle has an important role in the detoxification of ammonia. Recently Funahashi et al.[11] reported that Arg is mainly synthesized in the kidney and then transfered to other organs. These observations suggest that the urea cycle could begin detoxifying ammonia from just after birth, which would account for the marked decrease in the Arg level after birth and the maintenance of a low level in the liver.

The Arg levels in muscle were very high, being 140 nmol/g tissue at the lowest. It seems strange that the muscle needs such a high concentration of free Arg.

It is well known that GAA is synthesized by arginine-glycine transamidinase in the kidney. GAA in the kidney is transfered to

the liver and utilized to synthesize creatine phosphate which is
very important as a high energy source for other organs such as
the muscle. The GAA levels in kidney and liver increased after
birth and reached a maximum level at 3 weeks. They decreased
after that, and maintained a steady level up to 52 weeks. These
observations suggest that these changes in the GAA levels might
reflect the demand for creatine phosphate. The liver GAA levels
seem to effect the kidney GAA levels. The highest level of GAA in
brain was observed one week after birth, and the level rapidly decreased after that. This result suggests that a specific metabolic
pathway or regulatory mechanism for GAA may exist in the brain.

    It is reported that the pancreas contains high levels of
arginine-glycine amidinotransferase[12]. We did not study all the
changes in GAA levels in the pancreas throughout the life of mice,
however, the highest level in pancreas was not observed at 3 weeks,
which is different from kidney and liver.

    It is of interest that high HArg levels were detected in
fetuses and sucklings except in pancreas, while the levels in
adult mice were very low in all organs. It was reported that a
high level of HArg was excreted by citrullinemic[13] and hyper-
lysinemic[14] patients. There are also reports that the HArg level
in liver, kidney and serum increased after chronic alcohol
administration to rats[6,7]. Thus HArg synthesis seems to be activated under some pathological conditions. In the present study,
we showed that normal mice had higher concentrations of HArg in
the fetal and suckling periods than in other stages. Moreover it
was found that the HArg level changed dramatically before and just
after birth. It is not known why there are high HArg levels in
the early period of life. It is possible, however, that high HArg
levels might be caused by diet, i.e. the HArg metabolic activity
could change as nitrogen metabolism changes during life.

    GSA, GPA and GBA existed in the liver, although their concen-
trations were low. The levels of these guanidino compounds
changed during maturation differently from those of other guanidino
compounds. the levels of GSA, GPA and GBA were very low before
and just after birth and increased to high levels at 3 to 6 weeks,
with the relatively high levels being retained during adulthood.
The roles of these guanidino compounds and synthetic pathways are
still obscure. GSA is, however, considered to be a uremic toxin
and to be synthesized from urea through the guanidine cycle[15,16]
The changes observed during maturation suggest that these guanidino
compounds increase according to an increase in nitrogen intake.

    In the present study we showed that the development of organs
might influence the metabolism of guanidino compounds, and that
each guanidino compound might have some important physiological
role in mouse organs.

REFERENCES

1. A. Mori, Natural occurrence and analyses of guanidino compounds, Rinshokagaku, 9:232 (1980).
2. S. Giovannetti, P. L. Balestri and G. Barsotti, Methylguanidine in uremia, Arch. Intern. Med., 131:709 (1973).
3. I. M. Stein, B. D. Cohen and R. S. Kornhauser, Guanidinosuccinic acid in renal failure, experimental azotemia and
   in born errors of the urea cycle, New. Engl. J. Med.,
   280:926 (1960).
4. B. Marescau, J. Pintens, A. Lowenthal, E. Esmans, Y. Luyten,
   G. Lemiere, R. Dommisse, F. Alderweireldt, and
   H. G. Terheggen, Isolation and identification of 2-oxo-5-
   guanidinovaleric acid in urine of patients with hyperargininaemia by chromatography and gas chromatography/mass
   spectrometry, J. Clin. Chem. Clin. Biochem., 19:61 (1981).
5. A. Mori, Y. Watanabe and M. Akagi, Guanidino compound
   anomalies in epilepsy, in:"Advances in Epileptology:
   XIIIth Epilepsy International Symposium," H. Akimoto,
   H. Kazamatsuri, M. Seino, and A. Ward, eds., Raven Press,
   New York (1982).
6. I. Yokoi, J. Toma and A. Mori, Effects of taurine on guanidino compounds in the brain and serum of rats administered
   alcohol chronically, Neurosciences, 9:177 (1983).
7. I. Yokoi, J. Toma and A. Mori, The effect of chronic ethanol
   administration on the guanidino compounds in rat organs,
   (in this book).
8. A. Mori, Y. Watanabe and N. Fujimoto, Fluorometrical analysis
   of guanidino compounds in human cerebrospinal fluid,
   J. Neurochem., 38:448 (1982).
9. H. Kato, I. Oyamada, M. Mizutani-Funahashi and H. Nakagawa,
   New radioisotopic assays of argininosuccinic synthetase
   and argininosuccinase, J. Biochem., 79:945 (1976).
10. J. G. Morris and Q. R. Rogers, Arginine: An essential amino
    acid for the cat, J. Nutr., 108:1944 (1978).
11. M. Funahashi, H. Kato, S. Shiosaka and H. Nakagawa, Formation
    of arginine and guanidinoacetic acid in the kidney in vivo.
    Their relations with the liver and their regulation,
    J. Biochem., 89:1347 (1981).
12. J. B. Walker, Formamidine group transfer in extracts of human
    pancreas, liver, and kidney, Biochim. Biophys. Acta,
    73:241 (1963).
13. A. Scott-Emuakpor, J. V. Higgins and A. F. Kohrman,
    Citrullinemia: A new case, with implications concerning
    adaptation to defective urea synthesis, Pediat. Res.,
    6:626 (1972).
14. N. C. Woody and Eng Bee Ong, Paths of lysine degradation in
    patients with hyperlysinemia, Pediatrics, 40: 986 (1967).
15. S. Natelson, A. Koller, H. Tseng and R. F. Dods, Canaline
    carbamoyltransferase in human liver as part of a metabolic

cycle in which guanidino compounds are formed, <u>Clin. Chem.</u>, 23:960 (1977).

16.   B. D. Cohen and H. Patel, Guanidinosuccinic acid and the alternate urea cycle, <u>in</u>:"Urea cycle diseases" A. Lowenthal, A. Mori, and B. Marescau, eds., Plenum Press, New York (1982).

DEVELOPMENTAL CHANGES OF GUANIDINO COMPOUNDS IN CHICK EMBRYO

Yoko Watanabe, Takashi Kadoya, Masayoshi Fukui,
Rei Edamatsu, Akitane Mori, and *Sonoko Seki

Department of Neurochemistry
Institute for Neurobiology
Okayama University Medical School
2-5-1 Shikatacho,
Okayama 700, Japan

*Department of Physiology
Kanagawa Dental College
Inaokacho
Yokosuka, Kanagawa 238, Japan

INTRODUCTION

It is well known that many kinds of guanidino compounds are observed in invertebrates and vertebrates[1,2]. The appearance of guanidino compounds in animals is thought to be related to nitrogen metabolism but there are no reports concerning guanidino compounds and embryogenesis. The present study was undertaken to learn about the appearance and changes in concentrations of guanidino compounds in the whole chick embryo, yolk and embryonic organs at various stages of embryonic development. It is hoped that this study might contribute to our basic understanding of guanidino compound metabolism during embryogenesis.

METHODS

Fertile and unfertile eggs from white Leghorn hens were obtained from the Kanagawa Prefectural Life Stock Experiment Station (Atsugi, Kanagawa, Japan). The eggs were incubated at 37 °C and 70 % humidity. On the indicated days of incubation the embryo and yolk were harvested, and the liver, kidney and brain were removed. Three of each organ were pooled for one experiment.

Each of the samples was homogenized with 10 volumes of 1 % picric acid for deproteinization, and then centrifuged at 1100 x g for 20 minutes. The supernatant was passed through a column of Dowex 2x8 (Cl⁻), and the colorless eluate was dried in vacuo. The guanidino compounds concentrations were determined with a high performance liquid chromatograph, Guanidino compound analyzer JASCO G-520 (Japan Spectroscopic Co., Hachioji, Tokyo, Japan)[3]. Standard guanidino compounds were obtained as previously described[3]. The data were analyzed according to the Student's t-test.

RESULTS

Guanidino compounds in the whole embryo and yolk

Guanidino compounds in the whole embryos were studied from the 8th to 20th day of incubation. Arginine (Arg), creatinine (CRN), guanidinoacetic acid (GAA), homoarginine (HArg), guanidine (G), β-guanidinopropionic acid (GPA) and methylguanidine (MG) were detected in the embryo on all the days studied. Guanidino compounds in the yolk, which were obtained from both the fertile and unfertile eggs, were studied from the 2nd to 20th day of incubation. The same kinds of guanidino compounds as those in the embryo were detected in the yolk.

The concentration of Arg in the embryo gradually increased throughout the embryonic development from 250 nmole/g wet weight to 630 nmole/g. High concentrations of Arg were observed in the yolk and the levels in the yolk from the fertile eggs increased ($p<0.005$) on the 18th day of incubation (Fig. 1).

Fig. 2 shows the changes in the concentrations of GAA in the embryo and yolk. The embryo contained low levels of GAA (below 6 nmole/g) up to the 14th day of incubation, increasing ($p<0.05$) on the 16th day and remaining at a high level throughout the next 4 days. GAA concentrations in the yolk were very low (below 0.5 nmole/g) from the 2nd to 10th day of incubation, and then the level in the yolk from the fertile eggs increased to 5.7 nmole/g ($p<0.005$) on the 12th day and retained that level up to the 20th day of incubation.

The concentration of CRN in the embryo gradually increased according to the stage of development. The yolk contained a steady CRN level (from 14 to 34 nmole/g) up to the 16th day of incubation. Thereafter the level of CRN in the yolk from the fertile eggs increased ($p<0.001$) to 75 nmole/g, and stayed at that level (Fig. 3).

The changes in HArg concentrations in the embryo and yolk are shown in Fig. 4. Low concentration of HArg in the embryo was seen

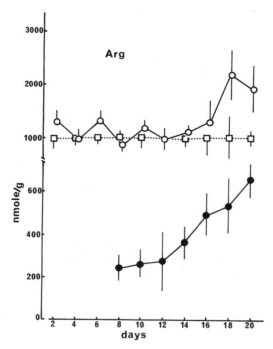

Fig. 1.   Changes of Arg in the embryo and yolk (●); embryo, (○);
          yolk (fertile egg), (□); yolk (unfertile egg).
          Each point and vertical bars represent the mean and the
          SD of five to eight experiments.

up to the 14th day with a slight increase thereafer.   The HArg
concentration in the yolk from the fertile eggs, however, began to
increase on the 12th day and continued to increase throughout the
next 8 days, attaining a level of 15 nmole/g on the 20th day.

     As shown in Fig. 5, G was detected on the 8th day.   However,
the level was very low and increased somewhat after embryonic
development.   The concentration of G in the yolk from the fertile
eggs slowly increased up to the 16th day of incubation and then
markedly increased (p<0.001) on the 18th day of incubation.

     MG and GPA were detected in the embryo and yolk on all the
days studied, however, their levels were always very low (below
0.4 nmole/g) as shown in Fig. 6.

Guanidino compounds in the embryonic liver, kidney and brain

     The concentrations of guanidino compounds in the liver,
kidney and brain were studied from the 14th to the 20th day of

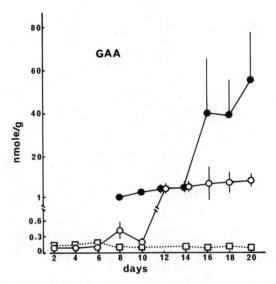

Fig. 2.   Changes of GAA in the embryo and yolk

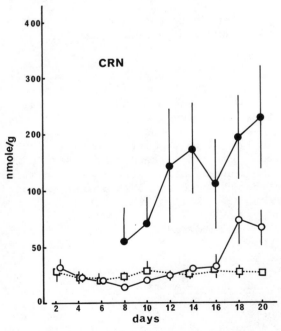

Fig. 3.   Changes of CRN in the embryo and yolk

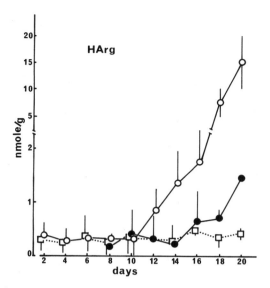

Fig. 4.   Changes of HArg in the embryo and yolk

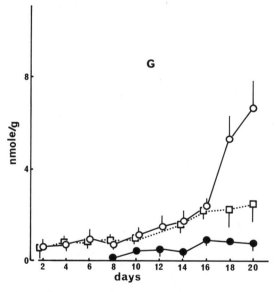

Fig. 5.   Changes of G in the embryo and yolk

incubation.   The Arg levels in the organs are shown in Table 1.
The Arg levels in the liver and brain did not show any change,
whereas Arg level in the kidney increased (p<0.01) on the 16th day
and retained the elevated level up to the 20th day.   As shown in

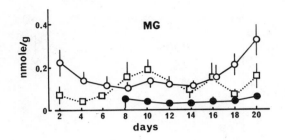

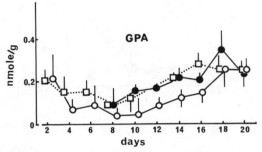

Fig. 6.   Changes of MG and GPA in the embryo and yolk

Table 1.   Arginine levels in liver, kidney and brain

Days of incubation	Liver	Kisney	Brain
14	373± 36 (8)	546± 75 (5)	198±21 (8)
16	415±105 (8)	635±152 (8)	181±28 (8)
18	429±103 (8)	775±156 (8)	180±14 (8)
20	381±144 (8)	717±603 (8)	186±31 (8)

The values in Table 1 to 7 are expressed as nmol/g wet weight, mean±SD with the number of experiments in parentheses.

Table 2, GAA levels in the liver gradually decreased from the 14th to the 20th day, whereas the level in the kidney was the highest on the 16th day of incubation and decreased after that ($p < 0.001$). The CRN levels in the brain and kidney did not change throughout all the days studied, whereas the level in the liver proved to be higher on the 18th day ($p < 0.05$) compared with other days (Table 3).

Table 2. Guanidinoacetic acid levels in lever, kidney and brain

Days of incubation	Liver	Kidney	Brain
14	26.1±6.8 (8)	53.3± 6.9 (5)	1.7±1.2 (8)
16	21.1±3.8 (8)	98.2±26.1 (8)[c]	2.0±0.7 (8)
18	14.0±5.3 (8)[a]	26.9± 8.6 (8)[a]	2.0±1.0 (8)
20	4.6±1.7 (8)[b]	23.4± 9.6 (8)	0.9±0.3 (8)[e]

a; p<0.05 vs. the 16th day, b; p<0.005 vs. the 18th day,
c; p<0.005 vs. the 14th day, d; p<0.001 vs. the 16th day,
e; p<0.05 vs. the 18th day.

Table 3. Creatinine levels in liver, kidney and brain.

Days of incubation	Liver	Kidney	Brain
14	53.9± 12.7 (8)	62.7±31.3 (5)	262± 86 (8)
16	85.2± 34.1 (8)	67.8±35.2 (8)	273±128 (8)
18	266.6±136.7 (8)[a]	94.9±57.7 (8)	438±214 (8)
20	96.4± 51.8 (8)[b]	71.8±48.0 (8)	272± 36 (8)

a; p<0.05 vs. the 16th day, b; p<0.05 vs. the 18th day.

As shown in Table 4, GPA was detected in all the organs studied, although their levels were low. The GPA level in the liver showed a decrease on the 20th day (p<0.005). Its level in the kidney also tended to decrease according to the stage of development, whereas the brain GPA level increased (p<0.05) on the 16th day and the increased level was retained up to the 20th day of incubation. HArg, G and MG were also detected in the liver, kidney and brain (Table 5 to 7), although their levels were very low. We did not observe any change in the HArg and G levels in the liver. The levels of HArg and G in the kidney and brain were somewhat high on the 20th day.

Table 4. β-Guanidinopropionic acid level in liver, kidney and brain

Days of incubation	Liver	Kidney	Brain
14	1.28±0.29 (8)	1.65±0.31 (5)	0.29±0.07 (8)
16	1.71±0.70 (8)	2.34±0.79 (8)	0.51±0.22 (8)[d]
18	1.48±0.53 (8)	1.54±0.29 (8)[b]	0.69±0.08 (8)
20	0.61±0.30 (8)[a]	0.96±0.23 (8)[c]	1.20±0.55 (8)

a; p<0.005 vs. the 18th day, b; p<0.05 vs. the 16th day,
c; p<0.005 vs. the 18th day, d; p<0.05 vs. the 14th day.

Table 5. Homoarginine levels in liver, kidney and brain

Days of incubation	Liver	Kidney	Brain
14	1.63±0.54 (8)	ND – 4.40 (5)	ND – 0.36 (8)
16	2.57±1.09 (8)	3.17±0.68 (8)	ND – 0.77 (8)
18	1.68±0.42 (8)	3.00±1.78 (8)	0.70±0.21 (8)
20	2.19±0.79 (8)	7.09±4.54 (8)	1.80±0.66 (8)[a]

a; p<0.005 vs. the 18th day.

DISCUSSION

We show that Arg, CRN, GAA, HArg, G, MG and GPA are detectable in the embryo, yolk and embryonic organs of chicks by high performance liquid chromatography. Most of the synthetic pathways of guanidino compounds are still obscure but nitrogen metabolism in the chick embryo must be controlled in a different manner from that of mammals since birds excrete waste nitrogen as uric acid.

In this study, we show that concentrations of guanidino compounds in the whole embryo increase during embryogenesis, especially GAA, Arg and CRN. This increase appears at the middle stage, from the 8th to 14th day, during which ossification occurs. The increase of GAA, the precursor of creatine, could reflect the need for creatine phosphate for storage of energy. Mellanby[4] reported an increased creatine in eggs during this period.

Table 6. Guanidine levels in liver, kidney and brain

Days of incubation	Liver	Kidney	Brain
14	2.64±0.62 (8)	ND – 4.25 (5)	ND – 1.05 (8)
16	2.16±0.31 (8)	ND – 4.00 (8)	ND – 0.85 (8)
18	2.25±0.36 (8)	2.57±1.68 (8)	ND – 0.48 (8)
20	1.86±0.74 (8)	3.17±0.90 (8)	0.75±0.18 (8)

Table 7. Methylguanidine levels in liver, kidney and brain

Days of incubation	Liver	Kidney	Brain
14	0.05±0.03 (8)	ND – 0.10 (5)	0.06±0.02 (8)
16	0.04±0.02 (8)	ND – 0.23 (8)	0.06±0.03 (8)
18	0.16±0.11 (8)[a]	ND – 0.11 (8)	0.08±0.03 (8)
20	0.11±0.02 (8)	0.17±0.05 (8)	0.08±0.01 (8)

a; p<0.05 vs. the 16th day.

We[5,6] have reported elsewhere that guanidinosuccinic acid (GSA) and γ-guanidinobutyric acid (GBA) were detectable in mouse liver and that GPA was found only in mouse liver, but not in kidney and brain. Moreover, GSA, GPA and GBA levels in liver were very low in fetal and suckling mice compared with adult mice. In the present study, we could not detect GSA and GBA in chick embryonic liver or in kidney and brain. On the other hand, GPA was detected not only in embryonic liver but also in embryonic kidney and brain. These results suggest that the metabolism of guanidino compounds is different in the chick embryo and mouse fetus.

The high level of Arg in the yolk was detected at all stages. This result confirms the findings of Arnold et al.[7] who reported that Arg is an essential amino acid for the chick during the rapid growth period after hatching.

The concentrations of guanidino compounds in the yolk from fertile eggs showed an increase during the later period of embryogenesis. The increase in HArg and G levels in the yolk were very large which suggest that HArg and G are synthesized during embryogenesis. However, it is difficult to remove the vitelline artery and vein from the yolk. So there is a possibility that this system might be responsible for the increased guanidino compounds seen at that stage.

We also detected guanidino compounds in the embryonic liver, kidney and brain, but these levels, especially GAA, CRN and GPA, fluctuated during late embryonic stages. Moreover the change in guanidino compounds in each organ did not necessarily reflect the change in the whole embryo. This suggests that metabolic rates might be too fast to determine the dynamic state of guanidino compound metabolism during this period.

CONCLUSION

1. Arg, CRN, GAA, HArg, G, MG and GPA were detectable in the embryo, yolk and embryonic organs by a high performance liquid chromatography. GSA and GBA, which are commonly detectable in the mammalian liver, were not observed in the chick embryonic liver.

2. The levels of guanidino compounds, especially GAA, Arg and CRN, in the whole embryo increased according to embryogenesis.

3. The levels of guanidino compounds, especially HArg and G, in the yolk from the fertile egg increased at the late stage of embryogenesis.

4. The levels of guanidino compounds, especially GAA, CRN and GPA, in the embryonic organs fluctuated at the late stage of embryogenesis. These changes in guanidino compound levels in each organ did not reflect the changes in guanidino compound levels in the whole embryo.

REFERENCES

1. A. Mori, Natural occurrence and analyses of guanidino compounds, Rinshokagaku, 9:232 (1980) (in Japanese).
2. Y. Robin, Metabolism of arginine in invertebrates: relation to urea cycle and to other guanidine derivatives, in:"Urea cycle diseases," A. Lowenthal, A. Mori, and B. Marescau, eds., Plenum Press, New York (1982).
3. A. Mori, Y. Watanabe and N. Fujimoto, Fluorometrical analysis of guanidino compounds in human cerebrospinal fluid, J. Neurochem., 38:448 (1982).
4. E. Mellanby, Creatin and creatinin, J. Physiol., 36:447 (1908).
5. S. Shindo, Y. Katayama, Y. Watánabe and A. Mori, Metabolism of guanidino compounds in mouse brain, Neurosciences, Kobe, 8:78 (1982) (in Japanese).

6. Y. Watanabe, S. Shindo and A. Mori, Developmental changes in
    guanidino compounds levels in mouse organs, (in this book).
7. A. Arnold, O. L. Kline, C. A. Elvehjem and E. B. Hart,
    Further studies on the growth factor required by chicks,
    the essential nature of arginine, J. Chem. Biol.,
    116:699 (1936).

METABOLISM OF L-[AMIDINO-^{15}N]-ARGININE TO GUANIDINO COMPOUNDS

Shoichiro Shindo and Akitane Mori

Inst. for Neurobiol., Okayama Univ. Med. Sch.
Okayama 700, Japan

INTRODUCTION

Since analytical methods for guanidino compounds have recently been developed, it has become possible to get much information about the metabolism of guanidino compounds, especially in uremia[1]. In the past, stable isotopes were not used for analyses of such compounds as much as radio isotopes were, because the sensitivity was not very high. Recently, however, a new method for nitrogen-15 (^{15}N) analysis was developed, optical emission spectroscopy[2], which is several hundred times more sensitive than mass-spectrometry which was used to analyze ^{15}N until now. To examine the metabolism of guanidino compounds in the mouse, the incorporation of a ^{15}N into some guanidino compounds was observed after the administration of L-[amidino-^{15}N]-arginine.

MATERIALS AND METHODS

Six-week-old male ddY mice weighing 30 g were used for all experiments. The mice were given mouse food (Oriental Yeast Co. Ltd., Tokyo) and water ad libitum. One hundred mg/kg of L-[amidino-^{15}N]-arginine (^{15}N-ARG) (95 atom %, Hikari Tsusho Co., Ltd., Tokyo) was administered intraperitoneally to mice which were sacrificed by decapitation 1 min, 10 min, 30 min, 60 min, 3 h and 6 h after the administration. Blood was heparinized in a test tube which could be used for separating the plasma. The kidney, brain and liver were removed quickly and frozen at -80 °C until analysis.

The plasma was adjusted to pH 2.2 with 1N HCl and deproteinized by ultrafiltration (Centriflo CF-25, Amicon Corp., Mass.).

71

The organ was deproteinized with 1 % picric acid, the excess of which was removed by applying to Dowex 2x8 (Cl⁻) column. The effluent was evaporated and the residue dissolved in dilute HCl (pH 2.2). The samples were frozen at -20 °C until analysis. A part of each sample was utilized for quantification of guanidino compounds and the remainder was utilized for determination of ^{15}N-enrichment of guanidino compounds. Guanidino compounds were quantified fluorometrically with a guanidino compound analyzer(G-520, JASCO Co., Ltd., Tokyo)[3].

The ^{15}N enrichment, expressed as ^{15}N-atom %, of guanidino compounds was analyzed by an optical emission spectroscopic method in which all organic nitrogens were changed into the gaseous form in a discharge tube[2]. Each sample was fractionated into 0.3-0.5 ml fraction with the guanidino compound analyzer, and the guanidino compound in each fraction was verified by chromatography. The ^{15}N-atom % of the following compounds was determined: urea (in the plasma, brain, kidney and liver), guanidinoacetic acid (GAA; in the plasma, brain, kidney and liver), creatine (CR; in the plasma, brain, and liver), creatinine (CRN; in the plasma, brain, kidney and liver), arginine (ARG; in the plasma, brain, kidney and liver), guanidinosuccinic acid (GSA; in the liver), guanidino-butyric acid (GBA; in the liver), guanidinopropionic acid (GPA; in the liver) and methylguanidine (MG; in the liver).

Nitrogen of the urea and MG fractions was obtained as $(NH_4)_2SO_4$ by the Kjeldahl method, and concentrated to 0.5 µg/µl in 0.1 N HCl for 75 h at 40 °C. This solution was dried in a 2 x 10 mm glass capillary (PYREX) which was then put into a 4 x 200 mm glass tube (PYREX), along with a mixture of CuO, Cu_2O and CaO. The CaO was dried before use at 900 °C for 3 h and packed with powder of CuO and Cu_2O at 600 kg/cm². Finally, a mixture of Xe and He gas (2:98) was added to about 10 Torr, and the tube was sealed. This process was performed with a discharge tube preparing apparatus (Ouchi Rika Kogyo Co., Ltd., Tokyo). The glass tubes were heated at 560 °C for 30 min to change NH_4Cl to gaseous nitrogen. Other guanidino compounds were desalted first by Amberlite CG-120 (H⁺) and evaporated exclude excess NH_4OH, and dissolved in water. These samples was poured into 4 x 150 mm glass tubes (PYREX) which were sealed at one end and lyophilized. Then, powdery CuO, Cu_2O and solidified CaO were added. The rest of the procedure was performed in the same manner as with urea and MG except that the time required to change the nitrogen to gas was 3 h rather than 30 min. The contents of these discharge tubes were analyzed with a ^{15}N-analyzer (N-150, JASCO Co., Ltd., Tokyo).

Four to six samples of each guanidino compound were analyzed for ^{15}N-enrichment. The results were expressed as atom % excess, namely, the measured value less the natural abundance of [15]N (0.37 atom %). An incorporation of more than 0.1 atom % excess was defined as significant, and the confidence values of ^{15}N atom % of the same samples were under 5 %. Statistical analysis was performed with Student's t-test.

RESULTS

The free ARG level in the plasma increased rapidly to about twice the control level within 1 min after, and decreased gradually thereafter to the normal level within 3 h after the administration of ^{15}N-ARG (Fig 1-A). The ARG level in the brain increased slightly, but significantly, 1 h after, and decreased to the normal level 3 h after the administration (Fig 1-B). The ARG level in the kidney changed in almost the same way as in the plasma, i.e., it reached nearly twice the control level and then decreased quickly to the normal level 1 h after the administration (Fig 1-C). Although, the free ARG level in the liver was at most 1/10 of that in the plasma, brain and kidney, it increased slightly 1 min after the administration, but soon returned to the control level (Fig 1-D). Other free guanidino compounds did not increase quantitatively in the organs or plasma compared to the control groups.

The ^{15}N atom % of GAA, CR, urea and CRN was determined in the plasma after administering ^{15}N-ARG (Fig-2). The ^{15}N atom % excess of ARG reached a maximum 1 to 10 min after the administration, rapidly decreased up to 60 min, and then gradually decreased. The time course of the ^{15}N atom % excess of ARG showed two phases: one having a half life of about 18 min and the other, a half life of 85 min. This result shows that ^{15}N-ARG was diluted about two fold by non-labeled ARG. ^{15}N was incorporated markedly into GAA and CR 10 min after the administration of ^{15}N-ARG. The ^{15}N atom % excess of GAA decreased after that, but that of CR reached a maximum at 30 min and decreased after that. ^{15}N was incorporated significantly into urea 1 min after and reached a maximum 10 min after the administration, and then decreased gradually. No ^{15}N was incorporated into CRN.

In the brain, ^{15}N-ARG atom % excess reached a maximum rapidly (10 min) after the ^{15}N-ARG administration, stayed at the level up to 30 min, and then decreased (Fig 3). ^{15}N was incorporated significantly into GAA 10 min after the administration, and was incorporated significantly into urea 1 min after. The ^{15}N atom % excess of urea and GAA reached a maximal value 60 min after and decreased gradually. ^{15}N was not incorporated into CRN in the brain. The incorporation of ^{15}N into CR was undetectable because of a high back groud level.

In the kidney, the ^{15}N atom % excess of ARG reached a maximum promptly (1 min) after the administration of ^{15}N-ARG and then decreased rapidly (Fig 4). ^{15}N-ARG was diluted about two times by

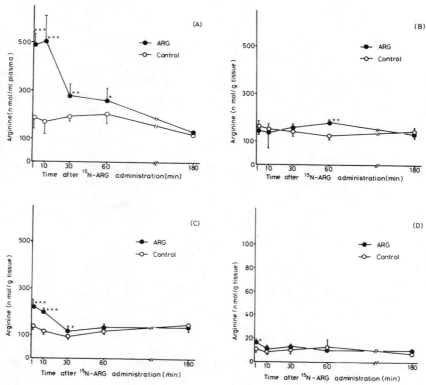

Fig. 1-A. The change of arginine level in plasma after ^{15}N-ARG
administration (Mean±S.D., N=5). The closed circles
refer to the group administered ^{15}N-ARG, and open
circles to the control group.
*:P<0.05, **:P<0.01, ***:P<0.001

B. The change of arginine level in brain after ^{15}N-ARG
administration (Mean±S.D., N=5). **;P<0.01.

C. The change of arginine level in kidney after ^{15}N-ARG
administration (Mean±S.D., N=5). *;P<0.05, ***;P<0.001.

D. The change of arginine level in liver after ^{15}N-ARG
administration (Mean±S.D., N=5). *;P<0.05

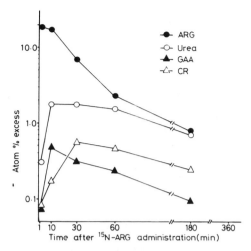

Fig. 2.   The change of ^{15}N enrichment in arginine (ARG), creatine
           (CR), urea and guanidinoacetic acid (GAA) in plasma after
           ^{15}N-ARG administration (Mean, N=4-6).

non-labeled ARG, a change which was two phased as in the plasma.
The half life of the first phase was about 10 min, and that of the
second phase was about 70 min. ^{15}N was incorporated in large
amounts into GAA in the kidney compared with other organs, and the
^{15}N atom % excess of GAA reached a maximum 10 min after administration, and then decreased quickly.   This change consisted of two
phases, just as ARG, with respective half lifes of about 20 min
and 50 min. ^{15}N was incorporated significantly into urea 1 min
after administration and reached a maximum at 30 min and then
decreased slowly.

        In the liver, the atom % excess of ARG reached a maximum 1
min after the administration of ^{15}N-ARG, and then decreased (Fig 5).
This time course was very rapid compared to the plasma and kidney,
and consisted of two phases, the half life of the first phase being
about 3 min and that of the second phase 60 min. ^{15}N was incorporated significantly into GAA and CR 1 min after administration.
The ^{15}N atom % excess of GAA reached a maximum from 10 min to 30 min
after administration and then decreased gradually.   However, the
^{15}N atom % excess of CR reached a maximum higher than that of GAA
60 min after administration, and then decreased gradually (Fig 5-A).
The ^{15}N atom % excess of urea in the liver, which was different
from that in the plasma, brain and kidney, reached a maximum 1 min
after administration and then decreased gradually. ^{15}N was incorporated into MG significantly 1 min after administration. ^{15}N was
also incorporated significantly into CRN 30 min after, and the ^{15}N
atom % excess reached a maximum between 30 and 60 min after admin-

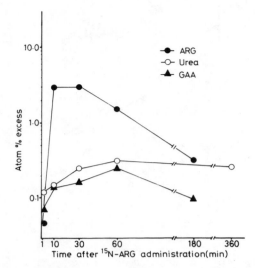

Fig. 3.   The change of ^{15}N enrichment in arginine (ARG), urea and
         guanidinoacetic acid (GAA) in brain after ^{15}N-ARG admini-
         stration (Mean, N=4-6).

istration, and then decreased gradually. ^{15}N was incorporated
significantly also into GSA, and the ^{15}N atom % excess reached a
maximum 60 min after administration, and then decreased gradually
(Fig 5-B).   No ^{15}N was incorporated into GPA and GBA.

DISCUSSION

   ^{15}N atom % has, until now, been analyzed mainly by mass-
spectrometry.   In the present study it was determined by a new
method, optical emission spectroscopy[2], in which the intensity of
the emission spectra of gaseous nitrogen is measured.   As little
as 0.2 µg nitrogen can be estimated by this method.

   ^{15}N-ARG administered intraperitoneally into mice was absorbed
rapidly and distributed via the circulation to the organs examined.   The net amount of free guanidino compounds did not increase,
except for ARG, indicating that ARG could be metabolized actively
to metabolites other than guanidino compounds.   It has been reported that guanidino compound levels, except for ARG, are changed
by continuous administration of ARG to mice for 3 to 6 days[4].
This observation suggests that the turnover rate of guanidino
compounds may be too fast to detect an increase in their concentration after one dose of ARG.

   In our experiment, the half life of ^{15}N-ARG in the plasma was

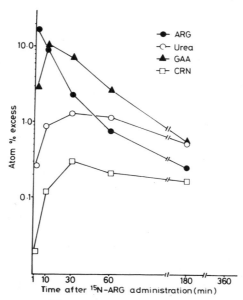

Fig. 4.   The change of ^{15}N enrichment in arginine (ARG), urea,
          guanidinoacetic acid (GAA) and creatinine (CRN) in kidney
          after ^{15}N-ARG administration (Mean, N=4-6).

about 18 min, though other investigators have reported that it is
6 to 8 min[5]. Since as the dose of ^{15}N-ARG we used was much higher
than in other experiments, both non-metabolized and recirculated
^{15}N-ARG were determined together.

The time course of the maximal incorporation of ^{15}N into ARG
was slower in the brain compared to kidney or liver, which suggests that ARG incorporation into the brain may be controlled by
the blood-brain barrier. ARG exists in relatively high amounts in
the brain and some ARG may pass through the blood-brain barrier
via an active transport system.

Recently, it was reported that ARG activates guanylate cycl-
ase in vitro[6], and therefore ARG could play a physiologically important role in brain. Nevertheless the metabolism of guanidino
compounds in brain is still obscure. In our experiment, whether
the ^{15}N in GAA and urea was derived from ^{15}N-ARG in the brain is
not clear, since part of ^{15}N may be a reflection of the change in
the plasma.

In the kidney, the ARG level and the ^{15}N atom % excess of ARG
dramatically increased and then decreased rapidly when ^{15}N was incorporated into GAA. This result suggests that exogeneous ARG was

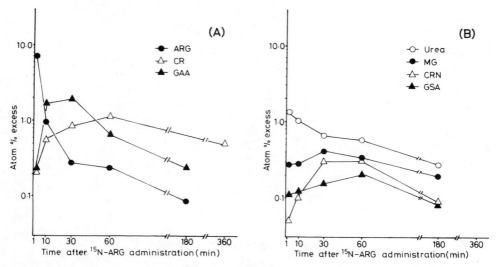

Fig. 5-A. The change of ^{15}N enrichment in arginine (ARG), guanidi-
          noacetic acid (GAA) and creatine (CR) in liver after
          ^{15}N-ARG administration (Mean, N=4-6).

      B. The change of ^{15}N enrichment in urea, methylguanidine
         (MG), creatinine (CRN) and guanidinosuccinic acid (GSA)
         in liver after ^{15}N-ARG administration (Mean, N=4-6).

metabolized quickly to GAA. The turnover rate of ARG could be very
high in the kidney, because the half life of ^{15}N-ARG was shorter
than in the plasma. According to a recent report, the kidney is
the organ responsible for synthesizing ARG from citrulline[7].
Together this data indicates that the kidney plays an important
role in ARG metabolism.

      The half life of ^{15}N-ARG in the liver was very short, and the
^{15}N atom % excess of urea in the liver reached its maximal value
1 min after administration. These findings suggest that exogenous
ARG may be rapidly metabolized to urea by arginase (EC 3.5.3.1.)
in the liver. The free ARG level increased significantly, but
very slightly, only 1 min after administration. These facts indicate that the ARG pool is small and the turnover rate is very high.

      Bloch and his colleagues, using ^{15}N-ARG, found that GAA was a
precursor of CR, and that the amidine group in CR was derived from
ARG and the sarcosine group was derived from glycine[8]. Also,
Walker and others, reporting on the properties of ARG-glycine
amidinotransferase (EC 2.1.4.1.), a GAA synthesizing enzyme[9,10],
noted that this enzyme was repressed by CR[10]. Others have reported

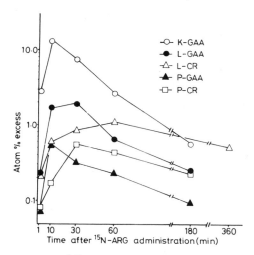

Fig. 6.   The change of ^{15}N enrichment in guanidinoacetic acid
          (GAA) and creatine (CR) after ^{15}N-ARG administration
          (Mean, N=4-6). K-GAA; kidney-GAA, L-GAA; liver-GAA,
          P-GAA; plasma-GAA, L-CR; liver-CR, P-CR; plasma CR.

that ARG-glycine amidinotransferase exists mainly in the kidney,
and GAA-methyltransferase (EC 2.1.1.2.), which is a CR synthesizing
enzyme, exists mainly in the liver, indicating an interesting liver-
kidney relationships[11,12]. Fig. 6 summarizes the time courses of
the ^{15}N atom % excess of GAA and CR in the plasma, liver and kidney
after the ^{15}N-ARG administration. ^{15}N was rapidly incorporated into
kidney GAA, liver GAA and liver CR at an early stage and continued
to be incorporated into liver GAA and CR while incorporation into
plasma GAA and kidney GAA rapidly decreased. The ^{15}N atom % excess
of plasma CR also increased late. These findings suggest that GAA
is synthesized mainly and very actively in the mouse kidney. The
GAA synthesized is transported to other organs via the circulation,
and CR, synthesized mainly in liver, is also transported to other
organs such as muscle.

Since GSA was first discovered by Bonas et al. in 1963 in the
urine of uremic patients[13], many authors have studied GSA metabolism in renal disorders[14]. Recently, GSA was found to be a normal
constituent of mouse liver, brain and urine, though the amount is
small[4]. However, the detailed metabolic pathway of GSA is still
unknown. Our experimental results suggest that ARG may be related
to GSA synthesis in liver, and synthesis of GSA from ARG is not
great in normal liver but increases dramatically in uremia.

MG is also considered to be a uremic toxin, and many papers
are concerned with MG metabolism in uremia[15]. In our experiment,

[15]N was incorporated into MG in the liver soon after ^{15}N-ARG administration, and faster than into CRN. These facts suggest that MG could be synthesized directly from ARG. MG has been thought to be synthesized from CRN, but the enzyme which catalyzes this reaction has not yet been determined[16].

GBA has been found in insects and mammalian tissues[17], and later in the rabbit brain during convulsive seizure induced by pentylenetetrazole[18]. GPA is reported to be present also in the rat liver[19]. In invertebrates, GBA and GPA are derived from ARG by an oxidation reaction[20], but, in vertebrates, they are synthesized by a transamidination reaction from ARG and γ-aminobutyric acid and β-alanine[21], respectively. Perez et al. reported that the guanidino-^{14}C of ARG was incorporated into GBA only in small amounts, but not at all into GPA[22]. In our experiment, ^{15}N was not incorporated into either GBA or GPA.

Recently, ARG was reported to be one of the essential amino acids for cats and young rats[23], and to play a physiologically important role in the urea cycle and guanidino compound metabolism in all animals. The metabolism, however, is still obscure.

REFERENCES

1. A. Ando, K. Kikuchi, H. Mikami, M. Fujii, K. Yoshihara,
    Y. Orita and H. Abe, Quantitative determination of gua-
    nidino compounds: The excellent preparation of biological
    samples, In:" Urea Cycle Disease," A. Lowenthal, A. Mori
    and B. Marescau, eds., Plenum Press, New York, (1983).
2. H. Kano, T. Yoneyama and K. Kumazawa, Emission specrto-
    metric ^{15}N analysis of the amino acids in plant tissues
    separated by thin layer chromatography, Analyt. Biochem.,
    67:327-331 (1975).
3. A. Mori, Y. Katayama, S. Higashidate and S. Kimura,
    Fluorometrical analysis of guanidino compounds in mouse
    brain, J. Neurochem., 32:643-644 (1979).
4. Y. Katayama, S. Shindo, A. Sawaki, Y. Watanabe, C. Hiramatsu
    and A. Mori, Biosynthesis of taurocyamine by trans-
    amidination reaction of taurine, Sulfur Amino Acids,
    2:297-304 (1979).
5. Q. R. Rogers, R. A. Freedland and R. A. Symons, In vivo
    synthesis and utilization of arginine in the rat,
    Am. J. Physiol., 233:236-240 (1972).
6. T. Deguchi and M. Yoshioka, L-Arginine identified as an
    endogenous activator for soluble guanylate cyclase from
    neuroblastoma cells, J. Biol. Chem., 257:10147-10151 (1982).
7. M. Funahashi, H. Kato, S. Shiosaka and H. Nakagawa, Formation
    of arginine and guanidinoacetic acid in the kidney in vivo.
    Their relations with the liver and their regulation,
    J. Biochem., 89:1347-1356 (1981).

8. K. Bloch and R. Shoenheimer, The biological precursors of creatine, J. Biol. Chem. 138:167-194 (1941).

9. J. B. Walker, Studies on the mechanism of action of kidney transamidinase, J. Biol. Chem. 2244:57-66 (1957).

10. J. B. Walker, Metabolic control of creatine biosynthesis. II. Restration of transamidinase activity following creatine repression, J. Biol. Chem., 236:493-498 (1961).

11. H. Borsook and J. W. Dubnoff, The formation of glycocyamine in animal tissues, J. Biol. Chem., 138:389-403 (1941).

12. G. L. Cantoni and P. J. Vignos Jr., Enzymatic mechanism of creatine synthesis, J. Biol. Chem., 209:647-659 (1954).

13. J. E. Bonas, B. D. Cohen and S. Natelson, Sepatation and estimation of certain guanidino compounds. Application to human urine, Microchem. J., 7:63-77 (1963).

14. B. D. Cohen and H. Patel, Guanidinosuccinic acid and the alternate urea cycle, In:"Urea Cycle Diseases," A. Lowenthal, A. Mori and B. Marescau, eds., Plenum Press, New York, pp. 435-441, (1983).

15. M. Mikami, Y. Orita, A. Ando, H. Fujii, T. Kikuchi, K. Yoshihara, A. Okada and H. Abe, Metabolic pathway of guanidino compounds in chronic renal failure, In:"Urea Cycle Diseases," A. Lowenthal, A. Mori and B. Marescau, eds., Plenum Press, New York, pp. 449-457, (1983).

16. G. Perez and R. Faluotico, Creatinine: A precursor of methyl-guanidine, Experientia, 29:1473-1474 (1973).

17. F. Irreverre, R. L. Evans, A. R. Hayden and R. Silber, Occurrence of gamma-guanidinobutyric acid, Nature, 180: 704-705 (1957).

18. D. Jinnai, A. Sawai and A. Mori, γ-Guanidinobutyric acid as a convulsive substance, Nature, 212:617 (1966).

19. H. Rosenberg, The occurrence of guanidinoacetic acid and other substituted guanidines in mammalian liver, Biochem. J., 72:582-585 (1959).

20. H. Hasegawa, Enzymological studies on arginine metabolism in Cambaloides Japonicus, Bull. Osaka Med. Sch., 9:135-150 (1963).

21. M. Shimoyama, Studies on arginine and other guanido-derivatives. XVI The influence of the addition of several kinds of acceptors on enzymatic transamidination from glyco-cyamine by hog kidney preparation, Bull. Osaka. Med. Sch., 7:105-110 (1961).

22. G. O. Perez, B. Rieteberg, B. Owens and E. R. Shiff, Effect of acute uremia on arginine metabolism and urea and guanidino acid production by perfused liver, Pflugers Arch., 372:275-278 (1978).

23. J. G. Morris and Q. R. Rogers, Arginine: An essential amino acid for the cat, J. Nutr., 108:1944-1953 (1978).

BIOSYNTHESIS OF GUANIDINOACETIC ACID IN ISOLATED RAT HEPATOCYTES

Kazumasa Aoyagi, Shoji Ohba, Mitsuhiro Miyazaki,
Sohji Nagase, Satomi Iida, Mitsuharu Narita,
and Shizuo Tojo

Department of Internal Medicine, Institute of Clinical
Medicine, University of Tsukuba
Sakura-mura, Ibaraki-ken 305, Japan

INTRODUCTION

Guanidinoacetic acid (GAA), a precursor of creatine, is formed from arginine and glycine by transamidination[1]. GAA formation in rat is not detected in liver homogenates[2] or in isolated perfused livers[3]. Therefore, creatine synthesis in rat liver is regulated by the activity of transamidinase in the kidney[4]. However, recent improvements in GAA analysis reveal that the blood level is 1/4 of the value[6] reported previously[3]. Moreover, Natelson et al. reported that isolated rat hepatocytes had GAA synthetic activity[5]. The details of which are not clear. Therefore, GAA synthesis in isolated rat hepatocytes was investigated quantitatively to evaluate the role of its synthesis in liver. In addition, some regulatory mechanisms in this synthesis were also investigated.

MATERIALS AND METHODS

Preparation of cells

Cells were prepared essentially according to the method of Berry and Friend[7] modified by Zahlten[8] except for the addition of 2.5 mM $CaCl_2$ and 3 % bovine serum albumin to replace 1.5 % gelatin in the medium for cell washing. Male Wistar rats (300–350 g) were fed on laboratory chow containing 25 % protein. Cell preparation was started at 10:00 A.M. by the injection of sodium pentobarbital (60 mg/kg) in the non-fasting animal. The viability of the cells, judged from the trypan blue exclusion test, was about 90 %. We calculated that 9.8×10^7 cells corresponded to 1 g wet liver as reported by Zahlten et al.[8].

Incubation of cells

Cells were incubated in 6 ml of Krebs-Henseleit bicarbonate buffer containing 3 % bovine serum albumin, 10 mM sodium lactate and indicated substances with shaking at 60 cycle/min in a 30 ml conical flask with a rubber cap under an atomosphere of mixed gases[5] (95 % O_2 + 5 % CO_2) at 37 °C. The equilibration of the buffer was repeated every hour.

The amount of cells used for each experiment is indicated in the results section. The incubation was stopped by the addition of 0.6 ml of 100 % (w/v) trichloroacetic acid. After sonication, the supernatant of cells and medium was obtained by centrifugation at 1,700 x g for 15 min at 4 °C. The pH of the extract was adjusted to 2.2 by 2 N NaOH, and 0.3 ml of the neutralized extract was used for GAA determination.

Assay of GAA

GAA was determined by high performance liquid chromatographic analysis using 9,10-phenanthrenequinone for the post-labelling method essentially according to the method of Yamamoto et al.[9]. We used 2 systems (A and B system) for GAA analysis. GAA was separated on a stainless steel column (4.0 x 125 mm) packed with Hitachi 2619 cation-exchange resin with a mean particle size of 6 μm (Hitachi Co., Tokyo, Japan). The buffer for column equilibration and elution was Na-citrate buffer (pH 3.0, 0.2 mole/liter based on Na concentration) and the pH was adjusted by HCl. GAA was eluted by buffer at a flow rate of 0.4 ml/min for 60 min. The column was washed with 0.2 N NaOH at a flow rate 0.8 ml/min for 20 min and then, equilibrated by the buffer (pH 3.0) at 0.8 ml/min for 40 min and at 0.4 ml/min for 20 min (System A). GAA was separated on a 4.6 x 125 mm column packed with TSK gel 1EX 210 SC (12 μm) (Toyo Soda Co., Ltd) and eluted with 0.2 N Na-citrate buffer pH 3.3 at 1.0 ml/min for 20 min and the column washed with 1 N NaOH at 1.0 ml/min for 10 min. The column was equilibrated with 0.2 N Na-citrate buffer pH 3.0 at 1.0 ml/min for 20 min (System B). Fluorescence was detected at 495 nm for emission and 365 nm for excitation. Phenanthrenequinone (special grade, Wako chemicals Co., Ltd, Osaka, Japan) was recrystallized from ethanol. These analyses were done automatically using a Tri-Rotor 111 pump (Japan Spectrascopic Co., Tokyo, Japan) and an autosampler.

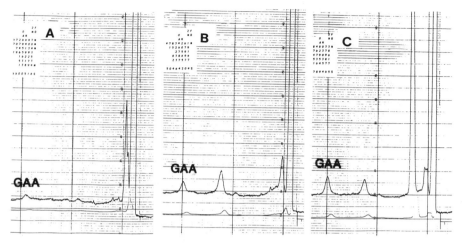

Fig. 1.   Chromatograms obtained from TCA extracts of cells and
          media incubated with (A) none, (B) 0.014 mM canaline
          and (C) 0.014 mM canaline and 10 mM glycine.

RESULTS

GAA systhesis from various amidine donors

     Isolated rat hepatocytes (0.088 g wet liver) were incubated
as described for 4 hours with various amidine donors with or
without 10 mM glycine[2].   Chromatograms obtained by system A from
the specimens of trichroloacetic acid extract of cells and media
incubated with 0.12 mM canaline with or without 10 mM glycine
(Fig. 1), 1 mM canavanine with or without glycine (Fig. 2) and 1
mM arginine with or without 10 mM glycine (Fig. 3).   The amount of
GAA present in the TCA extract are shown in Fig. 4.   GAA analysis
of the other experiments were done by system B.

Time course of the amount of GAA in the TCA extracts

     Cells (0.14 g wet liver) were incubated with 10 mM glycine
plus 1 mM canavanine or 5 mM arginine for 1, 2 and 4 hours.   The
amount of GAA in the TCA extracts increased up to 2 hours, and no
significant increase of GAA was observed by the incubation from 2
to 4 hours (Fig. 5).

Effect of ornithine and D,L-norvaline on the amount of GAA

     GAA formation in kidney is catalyzed by transamidinase[2].   It
is known that transamidinase is inhibited by ornithine and L-norvaline which is an ornithine analogue[2].   Cells (0.17 g wet liver)
were incubated with 10 mM glycine and 1 mM canavanine or 5 mM

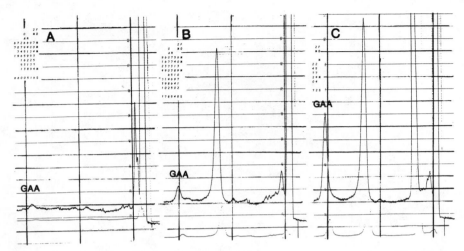

Fig. 2.   Chromatograms obtained from TCA extracts of cells and
          media incubated with (A) none, (B) 1 mM canavanine and
          (C) 10 mM glycine and 1 mM canavanine.

arginine in the presence of 1 mM and 5 mM ornithine and 5 mM
D,L-norvaline (Fig. 6). Ornithine decreased the amount of GAA
incubated with arginine and canavanine in the presence of glycine.
However, D,L-norvaline did not decrease the amount of GAA in the
presence of arginine and glycine. The inhibitory effect of
ornithine on transamidinase is considerably stronger than that of
L-norvaline[2]. The difference in the inhibitory effect on the two
agents on transamidinase may explain the difference in these
results.

Effect of methionine on the amount of GAA

      Creatine is formed from GAA and S-adenosylmethionine. The
amount of amino acids in isolated rat hepatocytes is small. So,
we tested the effect of methionine on the yield of GAA. The amount
of GAA was decreased by the addition of 1 mM methionine (Table 1).
These results suggest that the concentration of methionine in
isolated hepatocytes is insufficient and requires supplementation
under these conditions.

Effect of creatine on the amount of GAA

      It has been reported that creatine does not inhibit GAA syn-
thesis directly[3,10]. However, the high concentration (more than
20 mg/dl) that is found in rat liver following feeding a 2 %

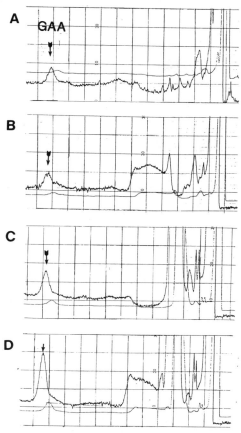

Fig. 3.   Chromatograms obtained from TCA extracts of cells and
          media incubated with none (A), 1 mM arginine (B), 10 mM
          glycine (C) and 1 mM arginine and 10 mM glycine (D).

creatine diet[3] is not explained. The time course for the production of GAA was investigated in the cells (0.07 g wet liver) incubated with 5 mM arginine and 10 mM glycine with or without 10
mg/dl creatine. Creatine did not influence the amount of GAA up
to 1 h (Fig. 7) and no remarkable change was observed at 2 and 4
hour incubations (Table 1). However, higher concentrations of
creatine (10, 20 and 50 mg/dl) increased the amount of GAA in the
cells (0.16 g wet liver) and media incubated for 2 hours in the
presence of 5 mM arginine, 10 mM glycine and 1 mM methionine
(Table 2). These results suggest that creatine synthesis is
directly inhibited by high concentrations of creatine or that
creatine stimulates GAA synthesis.

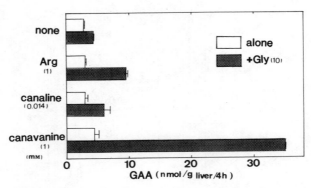

Fig. 4.   The amount of GAA present in the TCA extracts.

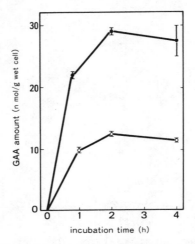

Fig. 5.   Time course of the amount of GAA in the TCA extracts.
Cells (0.14 g wet liver) were incubated with 10 mM
glycine and 1 mM canaline (•—•) and 10 mM glycine and
5 mM arginine (o—o). Each point represents the mean
of duplicate incubations. Bars indicate the range of
each determination.

DISCUSSION

GAA synthesis has not been detected in rat liver homogenates[2]
or in isolated perfused rat livers[5]. However we are able to show
GAA synthesis in isolated hepatocytes. The maximum rate of GAA
increase in cells and medium is about 10–15 nmol/g wet liver/h in
the presence of 10 mM glycine and 1 mM canavanine. The maximum
rate of GAA formation in rat liver is only 5 % of the amount needed
(5 mg) to maintain normal creatine metabolism. However, rat liver
has a considerable capacity to form creatine from GAA and methionine[3]. So, the regulation of GAA synthesis is probably a function

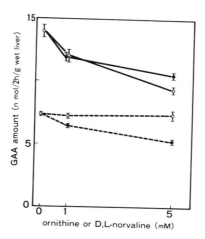

Fig. 6.   Effect of ornithine and D,L-norvaline on GAA levels in
          cells and media. Cells (0.17 g wet liver) were incubated
          for 2 hours with 10 mM glycine and 1 mM canavanine (——)
          or 10 mM glycine and 5 mM arginine (---) in the presence
          of ornithine (•) and 5 mM D,L-norvaline (○).
          Each point represents the mean of duplicate incubations.
          Bars indicate the range of each determination.

Table 1.   Effect of creatine and methionine on GAA levels in cells
           and media.

| substrate | GAA (nmol/g wet liver) | | |
| | Experiment 1 | Experiment 2 | |
	2 h	2 h	4 h
none	n.d.	–	–
Gly	37.5 ± 0.25	–	–
Arg	n.d.	–	–
Gly, Arg	10.0 ± 0	–	20.8 ± 0.5
Gly, Arg, creatine	8.3 ± 0.2	15.7 ± 0	21.1 ± 0.5
Gly, Arg, Met	7.4 ± 0.3	11.7 ± 0.3	11.1 ± 1.7
canavanine	4.3 ± 0.8	–	–
Gly, canavanine	31.6 ± 0.8	–	35.3 ± 0.3
Gly, canavanine, creatine	31.6 ± 1.2	23.1 ± 1.5	33.3 ± 0.9
Gly, canavanine, Met	17.0 ± 0.8	15.5 ± 0.1	17.2 ± 0.02

Each value is expressed as the mean ± range of a duplicate
incubations.

Table 2.   Effect of creatine on GAA levels in cells and media.

creatine (mg/dl)	Amount of GAA (nmole/g wet liver/2h)	P
0	3.09 ± 0.50	
10	3.97 ± 0.94	<0.10
20	4.21 ± 0.84	<0.025
50	5.51 ± 0.35	<0.001

Mean ± S.D. (n=5).   P; vs. the control value.

of creatine synthesis in isolated rat hepatocytes.

However, the amount of GAA increased up to at least 60 minutes. This increase in GAA suggests that creatine is not formed under these conditions. The addition of methionine decreased the amount of GAA at 2 and 4 hours by about 50 %. This suggests that methionine deficiency is the main cause of the decreased creatine synthesis.

Feedback inhibition of creatine synthesis by the high concentration of creatine is also possible. However, such inhibition occurs only at high concentrations of creatine as observed in rats fed a 2 % creatine diet.

The inhibitory effect of ornithine and D,L-norvaline suggests that transamidinase also reacts with amidine donors and glycine to form GAA.

The relationship between the concentration of GAA and the rate of creatine synthesis has not been priviously elucidated. It seems that the in vitro system described here can be helpful in standing this phenomenon.

SUMMARY

GAA synthesis was observed in isolated rat hepatocytes in the presence of 10 mM glycine and several amidine donors such as arginine, canaline and canavanine. The amount of GAA increased up to 2 hours in the presence of 10 mM glycine and 1 mM canavanine or 5 mM arginine. Ornithine and D,L-norvaline decreased the amount of GAA in the cells and media incubated with 10 mM glycine and 1 mM canavanine. Methionine (1 mM) decreased by 50 % the amount of GAA in the cells and media incubated with 10 mM glycine and 1 mM

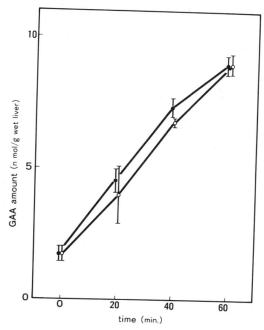

Fig. 7.   Effect of creatine on GAA level. Cells (0.07 g wet liver)
were incubated for the time indicated with 10 mM glycine
and 5 mM arginine (o—o) supplemented with 10 mg/dl
creatine (•—•). Each point represents the mean of
duplicate incubations. Bars indicated the range of each
determination.

canaline or 5 mM arginine. Creatine at high concentrations (more
than 10 mg/dl) increased the amount of GAA in the cells and media
incubated with 10 mM glycine, 5 mM arginine and 1 mM methionine.

These results indicate that GAA is synthesized in isolated
rat hepatocytes and also suggests that GAA is formed by transamidination and that high concentrations of creatine inhibit
creatine synthesis.

ACKNOWLEDGEMENTS

This study was supported in part by a Research Grant from the
Intractable Disease Division, Public Health Bureau, Ministry of
Health and Welfare, Japan and University of Tsukuba Project
Research.

REFERENCES

1. H. Borsook and J. W. Dubnoff, The formation of glycocyamine
   in animal tissues, J. Biol. Chem., 138: 389 (1941).
2. J. B. Walker, Formamidine group transfer in extracts of human
   pancreas, liver and kidney, Biochim. Biophys. Acta,
   73: 241 (1963).
3. G. B. Gerber, G. Garber, T. R. Koszalka and L. L. Miller, The
   rate of creatine synthesis in the isolated, perfused rat
   liver, J. Biol. Chem., 237: 2246 (1962).
4. J. B. Walker, Matabolic control of creatine biosynthesis,
   J. Biol. Chem., 236: 493 (1961).
5. S. Natelson, H-Y Tseng and J. E. Sherwin, On the biosynthesis
   of guanidinosuccinate, Clin. Chem., 24: 2108 (1978).
6. H. Mikami, Y. Orita, A. Ando, M. Fuji, T, Kikuchi,
   K. Yoshihara, A. Okada and H. Abe, Metabolic pathway of
   guanidino compounds in chronic renal failure in:"Urea cycle
   disease," A. Lowenthal, A. Mori and B. Marescau, eds.,
   Plenum, New York (1983).
7. M. N. Berry and D. S. Friend, High-yield preparation of isolated liver cells, J. Cell Biol., 43: 506 (1969).
8. R. N. Zahlten, W. S. Frederick and H. A. Lady, Regulation of
   glucose synthesis in hormonesensitive isolated hepatocytes,
   Proc. Nat. Acad. Sci. U.S.A., 70: 3213 (1973).
9. Y. Yamamoto, T. Manji, A. Saito, K. Maeda and K. Ohta, Ion-
   exchange chromatographic separation and fluorometric
   determination of guanidino compounds in physiological fluids,
   J. Chromatogr., 162:327 (1979).
10. J. B. Walker, Metabolic control of creatine biosynthesis,
    J. Biol. Chem., 235: 2357 (1960).

BIOSYNTHESIS OF GUANIDINOSUCCINIC ACID IN ISOLATED RAT HEPATOCYTES: EVALUATION OF GUANIDINE CYCLE AND ACIDOSIS

Kazumasa Aoyagi, Shoji Ohba, Mitsuhiro Miyazaki, Sohji Nagase, Satomi Iida, Mitsuharu Narita and Shizuo Tojo

Department of Internal Medicine, Institute of Clinical Medicine, University of Tsukuba Sakura-mura, Ibaraki-ken 305, Japan

INTRODUCTION

Guanidinosuccinic acid (GSA), a guanidine derivative, is implicated as a uremic toxin[1]. To clarify the synthetic pathway of GSA and the mechanism of its increased synthesis in renal failure, we investigated GSA synthesis in isolated rat hepatocytes, in vitro and obtained the following results[2]. 1) GSA synthesis increased as urea concentration rose. 2) Ornithine and arginine[3] which stimulated urea synthesis inhibited GSA synthesis. 3) D,L-Norvaline which is an inhibitor[4] of urea cycle enzymes: arginase, argininosuccinate synthetase and argininosuccinate lyase, inhibited GSA synthesis. These results support the theory that GSA is formed from urea via the guanidine cycle which consists of microsomal enzymes and urea cycle enzymes[5]. In this study, the guanidine cycle which is proposed as a synthetic pathway for GSA was examined in isolated rat hepatocytes. In addition, the effects of some basic conditions on GSA synthesis in isolated rat hepatocytes were investigated.

MATERIALS AND METHODS

Reagents

Guanidinosuccinic acid, L-canaline (free base) and L-canavanine were obtained from Sigma Chemical Co. (St. Louis, MO., U.S.A.). Collagenase CLS II was obtained from Worthington Biochemical Co. (Freehold, N.J.). Homoserine was obtained from Wako Junyaku Co..

Hydroxylamine was obtained from Nakarai Co.. 9,10-Phenanthrene-
quinone was recrystallized from ethanol.

Preparation of cells

Cells were prepared essentially according to the method of
Berry and Friends[6] modified by Zahlten[7] except for the addition of
2.5 mM $CaCl_2$ and 3 % bovine serum albumin to replace 1.5 % gelatin
in the medium for cell washing. Male Wister rats (300-350 g) were
fed on laboratory chow containing 25 % protein. Surgery for cell
preparation was started at 10:00 A.M. by the injection of sodium
pentobarbital (60 mg/kg) in the non-fasting state. The viability
of the cells, judged from the trypan blue exclusion test, was around
90 %. We calculated that 9.8×10^7 cells corresponded to 1 g wet
liver as reported by Zahlten et al[7].

Incubation of cells

Cells were incubated in 6 ml of Krebs-Henseleit bicarbonate
buffer containing 3 % bovine serum albumin, 10 mM sodium lactate,
10 mM NH_4Cl and the various substances indicated with shaking at
60 cycle/min in a 30 ml conical flask with a rubber cap under an
atmosphere of mixed gases (95 % O_2 + 5 % CO_2) at 37 °C. Equilibration of the buffer was repeated every hour. The amount of cells
used for each experiment is indicated in the results section. In
the pH dependent experiments when the pH of the medium needed
change, 1 N HCl or NaOH was added. Incubation was stopped by the
addition of 0.6 ml of 100 % (w/v) trichloroacetic acid. After
sonication, the supernatant of cells and medium was obtained by
centrifugation at 1,700 x g for 15 min at 4 °C. The pH of the
extract was adjusted to 2.2 by 2 N NaOH and 0.3 ml of the neutralized extract was used for GSA determination.

Assay of GSA

GSA was determined by high pressure liquid chromatographic
analysis using 9,10-phenanthrenequinone for the post-labelling
method according to the method of Yamamoto et al[8]. For the separation of GSA, we used three analytical system as indicated in
Table 1.

RESULTS

Effect of guanidine cycle intermediates on GSA synthesis in
isolated rat hepatocytes

The guanidine cycle was proposed as the pathway of GSA synthesis by Natelson et al.[6,9,10,11] (Fig. 1.). To clarify whether this
cycle is active or not, guanidine cycle intermediates were incu-

Table 1. Programs of GSA analysis

Program 1   (Column 4φ x 125 mm, 60 °C.)
                 (Resin: 2619 Hitachi cation exchange)

Min	Buffer				flow rate (ml/min)
0 - 5	pH 3.0	0.2 N	Na-citrate		0.4
5 - 65	pH 3.3	0.2 N	Na-citrate		0.4
65 - 85		0.2 N	NaOH		0.8
85 - 125	pH 3.0	0.2 N	Na-citrate		0.8
125 - 140	pH 3.0	0.2 N	Na-citrate		0.4

Program 2   (Column 7.5φ x 50 mm, 60 °C.)
                 (Resin: 1EX 210SC, Toyo Soda, 12 μm)

Min	Buffer			flow rate (ml/min)
0 - 5	pH 3.0	0.2 N	Na-citrate	0.8
5 - 75	pH 3.3	0.2 N	Na-citrate	0.8
75 - 85		0.2 N	NaOH	1.6
85 - 105	pH 3.0	0.2 N	Na-citrate	1.6
105 - 110	pH 3.0	0.2 N	Na-citrate	0.8

Program 3   (Column 4.6φ x 125 mm, 60 °C.)
                 (Resin: 1EX 210SC, Toyo Soda, 12 μm)

Min	Buffer			flow rate (ml/min)
0 - 20	pH 3.3	0.2 N	Na-citrate	1.0
20 - 30		0.2 N	NaOH	1.0
30 - 50	pH 3.0	0.2 N	Na-citrate	1.0

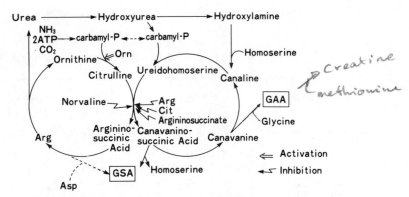

Fig. 1.   The proposed mechanisms of the inhibitory effect of
          various   intermediates of the urea cycle and D,L-norvaline
          on GSA synthesis based on the guanidine cycle hypothesis.

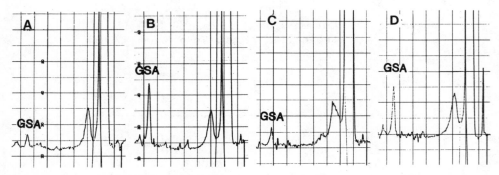

Fig. 2.   Chromatograms obtained from the TCA extracts incubated
          with (A) none, (B) 36 mM urea, (C) 40 mM hydroxyurea,
          (D) 36 mM urea and 5 mM hydroxyurea.

bated with isolated rat hepatocytes. Cells (0.18 g wet liver) were
incubated with 5 mM, 40 mM hydroxyurea, 5 mM hydroxyurea plus 36 mM
urea and 36 mM urea for 4 hours. Urea (36 mM) stimulates GSA
synthesis. However, hydroxyurea did not stimulate GSA synthesis
(Fig. 2.). Hydroxylamine (1 mM) plus homoserine (1 mM) did not
stimulate GSA synthesis. When cells (0.176 g wet liver cell) were
incubated with 1 mM L-canavanine, the peak, whose retention time
corresponded to that of GSA, appeared as shown in Fig. 3A using
Program 1, Table 1 for analysis. However, the peak had a notch as
shown by the arrow in Fig. 3B when valiable cells were incubated
with 36 mM urea plus 1 mM L-canavanine. Standard GSA and TCA
extracts of cells (0.22 g wet liver) and media incubated with 1 mM
L-canavanine or 36 mM urea were analized using Program 2 (Fig. 4).
The peak derived by the addition of canavanine (Peak C) separated

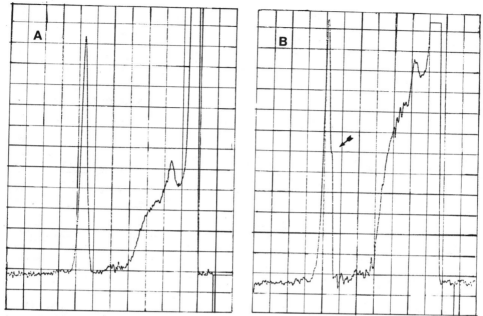

Fig. 3. Chromatograms obtained from the TCA extracts incubated
         with 1 mM canavanine (A) or 36 mM urea + 1 mM canavanine
         (B).

from the peak of standard GSA. The peak derived by the addition of
urea did not separate from standard GSA. Peak C derived by the
addition of L-canavanine increased as the incubation time increased.
A similar peak was observed by the addition of L-canaline to the
medium. However, this peak C decreased as the incubation time
increased (Fig. 5). Moreover, peak C was not observed by the
addition of L-canaline alone. These results suggest that peak C
is derived from the reaction of canaline and some substance in
cells or medium.

Effect of pentobarbital and lipoic acid on GSA synthesis in isolated rat hepatocytes

The first step of GSA synthesis from urea via the guanidine
cycle is oxidation of urea by liver microsomes[5]. Cohen et al
reported that phenobarbitol inhibited GSA synthesis by supressing
this step[12]. Pentobarbital (1 mg/6 ml) in our in vitro system
inhibited urea-stimulated GSA synthesis by 40 % (Table 2).

In the pathway for GSA synthesis, GSA formation from reductive
cleavage of canavaninosuccinic acid required the reduced form of
lipoic acid[13]. Addition of the reduced form of lipoic acid as well
as the oxidized form of lipoic acid (10 mg/6 ml) inhibited urea
stimulated GSA synthesis in our system by 75 and 56 %, respectively.

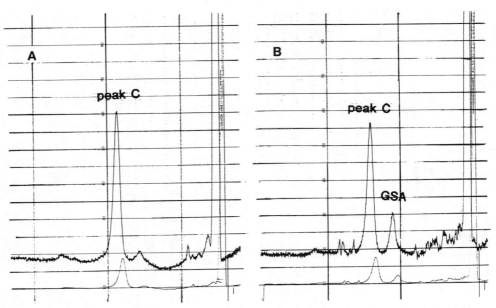

Fig. 4. Chromatograms obtained from TCA extracts incubated with
         1 mM canavanine (A) and TCA extracts incubated with 1 mM
         canavanine supplemented with standard GSA (B).

Effect of various carbon sources on GSA synthesis in isolated rat hepatocytes

     The effect of energy sources on GSA synthesis in isolated
hepatocytes was investigated. Lactate (10 mM) in the incubation
buffer was replaced by 10 mM pyruvate or 10 mM acetate. The
highest rate of GSA formation was observed in the presence of 10 mM
lactate (Fig. 6). In the presence of 10 mM pyruvate and 10 mM
acetate, the amount of GSA formed was 71 and 58 % of that found in
the presence of 10 mM lactate.

Effect of acidosis on GSA synthesis in isolated rat hepatocytes

     It has been thought that the acidosis observed in uremic
patients is one of the causes for stimulation of GSA synthesis.
Cells (0.076 g wet liver) were incubated in Krebs-Henseleit bicarbonate buffer at pH 6.8, pH 7.2 and pH 7.6 at the start of
incubation and pH 6.5, pH 6.95 and pH 7.15 at the end of 4 hours
(Fig. 7). The viability of cells, which was about 90 % as judged
from the trypan blue exclusion test, did not change at the end of
4 hrs incubation. In this experiment, GSA was determined using
program 3. The maximum rate of GSA synthesis was obtained with
the medium at pH 7.2 at the start of incubation. The rate of GSA
synthesis at pH 7.6 and at 6.8 decreased by 6 and 20 %, respec-

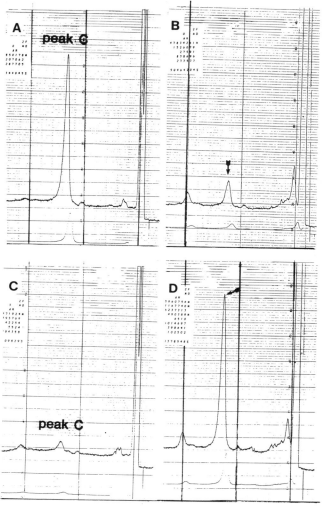

Fig. 5. Chromatograms obtained from TCA extracts incubated with
         10 µg/6 ml L-canaline at 0 time (A) and at 4 hours (B)
         or with 1 mM canavanine at 0 time (C) and at 4 hours (D).

tively. These results, in vitro, support the suggestion that
acidosis stimulates GSA synthesis.

DISCUSSION

     Among the proposed intermediates of the guanidine cycle,
neither hydroxyurea, canaline, canavanine, homoserine or hydroxyl-

Table 2.   Effect of pentobarbital on GSA synthesis in isolated rat
           hepatocytes

Conditions	GSA formed nmol/g wet liver/4h
none	8.0 ± 0.7
36 mM urea	35.1 ± 2.3
36 mM urea + 1 mg/6 ml pentobarbital	21.0 ± 0.0

Cells (0.078 g wet liver) were incubated as described
in Materials and Methods. Each value is expressed as
the mean ± range of   duplicate incubations.

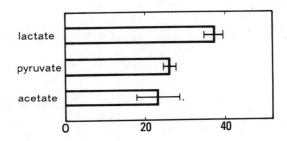

Fig. 6.   Effect of various carbon sources on GSA synthesis in
          isolated rat hepatocytes. Cells (0.076 g wet liver) were
          incubated in Krebs-Henseleit bicarbonate buffer containing
          10 mM lactate, 10 mM pyruvate or 10 mM acetate for 4 hours.
          Each column represents the mean of duplicate incubations.
          Bars indicate the range of each determination.

amine stimulated GSA synthesis in isolated hepatocytes. These
results do not support the theory that the guanidine cycle is the
pathway of urea stimulated GSA synthesis. However, there is a
possibility that by some metabolic deficiency in isolated rat
hepatocytes, the guanidine cycle could be inactivated in this
system. With regard to canaline and canavanine, the possibility
that they do not enter cells is denied since isolated hepatocytes
do form guanidinoacetic acid as shown in our other experiments.

    Pentobarbital inhibits GSA synthesis suggesting that micro-
somal function influences GSA synthesis as reported by Cohen
et al[13].

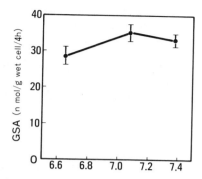

Fig. 7.   Effect of pH on GSA synthesis in isolated hepatocytes.
          The pH of the incubation medium is indicated as the mean
          value of the pH at the start and at the end of the incuba-
          tions. Each point represents the mean of duplicae incuba-
          tions. Bars indicate the range of each determination.

     Peak C was probably derived from canaline and some substance
in the cells and medium. This peak appeared very close to the
peak of GSA. The similar behavior of peak C and GSA on ion
chromatographic analysis may be the cause of the different results
between our findings and those of others[12].

     A differnt rate of GSA formation was obtained using various
carbon sources. One of the differences between lactate, pyruvate
and acetate is the capacity to form reduced NAD and NADP[14]. In
the presence of NH_4Cl and lactate, more aspartate is formed than
in the presence of other carbon sources. However, reductive
cleavage of canavaninosuccinic acid is inhibited by NADH and
NADPH[12]. We observed that ethanol, which increases NADH, inhibits
GSA synthesis in isolated rat hepatocytes.

     In our in vitro system, GSA synthesis is inhibited in acidosis.
However, the viability of cells, judged by the trypan blue exclusion test, did not change. The maximum rate of GSA formation is
obtained at a mean pH of 7.1 which is slightly more acid than
pH 7.3. These results suggest that in mild acidosis, GSA formation
reaches its maximum rate.

SUMMARY

     To clarify the metabolic pathway, the guanidine cycle was
evaluated in isolated hepatocytes. Pentobarbital which inhibits
microsomal function inhibited GSA synthesis in cells. Hydroxyurea,
hydroxylamine plus homoserine, canaline and canavanine, proposed
intermediates of the guanidine cycle, did not stimulate GSA synthesis. However, an unknown peak that might be derived from canaline

and some substance in the cells or the medium appeared close to the peak of GSA and may interfere with interpretation of these results. Lipoic acid did not stimulate GSA synthesis. Among the carbon sources lactate, pyruvate and acetate, GSA was formed at a muximum rate with lactate. The rate of GSA synthesis in pyruvate and acetate is 71 and 58 % of that of lactate. GSA was formed at a maximum rate in the incubation medium with a mean pH of 7.1. At the mean pH of 7.37 and 6.65, the amount of GSA formed was 94 and 80 % of that of pH 7.1.

These results suggest that the guanidine cycle is not active in isolated rat hepatocytes except for microsomal function, which might related to GSA synthesis. They also suggest that the redox state affects GSA synthesis, and that acidosis is low degree stimulates GSA synthesis.

ACKNOWLEDGEMENTS

This study was supported in part by a Research Grant from the Intractable Disease Division, Public Health Bureau, Ministry of Health and Welfare, Japan and University of Tsukuba Project Research.

REFERENCES

1. B. D. Cohen, Guanidinosuccinic acid in uremia, Arch. Intern. Med., 126:846 (1970).
2. K. Aoyagi, S. Ohba, M. Narita and S. Tojo, Regulation of biosynthesis of guanidinosuccinic acid in isolated rat hepatocytes and in vivo, Kidney International, in press.
3. S. Kim, W. I. Paik and P. P. Cohen, Ammonia intoxication in rats: protection by N-carbamoyl-L-glutamate plus L-arginine, Proc. Nat. Acad. Sci. USA, 69:3530 (1972).
4. T. Saeki, Y. Sato, S. Takada and T. Katsunuma, Regulation of urea synthesis in rat liver, J. Biochem., 86:745 (1979).
5. S. Natelson and J. E. Sherwin, Proposed mechanism for urea nitrogen re-utilization: relationship between urea and proposed guanidine cycles, Clin. Chem., 25:1343 (1979).
6. M. N. Berry and D. S. Friend, High-yield preparation of isolated liver cells, J. Cell. Biol., 43:506 (1969).
7. R. N. Zahlten, W. S. Frederick and H. A. Lady, Regulation of glucose synthesis in hormone-sensitive isolated hepatocytes, Proc. Nat. Acad. Sci. USA, 70:3213 (1973).
8. Y. Yamamoto, T. Manji, A. Saito, K. Maeda and K. Ohta, Ion-exchange chromatographic separation and fluorometric determination of guanidino compounds in physiological fluids, J. Chromatogra., 162:327 (1979).
9. S. Natelson, H. Tseng and J. E. Sherwin, On the biosynthesis

of guanidinosuccinate, Clin. Chem., 24:2108 (1978).

10. S. Natelson, A. Koller, H. Tseng and R. F. Dods, Canaline carbamoyl-transferase in human liver as part of a metabolic cycle in which guanidino compounds are formed, Clin. Chem., 23:960 (1977).

11. A. Koller, L. Aldwin and S. Natelson, Hepatic synthesis of canavaninosuccinate from ureidohomoserine and aspartate, and its conversion to guanidinosuccinate, Clin. Chem., 21:1777 (1975).

12. K. Takahara, S. Nakanishi and S. Natelson, Studies on the reductive cleavage of canavanine and canavaninosuccinic acid, Arch. Biochem. Biophys., 145:85 (1971).

13. H. Patel and B. Cohen, The guanidine cycle, Third international congress on nutrition and metabolism in renal diseases (France) 1982.

14. M. Stubbs and H. A. Krebs, The accumulation of aspartate in the presence of ethanol in rat liver, Biochemical J., 150:41 (1975).

THE ROLE OF THE LIVER AS A SIGNIFICANT MODULATOR

OF THE SERUM GUANIDINOACETIC ACID LEVEL IN MAN

Hiroaki Muramoto, Yohei Tofuku, Mitsuhiko Kuroda
and Ryoyu Takeda

Department of Internal Medicine (II), School of
Medicine, Kanazawa University
13 - 1 Takaramachi, Kanazawa City, PC-920, Japan

INTRODUCTION

The metabolism of guanidinoacetic acid (GAA) may be affected by the status of both the kidney and the liver, as GAA is synthesized in the kidney and metabolized to creatine in the liver[1]. Little attention has been focused on GAA metabolism in patients with liver damage, although there have been many studies on GAA metabolism in uremic patients[2,3,4].

To examine the role of the liver in modulating the serum GAA level, we studied changes in the serum levels of guanidine compounds in renal failure patients with liver damage in comprison with renal failure patients without liver damage and patients with liver cirrhosis.

MATERIALS AND METHODS

Blood samples were obtained from 38 patients with acute renal failure (ARF), 111 with chronic renal failure (CRF), 17 with liver cirrhosis and 60 healthy subjects in the fasting state or before hemodialysis (HD) sessions. For measurements of serum concentrations of GSA, methylguanidine (MG) and GAA, each serum was pretreated for deproteinization with trichloroacetic acid at a 10 % final concentration[5], and then, 0.5 ml aliquots of the supernatant was applied to an automated guanidine analyzer, JASCO model G-520 (Japan spectrophotometric Co., Tokyo, Japan)[6]. Serum concentrations of urea-N and creatinine were determined by an autoanalyzer using a part of the same blood samples described above.

The ARF group included 38 patients, all of whom had such severe renal failure as to require HD therapy. Although 27 of these ARF patients had no liver dysfunction, 11 patients had liver damage, which was deduced from the following findings: an elevation of the serum transaminase level above two times the normal value, or marked jaundice giving serum bilirubin levels above 4 mg/dl, both or either of which were combined with an abnormality of the hepaplastin test or thrombo test below 40 % of the normal value. No patients with liver damage were included among the CRF patients. The CRF patients were divided into three subgroups according to the stage of renal disease:CRF-nondialyzed included 28 patients under conservative treatment; CRF-initial HD, 24 patients at the initiation of HD therapy; and CRF-regular HD, 59 patients undergoing regular dailysis for more than three months. None of the 17 patients with liver cirrhosis had renal failure.

RESULTS

Serum levels of urea-N, creatinine, GSA, MG, GAA and arginine in patients with ARF and CRF

The mean serum values of urea-N, creatinine, GSA, methylguanidine (MG), GAA, and arginine in patients with ARF and CRF in various stages are summarized in Table 1.

The mean serum urea-N level in ARF patients was almost the same as that in CRF-initial HD, and was higher than that in CRF-nondialyzed and CRF-regular HD. The mean serum creatinine level in ARF patients was significantly lower than CRF-initial HD and CRF-regular HD.

The mean serum GSA level in the ARF patients was 274.4 ± 43.5 μg/dl, and was significantly lower than that in CRF-initial HD and CRF-regular HD, although it was higher than that in CRF-nondialyzed. Although the GAA level in ARF patients seemed to be the lowest among the four groups, no significant difference was found except for CRF-regular HD. The serum MG level in the same group was significantly lower than that of CRF-initial HD and CRF-regular HD, with same tendency as seen in the serum GSA level. The mean serum arginine level in ARF patients was the lowest among these groups.

The serum profile of nitrogen metabolites and the guanidino compounds in ARF patients with liver damage in comparison with ARF patients without liver damage and patients with liver cirrhosis:

Serum levels of these measurements in the ARF patients with liver damage are summarized in Table 2 with reference to the presence or absence of liver damage.

Table 1. Serum levels of urea-N, creatinine, GSA, MG, GAA and arginine in patients with ARF and CRF.

	ARF	CRF Nondialyzed	Initial HD	Regular HD
No. (cases)	38	28	24	59
Urea-N (mg/dl)	97.1± 6.7	48.8± 4.5	117.3± 7.0	83.8± 3.2
Creatinine (mg/dl)	7.2± 0.6	4.2± 0.4	13.6± 0.9	14.2± 0.5
GSA (μg/dl)	274.4±43.5	118.7±19.6	1146.0±128.0	434.0±40.0
MG (μg/dl)	10.0± 1.9	4.6± 2.1	45.4± 7.8	78.6± 6.0
GAA (μg/dl)	16.0± 2.7	19.5± 1.6	20.4± 3.3	27.3± 2.1
Arginine (μmol/l)	99.5± 9.6	131.9± 8.9	127.0± 6.4	154.4± 6.5

mean ± SE; *,p<0.05; **,p<0.01; ***,p<0.001

No differences in serum levels of urea-N and creatinine, were seen between ARF patients with and without liver damage. As expected, the serum levels of urea-N and creatinine were normal in the liver cirrhosis patients, as none of them had renal failure.

No difference of serum GSA level was noted in the ARF patients, regardless of the presence or the absence of liver damage. No difference was seen either in the serum MG levels within these patient groups. In contrast, the serum GAA level was found to be

Table 2. The serum profiles of nitrogen metabolites and guanidino compounds in ARF patients with liver damage in comparison with ARF patients without liver damage and patients with liver cirrhosis.

	ARF Liver Damage(−)	Liver Damage(+)	Liver Cirrhosis	Normal Controls
No. (Cases)	27	11	17	60
Urea-N (mg/dl)	93.8± 7.6	94.4±14.3 ***	15.9± 1.6 ***	
Creatinine (mg/dl)	7.9± 0.7	5.5± 1.1 ***	1.0± 0.1 ***	
GSA (µg/dl)	305.8±56.3	197.4±55.7 *** ***	2.5± 1.1 *** ***	0.5±0.2
MG (µg/dl)	10.8± 2.1	8.1± 4.2 *** ***	n.d. *** ***	n.d.
GAA (µg/dl)	10.7± 1.5	29.1± 7.5 *** **	59.7±13.2 ***	28.8±1.1
Arginine (µmol/l)	96.7±10.7	106.7±21.0 *	101.1± 7.5 *	122.0±4.8

mean ± SE; *,p<0.05; **,p<0.01; ***,p<0.001; n.d., not detectable

markedly affected by the presence of liver damage. The serum GAA level in patients with liver cirrhosis was significantly higher than that of the normal controls (p<0.001). The GAA level in the ARF patients with liver damage was significantly higher than that in ARF patients without liver damage (p<0.01), and was found to be as high as the normal controls, despite their having such severe renal damage as to require HD therapy. GAA synthesis in their kidneys might be defective. Concerning serum arginine levels, little difference was noted among the groups except for the normal controls who showed a relatively high level.

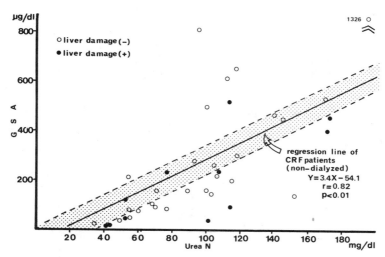

Fig. 1.   Relationship between serum levels of urea-N and GSA in
          patients with ARF.

Relationship between serum levels of urea-N and GSA or GAA

The relationship between serum levels of urea-N and GSA is
shown in Fig. 1.   The correlation between these two measurements
in nondialyzed CRF patients is depicted by a regression line and
a dotted zone which represents the area of one standard deviation
of the regression analysis.   The results imply that the serum GSA
levels of the ARF patients with liver damage were relatively low
in relation to the degree of the urea-N levels, as the value in the
area under the dotted zone indicates relatively low value of the
one in relation to the values of the other.   In the area under the
dotted zone, 8 of 11 (approximately 73 %) ARF patients with liver
damage depicted by closed circles were seen, whereas 12 of 28
(approximately 43 %) of the ARF patients without liver damage depicted by open circles were fonnd in this area.

The relationship between serum levels of urea-N and GAA is
shown in Fig. 2.   The regression line indicates a significant
inverse correlation between these two measurements in CRF-nondialyzed patients.   The dotted zone represents the area of one
standard deviation of the regression analysis, and the symbols
used are the same as in Fig. 1.   Although both symbols on the
whole seem to show the same distribution as the regression line, 6
of 11 (approx. 55 %) ARF patients with liver damage depicted by
closed circles were noted in the area over the dotted zone, whereas
only 5 of 28 (approx. 18 %) ARF patients without liver damage
depicted by open circles were seen in this area.

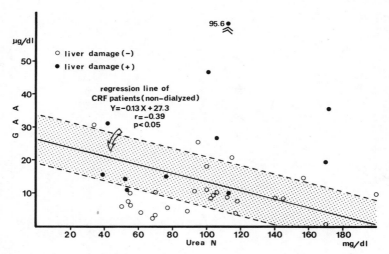

Fig. 2. Relationship between serum levels of urea-N and GAA in
         patients with ARF.

        These findings in the ARF patients with liver damage indicate
that serum GAA levels tend to be relatively higher and serum GSA
levels tend to be relatively lower, in relation to the degree of
serum urea-N levels in ARF patients with liver damage.

DISCUSSION

        The synthesis of GAA is considered to be carried out mainly
in the kidney and metabolized to creatine in the liver[1,2].
Although the metabolic pathway of GSA has not been fully elucidated
and it is unclear whether it is synthesized from arginine[2,7] or from
canavanine[8,9], there is no controversy concerning the fact that the
liver participates in GSA synthesis. As many studies on the metabolism of these guanidino compounds have been directed toward renal
failure, the present study was undertaken to focus on changes in
the serum profile of these guanidino compounds in patients with
liver damage.

        In the study of CRF-nondialyzed patients, changes in the
serum levels of GSA and GAA were characteristic of those previously reported[3,10]: the serum GSA levels were significantly higher
with a close correlation to the serum urea-N level, whereas the
serum GAA level was significantly lower with an inverse correlation
to the serum urea-N level. As in this group of patients, metabolism of these guanidino compounds is not affected by HD therapy.
The results reflect the role of the kidney as a key modulator of
the serum level of these compounds.

In ARF patients without liver damage, the serum profile of these compounds was found to have the same characteristics as seen in CRF-nondialyzed patients. In contrast to these results, in ARF patients with liver damage, the serum GAA level was shown to be significantly higher than that in ARF patients without liver damage, and was as high as the normal controls. Furthermore, the serum GAA level in patients with liver cirrhosis was found to be significantly higher than the normal controls. These results indicate that the serum GAA level may be intensely affected by liver damage in the opposite direction from that seen in renal failure. These changes in serum GAA are probably caused by GAA retention in the blood resulting from a defect in converting GAA to creatine in the liver.

The serum GSA level in ARF patients with liver damage was found to be relatively low in relation to the urea-N level, and no difference in serum urea-N levels was found between the ARF patients with or without liver damage. As a supplemental comment, it seems likely that liver damage affects GSA synthesis through a metabolic process designed to reutilize over-flow urea-N[9] rather than through the urea cycle[3].

SUMMARY

Changes in serum guanidino compound levels in renal failure patients with liver damage were evaluated by comparing them to renal failure patients without liver damage and patients with liver cirrhosis. In nondialyzed CRF patietns, the mean serum levels of GSA and MG were significantly higher and the serum GAA level was significantly lower. Although there were no differences in the mean serum levels of urea-N, creatinine, GSA, MG and arginine between ARF patients with and without liver damage, the mean serum GAA level in the former group was significantly higher than that in the latter group. In patients with liver cirrhosis, the mean serum GAA level was significantly higher than that of the normal controls.

These results indicate that the serum GAA levels are significantly affected by liver damage in the opposite direction from that seen in renal failure, probably caused by GAA retention in the blood which results, in turn, from a defect in converting GAA to creatine in the liver.

REFERENCES

1. J. F. van Pilsum, G. C. Stephens and D. Taylor, Distribution of creatine, guanidinoacetic and the enzymes for their biosynthesis in the animal kingdom, Biochem. J., 126:325 (1972).

2. B. D. Cohen, I. M. Stein and J. E. Bonas, Guanidinosuccinic aciduria in uremia: A possible alternate pathway for urea synthesis, Am. J. Med., 45:63 (1968).
3. B. D. Cohen, Guanidinosuccinic acid in uremia, Arch. Intern. Med., 126:846 (1970).
4. M. Sasaki, K. Takahara and S. Natelson, Urinary guanidino-acetate/guanidinosuccinate ratio: An indicator of kidney dysfunction, Clin. Chem., 19:315 (1973).
5. A. Ando, T. Kikuchi, H. Mikami, M. Fujii, K. Yoshihara, Y. Orita and H. Abe, Quantitative determination of guanidino compounds: the excellent preparation of biological samples, In:"Urea Cycle Diseases," A. Lowenthal, A. Mori and B. Marescau, eds., Plenum Publishing Corporation (1983).
6. Y. Yamamoto, A. Saito, T. Manji, H. Nishi, K. Ito, K. Maeda, K. Ohta and K. Kobayashi, A new automated analytical method for guanidino compounds and their cerebrospinal fluid levels in uremia, Trans. Am. Soc. Artif. Intern. Orgns., 24:618 (1978).
7. G. Perez, A. Rey and E. Schiff, The biosynthesis of guanidino-succinic acid by perfused rat liver, J. Clin. Invest., 57:807 (1976).
8. A. Koller, L. Aldwin and S. Natelson, Hepatic Synthesis of canavaninosuccinate from ureidohomoserine and aspartaté, and its conversion to guanidinosuccinate, Clin. Chem., 21:1777 (1975).
9. S. Natelson and J. E. Sherwin, Proposed mechanism for urea nitrogen re-utilization: Relationship between urea and proposed guanidine cycle, Clin. Chem., 25:1343 (1979).
10. M. Shitomoto and S. Otsuji, Changes of guanidino compounds in chronic renal failure, Japan. J. Nephrol., 21:33 (1979) in Japanese .

CONVERSION OF UREA NITROGEN INTO PROTEIN NITROGEN BY A FRESHWATER FISH, SAROTHERODON MOSSAMBICUS

K.C. Chitra, K.S. Swami and K.S. Jagannatha Rao

Department of Zoology
Sri Venkateswara University
Tirupati- 517 502, India

INTRODUCTION

Urea is used as a source of non-protein nitrogen in the feed of ruminants[1-3]. It is included in some of the compound feeding formulae for dairy and beef cattle[4]. In view of the non palatability of urea, it is compounded with other palatable food components so that it is accepted by the cattle. The cattle have been acclimated to urea and it has been demonstrated by in vitro techniques that rumen fluid has high urease activity at all times[5].

Since urea is extensively used in agriculture, fish in nearby waters are constantly exposed to urea stress, thereby altering their metabolism[6]. However there is no information regarding the metabolic conversion of this non-protein nitrogen into tissue protein contributing to the growth of the animal. This aspect has been investigated in the present study by exposing the fish, Sarotherodon mossambicus to urea stress and analysing the tissue protein and associated metabolism.

MATERIALS AND METHODS

The teleost fishes Sarotherodon mossambicus weighing about 16 ± 2 g have been procured from the local fish ponds and have been acclimated to laboratory conditions in large cement aquaria. One batch of fishes were exposed to 40 ppm urea (Technical grade 99.5 % M.W. 60.5) and another, kept in a similar situation without urea, served as controls. The fishes were exposed for a period of 24 hours and the following biochemical parameters were investigated

113

in the brain, liver, muscle and gill tissues after quick excision and chilling to 5 °C.

The protein and free amino acid level was estimated by the methods of Lowry et al.[7] and Moore and Stein[8] respectively. The free ammonia and urea concentration was determined by method of Bergmeyer[9] and the diacetylmonoxime method of Natelson[10]. The glutamate dehydrogenase and aspartate aminotransferase and alanine aminotransferase enzymes were estimated by the method of Lee and Lardy[11] and Reitman and Frankel[12]. The arginase and succinate dehydrogenase activity was estimated by the methods of Campbell[13] and Nachlas et al.[14].

RESULTS AND DISCUSSION

The biochemical parameters determined in the control fish showed tissue specificity. The protein level was 90.68 mg/g wet wt in the brain tissue while the other tissues had a higher level. The free amino acid concentration was low in the brain and gills (47.23 and 32.93 µmol of tyrosine/g wet weight of tissue respectively) while it is highest in the liver (103.2 µmol of tyrosine/g wet weight of tissue). The free ammonia level showed a variation corresponding with that of the amino acid level in all tissues. The brain and gills contained low ammonia while liver and muscle had higher ammonia levels (brain 3.28, gill 4.57 and muscle 8.9 µmol of ammonia/g wet tissue). The level of urea did not correspond to the ammonia level indicating that ureogenesis in general is not based on ammonia production. The high urea level in the liver and the low ammonia is an indication of the operation of ornithine cycle. The muscle on the other hand, had high ammonia (8.9 µmol of ammonia/g wet tissues) and urea (1.84 µmol of urea/g wet tissues) indicating an unlinking of these two parameters and presumed non operation of the ornithine cycle.

The level of activity of glutamate dehydrogenase (GDH), an ammonia secreting enzyme, was highest in the liver (0.148 µmol of formazan formed/mg protein/hour) and lowest in the gills (0.003 µmol of formazan formed/mg protein/hour). The high activity of the enzyme in liver corresponds to the free ammonia level but the situation is not similar in the case of other tissues, suggesting that other modes of ammonia secretion are operative[15].

The aspartate and alanine aminotransferase activities are highest in the liver (0.52 and 0.35 µmol of pyruvate/mg protein/ hour respectively). The high aminotransferase and glutamate dehydrogenase activity in the liver are indicative of prevalent transamination and deamination reactions leading to ammonia secretion which consequently give rise to urea[16,17]. The arginase activity is highest in the liver (0.99 µmol of urea/mg protein/hour)

Table 1.   Biochemical components in the different tissues of urea
           exposed (40 ppm conc) fishes Sarotherodon mossambicus.

Name of the bio-chemical parameter	Concentration of urea	Brain	Liver	Muscle	Gill
Proteins	'0'	90.68	134.7	183.5	117.8
	SD	4.42	4.09	6.94	6.33
	40 ppm	109.6	121.7	292.6	127.9
	SD	22.62	11.25	14.64	7.62
%difference over control		+ 20.96	# -9.65	* 59.49	# 8.56
Free Amino acids	'0'	47.23	103.2	97.40	32.93
	SD	5.13	12.72	11.19	5.34
	40 ppm	164.7	165.4	186.7	133.1
	SD	10.93	12.32	6.44	12.33
%difference over control		* 248.8	* 60.33	* 91.65	* 304.2
Ammonia	'0'	3.28	5.68	8.90	4.57
	SD	0.16	0.41	0.27	0.49
	40 ppm	4.78	6.46	8.79	3.38
	SD	0.92	0.73	0.70	0.51
%difference over control		* 45.94	# 13.74	-1.24	* -26.08
Urea	'0'	4.31	5.10	1.84	3.38
	SD	0.37	0.50	0.50	1.27
	40 ppm	7.17	12.81	4.40	4.91
	SD	0.79	1.93	0.27	1.04
%difference over control		* 66.34	* 151.2	* 138.5	* 45.10

+; $p < 0.05$,   #; $p < 0.01$,   *; $p < 0.001$.

and lowest in the gills (0.014 µmol of urea/mg protein/hour). This agrees with the urea level in the case of liver but not so in other tissues. The succinate dehydrogenase activity is highest in the liver (1.59 µmol of formazan/mg protein/hour) and lowest in gill (0.0461 µmol of formazan/mg protein/hour). These studies in general indicate that liver is an efficient organ in protein degradation, oxidative deamination and urea synthesis, while other tissues possess incomplete metabolic segments concerned with protein degradation and urea synthesis.

Table 2. Different enzyme levels of some of the tissues of urea
         (40 ppm conc) exposed fishes.

Name of the enzyme	Concen-tration of urea	Brain	Liver	Muscle	Gill
	'0'	0.0262	0.1477	0.0065	0.0030
Glutamate	SD	0.0031	0.0061	0.0008	0.0006
dehydrogenase (GDH)	40 ppm	0.0434	0.0534	0.0022	0.0030
	SD	0.0068	0.0046	0.0002	0.0003
%difference over control		65.65*	-63.85*	-66.85*	0.00
	'0'	0.2668	0.5225	0.1752	0.1367
Aspartate	SD	0.0090	0.0200	0.0060	0.0070
aminotransferase (ATT)	40 ppm	0.0165	0.1877	0.0497	0.1197
	SD	0.0020	0.0400	0.0090	0.0060
%difference over control		-93.82*	-64.07*	-71.63*	-12.44*
	'0'	0.1375	0.3533	0.1406	0.1048
Alanine	SD	0.0090	0.0600	0.0200	0.0200
aminotransferase (AlAT)	40 ppm	0.1772	0.2544	0.0855	0.1284
	SD	0.0060	0.0090	0.0020	0.0100
%difference over control		28.87*	-27.71*	-39.19*	22.52*
	'0'	0.0313	0.0985	0.0053	0.0140
Arginase	SD	0.0060	0.0100	0.0020	0.0020
	40 ppm	0.0551	0.0704	0.0077	0.0232
	SD	0.0080	0.0100	0.0010	0.0020
%difference over control		76.04*	-28.53*	45.28*	65.71*
	'0'	0.2573	1.5882	0.1996	0.0461
Succinate	SD	0.0800	0.2000	0.0200	0.0200
dehydrogenase (SDH)	40 ppm	1.5499	1.4710	0.1935	0.2310
	SD	0.4100	0.1600	0.0100	0.0200
%difference over control		502.4*	-7.38	-30.06*	401.1*

*; $p < 0.001$.

When these fishes are exposed to 40 ppm of urea for a 24 hour
period the liver seems to be the primary target of fertilizer toxicity[18]. The protein content decreased in the liver (-9.65 %)
while the amino acid content showed a slight increase unlike that
seen in the other tissues (60.33 %). The ammonia content in liver
is also less than that in the brain but more than the muscle and
gill. Though the GDH activity showed 63.85 % decrement, the 13.74 %

increase in ammonia in the liver indicates ammonia secretion from other metabolites. Though the urea content was the highest in the liver, the arginase activity levels was lowest. The apparent anomaly between urea production and arginase activity may be caused by the fact that urea produced in the liver is at a low rate but it accumulates in tissue because of circulatory failure. The low ammonia secretion from the amino acid sources could also be due to low transaminase activities (AAT is 64.07 % and AlAT -27.71 %). The low succinate dehydrogenase activity in the liver is an indication of low energy production.

The organ which is most affected next to liver is the brain where there is extensive free amino acid accumulation (248.8 %). The high amino acid could be due to low transaminase activities (AAT -93.82 %) since amino acids are not transaminated to the same extent as in liver. There is some amount of urea produced from arginase and some ammonia from GDH. The SDH activity increased considerably (502.4 %) indicating violent neural activity which requires energy. The muscle and gill are also affected to some extent by urea stress but there is no degradation of protein and free amino acids. Though there is some transaminase and low GDH activity, ammonia production is low and there is arginase activity with urea production.

These studies in general indicate that the liver is the primary target of urea stress with diminished ureogenesis while brain and other tissues have acquired urea synthesis probably from guanidino compounds. There is a possibility of the activation of metabolism concerned with guanidino compounds which might have resulted in a stepped up protein synthesis in all tissues except liver.

REFERECES

1. K. Singhal and V. D. Mudgal, Comparative study of urea and Biuret feeding on nutritive utilization and milk production in Goats, Indian J. Dairy Sci., 332 (1980).
2. R. R. Freitag, W. H. Smith and W. M. Beeson, Factors related to the utilization of urea Vs protein nitrogen supplemented diets by theruminant, J. Anim. Sci., 27:478 (1968).
3. E. E. Bartley and C. W. Deyoe, in:"Recent advances in Animal nutrition," Butterworth, London (1977).
4. J. Bond and R. R. Oltjin, Growth and reproduction performance of beef females fed high urea containing diets, J. Anim. Sci., 37:1040 (1973).
5. R. M. Pearson and J. A. B. Smith, The utilization of urea in the bovine rumen II. The conversion of urea to ammonia, Biochem. J., 37:148 (1943).

6. V. M. S. Srivastava, S. Lata, D. Srivastava and R. S. Tripati,
    Prospective ecological impact of urea and carpet chemicals
    in Gyanpur in relation to human values, Proc. 69th session
    of Indian Sci. Congress. Association, Mysore (1982).

7. O. H. Lowry, N. J. Rose Brough, A. L. Farr and R. J. Randall,
    Protein measurement into the Folin phenol reagent, J. Biol.
    Chem., 193:265 (1951).

8. S. Moore and W. H. Stein, Amodified ninhydrin reagent for the
    photometric determination of amino acids and related com-
    pounds, J. Biochem., 211:907 (1954).

9. H. O. Bergmeyer, in:"Methods of enzymatic analysis,"
    Academic press, New York (1965).

10. S. Natelson, in:"Techniques of clinical chemistry,"
    Thomas C. C. Publi. Illionois (1971).

11. Y. L. Lee and H. A. Lardy, Influence of thyroid hormones on
    L-glycerophosphate dehydrogenase and other dehydrogenases
    in various organs of rat, J. Biol. Chem., 240:1427 (1965).

12. S. Reitman and S. Frankel, A colorimetric method for the de-
    termination of serum glutamic oxaloacetic and glutamic
    pyruvic transaminases, Am. J. Clin. Pathol., 28:56 (1957).

13. J. W. Campbell, Studies on tissue arginase and ureogenesis in
    the elasmobranch Mustelus canu, Arch. Biochem. Biophys.,
    93:448 (1961).

14. M. M. Nachlas, S. P. Margulins and A. M. Seligman, Sites of
    electron transfer to tetrazolium salts in the succinoxidase
    system, J. Biol. Chem., 235:2739 (1969).

15. R. T. Bogusky, Lowenstein, L. M. and J. M. Lowenstein, The
    purine nucleotide cycle. A pathway for ammonia production
    in the rat kidney, J. Clin. Invest., 58:326 (1976).

16. A. E. Braunstein, Transamination and the integrative func-
    tions of the dicarboxylic acids in nitrogen metabolism,
    Advan. Protein Chem., 3:1 (1951).

17. Robert R. Bidigare and Frederick D. King, GDH activity in P.
    Flexuosus, Comp. Biochem. and Physiol., 70B:3 (1981).

18. S. K. Sarkar and S. K. Konar, Acute toxicity of agricultural
    fertilizers to aquatic organisms, Geobios., 8:225 (1982).

TOPOGRAPHICAL DISTRIBUTION OF SHEEP BRAIN ARGINASE:

ITS RESPONSE TO SOME GUANIDINO COMPOUNDS

V. Mohanachari, K. Satyavelu Reddy, K. Indira
and K.S. Swami

Department of Zoology
Sri Venkateswara University
Tirupati- 517 502, India

INTRODUCTION

Despite several reports on the role of guanidines in the neurological symptoms of epilepsy, hyperargininemia and renal failure[1-3], many points are still unclear and a need for studies on the direct toxic effects of guanidines was indicated in the symposium on urea cycle diseases[2]. Though there is a clear metabolic interrelation between arginine and guanidines, the whole metabolic profile appears complicated in the mammalian brain due to a) the lack of a functional urea cycle b) the presence of complex regional differences and c) the unknown functional significance of arginase[4]. Hence, these metabolic uncertainities have prompted the authors to take up a study of the topographical distribution of mammalian brain arginase and its response to some guanidines, in vitro, in order to assess the direct neurotoxic nature of these compounds.

MATERIAL AND METHODS

Material

Brains were procured from healthy sheep of similar age after decapitation at the local slaughter house and transported in a dry beaker kept in a freezing mixture. The meningi were removed and brains were washed repeatedly with Krebs-Henseleit Ringer's solution. The different parts of the brain such as cerebellum, cerebrum and brain stem were separated and 10 % (w/v) homogenates were prepared in 0.1 % acetyltrimethylammonium bromide. The homogenates

were spun at 5000 x g for 20 min and the clear cell free extract
was employed as the enzyme source to estimate arginase activity as
described earlier[5]. Unless otherwise specified, all these steps
were carried out at 0-3 °C.

Dose versus response study

In addition to the reaction mixture, varied concentrations of
guanidine hydrochloride (GHCl) (10, 20, 30, 50, 100, 150, 200 and
250 mM) and guanidinoacetic acid (GAA) (2.5, 5, 7.5, 12.5 and 25
mM) were added separately in vitro to study the dose dependent
arginase response in different brain regions.

Influence of GHCl and GAA on substrate dependency of arginase

The influence of GHCl (20, 100 and 250 mM) and GAA (5 and 12.5
mM) on the substrate (L-arginine) dependency of arginase was
tested, varying the substrate concentration from 10 to 80 mM. The
Vmax and Km were determined using least squares as the best fit.
The protein concentration in the enzyme source was estimated by
the method of Lowry et al.[6].

RESULTS AND DISCUSSION

The results are summarised in tables 1-3.

The results clearly indicate that the neural arginase activity is variable topographically. The cerebellum registered the
highest activity followed by cerebrum and brain stem. Though
these regional differences follow the same pattern as that of rat
brain arginase[7], it is difficult to interpret, as these results
are attributable to the differential distribution of effectors
modulating regionally the arginase activity.

The dose dependent studies with GHCl have shown a consistent
inhibition of arginase activity with increasing concentrations of
GHCl in all the regions of brain. With GAA, the response was different in the sense that with lower concentrations of GAA, the
inhibition was more prominent and with increasing concentrations,
the inhibitory influence was reduced in the cerebellum and cerebrum,
while the brain stem arginase showed consistent inhibition.

Substrate dependent studies in the presence of GHCl (20, 100,
and 250 mM) and GAA (5, and 12.5 mM) revealed a decreased Vmax
with all concentrations of GHCl and GAA (except in 12.5 mM of GAA)
suggesting that the hydrolysis of L-arginine by arginase is attenu-
ated, perhaps by masking the enzyme active sites by these guanidines. Further, the increase in Km in all the regions (Table 3)
suggests decreased affinity between E and S, thereby reducing the

Table 1. Neural arginase activity as a function of GHCl concentration

Region of the brain	Control	GHCl in mM							
		10	20	30	50	100	150	200	250
Cerebrum	0.528	0.586	0.647	0.545*	0.490	0.452	0.364	0.282	0.207
S.D.	0.002	0.031	0.042	0.032	0.010	0.021	0.012	0.021	0.031
		(11.0)	(22.5)	(3.2)	(-7.2)	(-14.4)	(-31.1)	(-46.6)	(-60.8)
Cerebellum	0.691	0.647	0.545	0.490	0.452	0.417	0.347	0.282	0.236
S.D.	0.017	0.024	0.053	0.025	0.042	0.013	0.021	0.010	0.001
		(-6.4)	(-21.1)	(-29.1)	(-34.6)	(-39.7)	(-49.8)	(-59.2)	(-65.8)
Brain stem	0.417	0.417	0.417	0.382	0.347	0.315	0.282	0.250	0.207
S.D.	0.014	0.012	0.078	0.012	0.018	0.019	0.026	0.016	0.016
		–	–	(-8.4)	(-16.8)	(-24.5)	(-32.4)	(-40.0)	(-50.4)

Arginase activity is in x 10^{-6} M of urea per mg protein per hour. The values are mean ± S.D of 8 samples. Values in parentheses are % change over corresponding controls. Experimental value marked with asterisk is not significant, while values without asterisk are significant at $p < 0.001$.

Table 2.   Neural arginase activity as a function of GAA concen-
           tration

Region of the brain	Control	GAA in mM				
		2.5	5	7.5	12.5	25
Cerebrum	0.528	0.469	0.434	0.490	0.565*	0.606
S.D	0.002	0.002	0.016	0.017	0.035	0.009
		(−11.2)	(−17.8)	(−7.2)	(7.0)	(14.8)
Cerebellum	0.691	0.282	0.282	0.364	0.469	0.490
S.D	0.017	0.012	0.012	0.016	0.025	0.076
		(−59.2)	(−59.2)	(−47.3)	(−32.1)	(−29.1)
Brain stem	0.417	0.382	0.297	0.282	0.236	0.207
S.D	0.014	0.004	0.015	0.015	0.075	0.065
		(−8.4)	(−28.8)	(−32.4)	(−43.4)	(−50.4)

Arginase activity is in x 10^{-6} M of urea per mg protein per hour.
The values are mean ± S.D of 8 samples.   Values in parentheses are
% change over corresponding controls.   Experimental value marked
with asterisk is significant at p<0.01,   while values without
asterisk are significant at p<0.001.

chances and frequency of ES complex formation.   Though the inhibition is of mixed type, it is tending towards competitive inhibition as the Km is more affected than Vmax.

In general, the response was relatively more potent with GAA
than GHCl in all regions of brain.   The GHCl influence was greater
in the cerebellum, while the GAA effect was more in both the cerebellum and brain stem.   Since the neural arginase activity in different regions is inhibited by these guanidines, apparently due to
competitive inhibition, the opportunities for arginine diversion
for some other reactions appear to be greater.   Since this route
of arginine hydrolysis by arginase is blocked, the metabolic consequences associated with urea and ornithine are impaired in
neural tissue, particularly with ornithine, as it is an exclusive
necessity for polyamine synthesis and also for glutamate/proline
formation in terms of neurotransmission.

Though through investigations are necessary for conclusive
demonstration of the toxic effects of guanidines, the abnormal
neurological symptoms elicited in epilepsy, hyperargininemia and
renal failure might have originated in a basic biochemical lesion,

Table 3.   Substrate dependent neural arginase activity in the presence of different concentrations of GHCl and GAA

Conc of Guanidines	Cerebrum		Cerebellum		Brain stem	
	Vmax	Km	Vmax	Km	Vmax	Km
Control	0.528	8.0	0.691	6.0	0.417	13.0
GHCl 20 mM	0.647	3.5	0.545	7.9	0.417	6.0
	(22.5)	(−56.3)	(−21.1)	(31.7)	−	(−53.8)
100 mM	0.452	18.0	0.417	8.3	0.315	30.0
	(−14.4)	(−125)	(−39.7)	(38.3)	(−24.5)	(131)
250 mM	0.207	25.0	0.236	15.0	0.207	21.0
	(−60.8)	(213)	(−65.8)	(150)	(−50.4)	(61.5)
GAA 5 mM	0.434	10.0	0.282	14.39	0.297	10.0
	(−17.8)	(25.0)	(−59.2)	(140)	(−28.8)	(−23.1)
12.5 mM	0.565	4.9	0.469	8.1	0.236	18.0
	(7.0)	(−38.8)	(−32.1)	(35.0)	(−43.4)	(38.5)

The Vmax values are in x 10^{-6}M of urea per mg protein per hour. The Km values are in x 10^{-3}M of L-arginine. All the values are mean of 8 samples. S.D. values are not shown as all of them are below 5 %. Values in parentheses are percent change over corresponding controls.

such as the inhibition of arginase in different regions of brain, as the present study. These studies further warrant an in depth analysis of the consequences of accumulation of guanidines in different regions of the brain and a need for resolving threshold in vivo concentrations to trigger similar effects in rats and humans.

ACKNOWLEDGMENTS

     Dr. V. Mohanachari wishes to thank Prof. A. Lowenthal, Belgium for his help. This work was supported by Council of Scientific & Industrial Research, India through a fellowship to Dr. V. M.

REFERENCES

1. A. Mori, Y. Watanabe, S. Shindo, M. Akagi and M. Hiramatsu, γ-Guanidinoglutaric acid and epilepsy, in: "Urea Cycle Diseases," A. Lowenthal, A. Mori and B. Marescau, eds.,

Plenum Press, New York and London (1982).

2. B. Marescau, A. Lowenthal, H. G. Terheggen, E. Esmans and
    F. Alderweireldt, Guanidino compounds in hyperargininemia,
    in:"Urea Cycle Diseases," A. Lowenthal, A. Mori and
    B. Marescau, eds., Plenum Press, New York and London (1982).

3. H. Mikami, Y. Orita, A. Ando, M. Fujii, T. Kikuchi,
    K. Yoshihara, A. Okada and H. Abe, Metabolic pathway of
    guanidino compounds in chronic renal failure, in:"Urea
    Cycle Diseases," A. Lowenthal, A. Mori and B. Marescau, eds.,
    Plenum Press, New York and London (1982).

4. B. Sadasivudu and T. I. Rao, Studies on functional and metabo-
    lic role of urea cycle intermediates in brain,
    J. Neurochem., 27:785 (1976).

5. V. Mohanachari, P. Neeraja, K. Indira and K. S. Swami,
    Sciatectomic stimulation of muscle arginase and its implica-
    tions, in:"Urea Cycle Diseases," A. Lowenthal, A. Mori and
    B. Marescau, eds., Plenum Press, New York and London (1982).

6. O. H. Lowry, N. J. Rosebrough, A. L. Farr and R. J. Randall,
    Protein measurement with the Follin Phenol reagent,
    J. Biol. Chem., 193:265 (1951).

7. A. Vanella, R. Pinturu, P. D'Urso and I. Disiluestro, Arginase
    activity and polyamine content in different regions of rat
    brain during post natal development, Acta. Neurol (Naples).,
    34:452 (1979).

THE EFFECT OF SODIUM VALPROATE ON BLOOD AMMONIA LEVELS

AND THE CONCENTRATIONS OF GUANIDINO COMPOUNDS IN MOUSE TISSUE

Jium Shiou Chang, Yoko Watanabe and Akitane Mori

Department of Neurochemistry
Institute for neurobiology
Okayama University Medical School
2-5-1 Shikatacho
Okayama 700, Japan

INTRODUCTION

Sodium valproate (SVA) is an anticonvulsant drug which has been widely used for the treatment of epilepsy. Recently it was reported that SVA has hepatic toxicity[1,2,3,4] and induces hyper-ammonemia in humans[5,6,7]. These reports prompted us to suspect that SVA might affect guanidino compound metabolism as well as nitrogen metabolism. In the present study we investigated the effect of SVA on both blood ammonia levels and the concentrations of guanidino compounds in the liver, kidney, pancreas, brain, plasma and urine of mice who were administered SVA intraperitone-ally.

MATERIALS AND METHODS

Materials

SVA was kindly supplied by Kyowa Hakko Kogyo Co., Ltd. (Tokyo, Japan). Standard guanidino compounds were obtained as previously described[8]. All other chemicals were of reagent grade.

Animals

Male ddY mice were purchased from Shizuoka Agricultural Co-operative Association for Laboratory Animals (Shizuoka, Japan) and maintained on food and water ad libitium in a light (lights on 07:00-19:00) and temperature (25±2 °C) regulated room for a minimum of one week before the following experiments. Six to seven-week-old animals were killed between 13:00-14:00.

The mice were intraperitoneally administered 400 mg/kg of SVA. Control mice were treated the same as the experimental animals, but without the administration of SVA. Thirty and 60 minutes after SVA administration, the mice were sacrificed by decapitation. Blood from the carotid arteries was collected in a tube containing heparin sodium, and the liver, kidney, pancreas and brain were rapidly removed and iced, weighed and stored frozen at -80°C until the extraction procedure was carried out. The blood obtained from 3 mice was pooled, and plasma was obtained by centrifugation at 1100 x g.

To collect 24 hour specimens of urine, mice were housed in groups of 3 in metabolic cages. Urine collecting vessels containing 1 ml of 1N-HCl to control bacteria were changed at 13:00 everyday. At 13:00 on the third day the mice were intraperitoneally administered 400 mg/kg of SVA, and urines were collected for the following 6 days.

Extraction and determination of guanidino compounds

The extraction procedure for guanidino compounds from mouse tissues has been described elsewhere[9].

Plasma and urine were ultrafiltered by passing through Amicon centriflo CF25 (Amicon Corp., Mass. U.S.A.) filters under centrifugation at 800 x g for 20 min at 4 °C, after adjusting the samples to pH 2.2 with a dilute HCl solution.

The guanidino compound concentrations were determined with a high performance liquid chromatograph, guanidino compound analyzer JASCO G-520 (Japan Spectroscopic Co., Hachioji, Tokyo, Japan)[8].

Determination of blood ammonia levels

The blood ammonia levels were determined with a kit of Ammonia-Test Wako (Wako Pure Chemical Industries Ltd., Osaka, Japan) immediately after obtaining blood from 3 mice.

RESULTS

Blood ammonia levels

As shown in Table 1, the blood ammonia level of control mice was found to be 54±4 µM (mean±SE). In the SVA administered group, blood ammonia levels were 101±8 µM and 90±7 µM, 30 and 60 minutes, respectively, following administration, and were significantly higher than the level found in the control group ($p < 0.01$).

Table 1.   Ammonia levels in mice blood

	Ammonia level in the blood (μM)
Control group	54 ± 4
SVA administered group	
30 minutes after	101 ± 8*
60 minutes after	90 ± 7*

* Significantly different (p<0.01) from the control group.
Each value represents the mean±SE of 8 determinations.
In the SVA administered group, the blood was obtained 30
and 60 minutes after administration. Data were analyzed
according to the Student's t-test throughout Table 1 to 6.

Table 2.   Guanidino compounds in liver tissue

	Control group n=6	SVA administered group 30 minutes after n=6	SVA administered group 60 minutes after n=6
GSA	4.74 ± 0.82	4.26 ± 0.49	2.27 ± 0.63*
GAA	31.3 ± 2.9	25.0 ± 2.3	24.4 ± 1.7
GPA	1.70 ± 0.16	1.58 ± 0.10	1.73 ± 0.13
CRN	102 ±23	40.3 ± 4.1	43.3 ± 6.3
GBA	15.0 ± 2.1	16.6 ± 2.0	15.6 ± 2.7
Arg	12.6 ± 1.1	18.6 ± 0.9*	17.4 ± 2.2
HArg	1.60 ± 0.29	0.98 ± 0.11	1.25 ± 0.16
G	1.60 ± 0.29	2.91 ± 0.85	3.49 ± 1.26
MG	0.19 ± 0.03	0.08 ± 0.01	0.07 ± 0.01*

* Significantly different (p<0.05) from the control.
Values are expressed as nmol/g tissue (mean±SE).

Guanidino compound contents in tissues and plasma

As shown in Table 2, liver tissue levels of guanidinosuccinic acid (GSA), guanidinoacetic acid (GAA), β-guanidinopropionic
acid(GPA), creatinine (CRN), γ-guanidinobutyric acid (GBA),
arginine (Arg), methylguanidine (MG) and small amounts of homoarginine (HArg) and guanidine (G) were determined. SVA resulted

Table 3.   Guanidino compounds in kidney tissue

	Control group n=6	SVA administered group	
		30 minutes after n=6	60 minutes after n=6
GSA	0.49± 0.07	0.83± 0.07	0.15± 0.06*
GAA	396 ± 13	196 ± 22 **	149 ± 26 **
CRN	96.7 ± 13.8	63.5 ± 10.9	56.5 ± 11.7
Arg	181 ± 12	183 ± 13	178 ± 12
HArg	Trace	Trace	Trace
G	Trace	Trace	Trace
MG	0.21± 0.02	0.11± 0.02*	0.14± 0.02

* Significantly different (p<0.05) from the control.
** Significantly different (p<0.01) from the control.
Values are expressed as nmol/g tissue (mean±SE).

in a marked decrease in GSA levels in liver 60 minutes after its
administration (p<0.05 compared to the control group). The MG
content in liver was also found to be significantly decreased after
SVA administration (p<0.05). On the other hand, the Arg level in
liver increased 30 minutes after injection (p<0.05), but not sig-
nificanly 60 minutes post injection. CRN content in the liver
decreased somewhat after the administration. Other guanidino
compounds did not change in liver after SVA administration.

     In kidney, GAA, CRN, Arg, small amounts of GSA and MG as well
as a trace of HArg and G were observed (Table 3). SVA resulted in
a significant decrease in GAA content in the kidney 30 and 60
minutes after its administration (p<0.01). GSA was found to be
significantly decreased in the kidney at 60 minutes (p<0.05) and
MG decreased at 30 minutes (p<0.05). Other guanidino compounds
did not show any change after the administration of SVA.

     In the pancreas, besides GAA, CRN and Arg, trace amounts of
HArg and MG were discovered (Table 4). The Arg content in the
pancreas was found to be significantly increased 60 minutes after
SVA administration, whereas the content of the other compounds did
not change.

     In brain, GAA, CRN, Arg, MG and trace amounts of HArg and GSA
were found (Table 5). SVA administration affected only the Arg
levels in brain, which decreased at 60 minutes (p<0.05).

     Plasma guanidino compounds are shown in Table 6. No signifi-

Table 4. Guanidino compounds in pancreatic tissue

	Control group n=6	SVA administered group 30 minutes after n=6	60 minutes after n=6
GAA	319 ± 48	380 ± 49	431 ± 41
CRN	232 ± 42	284 ± 66	219 ± 35
Arg	169 ± 9	225 ± 31	233 ± 19*
HArg	Trace	Trace	Trace
MG	Trace	Trace	Trace

* Significantly different (p<0.05) from the control.
Values are expressed as nmol/g tissue (mean±SE).

Table 5. Guanidino compounds in brain tissue

	Control group n=6	SVA administered group 30 minutes after n=6	60 minutes after n=6
GSA	Trace	Trace	Trace
GAA	7.64± 0.46	7.47± 0.49	7.43± 0.61
CRN	571 ±73	671 ±85	376 ±62
Arg	147 ± 6	136 ± 7	119 ± 9 *
HArg	Trace	Trace	Trace
MG	0.13± 0.10	0.11± 0.04	0.07± 0.04

* Significantly different (p<0.05) from the control.
Values are expressed as nmol/g tissue (mean±SE).

cant changes were observed in the levels in plasma after SVA administration.

Urinary excretion of guanidino compounds

The amount of GSA, GAA, GPA, CRN, GBA, Arg, G and MG excreted in 24 hours in the urine of mice is shown in Fig. 1. The urinary excretion of Arg was found to have increased from 172 nmol/24 hours to 507 nmol/24 hours on the first day after SVA administration. The elevated Arg excretion fell to 282 nmol/24 hours on the second day after the administration. The urinary excretion of other

Table 6. Guanidino compounds in plasma

	Control group n=6	SVA administered group 30 minutes after n=6	60 minutes after n=6
GSA	Trace	Trace	Trace
GAA	3.89± 0.74	3.59± 0.21	3.21± 0.14
CRN	12.4 ± 0.66	16.1 ± 1.9	15.6 ± 1.8
Arg	226 ±37	167 ±11	157 ±14
HArg	Trace	Trace	Trace
MG	Trace	Trace	Trace

Values are expressed as nmol/ml (mean±SE).

guanidino compounds did not change after SVA administration. The urine excretion volume also increased, but only on the first day after administration as shown in Fig. 1.

DISCUSSION

There have been some reports that SVA induces hyperammonemia in humans[5,6,7]. In this study, we showed that SVA, intraperitoneally administered, can increase blood ammonia levels in mice within 30 min of administration. We also showed that SVA affects the guanidino compound concentrations in various mouse organs.

It is well known that most of the ammonia in mammals is excreted as urea by the kidney after the ammonia is converted to urea through the urea cycle in the liver. It is suspected that increases in blood ammonia levels are due to inhibition of ureagenesis. Coude et al.[10] reported that valproate inhibits citrullinogenesis in the rat. However, our results indicate that the elevated blood ammonia levels may have been caused by arginase inactivation, since the administration of SVA resulted in increases in both the liver and pancreatic Arg levels and the urinary excretion of Arg. Unlike the other organs, the Arg concentration in brain decreased. We reported previously that the administration of ammonium acetate to mice induced a decrease of Arg in brain tissues[11]. Thus, elevated ammonia levels seem to affect Arg metabolism in brain.

GSA is thought to be synthesized from urea through the guanidine cycle[12,13,15]. If it is true that arginase is inactivated by SVA administration, urea production would be decreased,

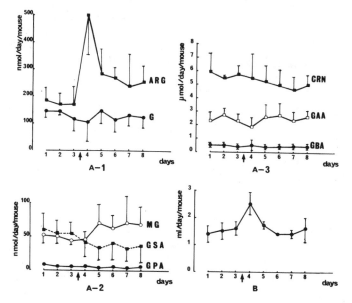

Fig. 1.   The effect of SVA administration on the urinary
excretion of guanidino compounds and urine volume.
The arrow indicates the time of the intraperitoneal
administration of SVA,
A : Guanidino compound excretion was expressed as
nmol or μmol/24 hours/mouse.
B : Urine volume was expressed as ml/24 hours/mouse.
Each point and bar represents the mean and S.E.,
respectively, of 3 experiments.

accounting for the decrease in the GSA concentration of the liver
and kidney.

It is well known that GAA is synthesized by Arg-Gly amidinotransferase in the kidney. The GAA concentration in kidney was
shown to be markedly decreased after SVA administration. Therefore SVA might affect not only the urea cycle but also other pathways of nitrogen metabolism. GAA was observed to be plentiful in
pancreatic tissue, but its concentration did not change after SVA
administration. It is suspected that there is a specific regulatory mechanism for GAA synthesis in the pancreas.

MG is considered to be synthesized mainly from CRN[14,15]. MG in kidney and liver tissue decreased after SVA administration which might be secondary to the decrease in GAA synthesis. In fact, CRN, which is synthesized from GAA, was shown to be somewhat decreased in liver, kidney and brain tissue.

GPA and GBA,whose synthetic pathways are still obscure, did not change in liver or urine. Synthesis of these substances seems to be regulated in a different way from that of other guanidino compounds such as Arg, GAA, GSA and MG.

In conclusion, we found that SVA directly or indirectly affects the metabolism of guanidino compounds, though the mechanism of this effect awaits clarification.

REFERENCES

1. J. F. Donat, J. A. Bocchini, E. Gonzalez and R. N. Schwendimann, Valproic acid and fatal hepatitis, Neurology, 29:273 (1979).
2. F. J. Suchy, W. F. Balistreri, J. J. Buchino, J. M. Soneimer, S. R. Bates, G. L. Kearns, J. D. Stull and K. E. Bove, Acute hepatic failure associated with the use of sodium valporate, N. Engl. J. Med., 300:962 (1979).
3. T. A. Bowdle, I. H. Patel, A. J. Wilensky and C. Comfort, Hepatic failure from valproic acid, N. Engl. J. Med., 301:435 (1979).
4. N. Gerber, R. G. Dickinson, R. C. Harland, R. K. Lynn, D. Houghton, J. I. Antonias and J. C. Schimschock, Reye-like syndrome associated with valproic acid therapy, J. Pediatr., 95:142 (1979).
5. D. L. Coulter and R. J. Allen, Hyperammonemia with valproic acid therapy, J. Pediatr., 99:317 (1981).
6. S. Rawat, W. J. Borkowski and H. M. Swick, Valproic acid and secondary hyperammonemia, Neurology, 31:1173 (1981).
7. M. L. Batshaw and S. W. Brusilow, Valproate-induced hyperammonemia, Ann. Neurol., 11:319 (1982).
8. A. Mori, Y. Watanabe and N. Fujimoto, Fluorometrical analysis of guanidino compounds in human cerebrospinal fluid, J. Neurochem., 38:448 (1982).
9. Y. Watanabe, S. Shindo and A. Mori, Developmental changes in guanidino compound levels in mouse organs (in this book).
10. F. X. Coude, D. Rabier, L. Cathelineau, G. Grimber, P. Parvy and P. Kamoun, A mechanism for valproate-induced hyperammonemia, in:"Urea Cycle Diseases," A. Lowenthal, A. Mori and B. Marescau, eds., Plenum Press, New York (1982).
11. S. Shindo, Y. Watanabe, Y. Katayama and A. Mori, Effects of electroconvulsive shock and ammonium acetate on guanidino compounds in the mouse brain, Neurosciences, 8:347 (1982).
12. S. Natelson, A. Koller, H. Tseng and R. F. Dods, Canaline corbamoyltransferase in human liver as part of a metabolic

cycle in which guanidino compounds are formed, Clin, Chem., 23:960 (1977).

13. B. D. Cohen and H. Patel, Guanidinosuccinic acid and the alternate urea cycle, in:"Urea Cycle Diseases," A. Lowenthal, A. Mori and B. Marescau, eds., Plenum Press, New York (1982).

14. G. Perez and R. Faluotico, Creatinine: A precursor of methyl-guanidine, Experientia, 29:1473 (1973).

15. H. Mikami, Y. Orita, A. Ando, M. Fujii, T. Kikuchi, K. Yoshihara, A. Okada and H. Abe, Metabolic pathway of gua-nidino compounds in chronic renal failure, in:"Urea Cycle Diseases," A. Lowenthal, A. Mori and B. Marescau, eds., Plenum Press, New York (1982).

THE GUANIDINE CYCLE

Harini Patel and Burton D. Cohen

Bronx-Lebanon Hospital
Bronx, N.Y., U.S.A.

INTRODUCTION

Disturbances in nitrogen balance are the hallmark of renal insufficiency. More sophisticated studies indicate that this is not simply a matter of balance. While failure to excrete adequate quantities may be the initiating event, a variety of adaptions and compensatory phenomena complicate the picture.

Some twenty-five years ago it was shown, by incubating rat liver slices in plasma from uremic animals, that urea formation increase in renal insufficiency[1]. Subsequent studies confirm this finding and these are recently reviewed[2] with the conclusion that loss of renal function increases urea production by some unknown but possibly adaptive, hormonal mechanism.

Central to these findings and the hypotheses they generate is the time honored thesis that urea is metabolically inert. Except for a small quantity which is degraded by bacterial ureases, a process unaltered in uremia, the bulk of urea is presumably synthesized solely for the elimination of excess nitrogen and enters into no further biochemical activities.

Recently, Natelson and Sherwin[3] have challenged this hythesis with the introduction of the guanidine cycle, a pathway for the conversion of urea nitrogen to the more stable storage form, creatine (Figure 1). The pathway begins via the enzymatic oxidation of urea to hydroxyurea, a reaction which is known to occur in reverse[4]. It proceeds by a series of steps analgous to the urea cycle to produce creatine from canavanine with guanidinosuccinic acid (GSA) as an inert, overflow byproduct. Production of the

135

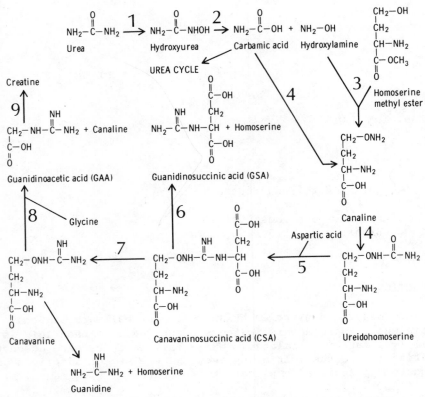

Fig. 1.   The Guanidine Cycle adapted from Natelson and Sherwin[3].
Note that carbon derived from urea reenters the carbon
pool via carbamyl phosphate derived from carbamic acid.
Hence, the failure of previous efforts[4] to demonstrate
an increased specific activity of GSA following the injection of C-14 labelled urea. The enzymes involved are as
follows: (1) mixed function oxidase (2) hydrolase (3) non-
enzymatic condensation, (4) canaline carbamyl transferase
(5) CSA synthetase (6) CSA reductase (7) CSA lyase (8)
transamidinase (9) transmethylase.

latter should increase whenever there is an excess of urea[5,6] and
decrease in states where urea production is inhibited[7].

In the following studies we present data in support of this
hypothesis. The first is evidence that, contrary to previous
reports[8], GSA is, indeed, inert and participates in no other biochemical or physical reactions. The second presents data confirming the general observation that the liver is the source of
this substance[7,9] and, finally, we offer a study showing the

relationship between GSA, urea and hydroxyurea suggesting some of the conditions for their interaction.

METHODS

Analysis of GSA and guanidinoacetic acid (GAA)

Separation and measurement were carried out by a method[10] involving gas-liquid chromatography (GLC) modified for micro-analysis as follows:

a. Purification. One ml of plasma or serum was passed through a mini Dowex-50 (H^+) column (7 x 25 mm) and washed with 10 ml of distilled water. This was eluted with 5.0 ml of 7N NH_4OH and partially evaporated at 100°C. Reduction to dryness is carried out at 70°C. The mean recovery of GSA by this method is 101 % and that of GAA is 85 %. Samples such as urines and tissue homogenates were initially chromatographed through Dowex-1-acetate columns[7] and portions of the eluate having the Sakaguchi color reaction were pooled. This was lyophilized and dissolved in 5.0 ml distilled water. One ml was evaporated to dryness and derivatized.

b. Derivatization. The evaporated samples were esterified with 0.5 ml of 3N HCl in n-butanol for 25 minutes at 110°C. The acylation was carried out at the same temperature with 0.5 ml of 33 % trifluoroacetic anhydride in methylenechloride, cooled to room temperature and evaporated with a stream of nitrogen. After reconstitution in 100 µl of methylenechloride, 5 to 10 µl were injected into packed column consisting of a mixture of 1.5 % OV-210 and 1.5 % OV-17 on Chromosob-W H.P. The mean recovery of GSA by this procedure is 95.3 %.

Binding studies

Physical reactions of GSA and GAA with various substrates were carried out as follows:

a. Protein binding. Plasma was obtained from heparinized blood. Red cell protein was prepared by hemolysis after washing with physiologic saline. To 5.0 ml of each solution GSA and GAA was added to achive a concentration of 0.25 to 2.0 mg/dl and 0.1 to 0.5 mg/dl respectively. Incubation was then carried out overnight at 37 °C and physiologic pH. The solutions were ultrafiltered through Centricone C-25 (Amicone Corp.) to yield sufficient filtrate for the analysis of GSA and GAA by gas-liquid chromatography. An aqueous solution of GSA and GAA was used as control and analyzed simultaneously.

b. Diffusion studies. Pooled, washed red cells were sus-

pended in pH 7.4 buffer and the hematocrit noted. These were
incubated overnight with GSA and GAA as above and the concentrations of the latter determined before and after incubation to
calculate the percent diffusion.

Reactivity studies

The biochemical activity of GSA was compared with that of GAA
as follows:

A 33 % guinea pig liver homogenate was prepared and incubated
with GSA and GAA after the method of Borsook and Dubnoff[11]. After
incubation, each tube was acidified and centrifuged. An aliquot of
the supernatant was analyzed by GLC and reduction in the concentration of reactants was compared with that before incubation.

Tissue analyses

Tissue from various organs were obtained at post mortem examination from patients dying with uremia as well as a variety of
non-uremic conditions. These were promptly frozen along with a
serum sample obtained immediately prior to death. Fifteen to 20 g
of the tissues were blended in the cold with 30 ml of 8 % TCA and
centrifuged at 10,000 rpm[12]. The supernatant was extracted with
ether and applied to the Dowex-1-acetate anion exchange column[7].
The tubes showing GSA by Sakaguchi color reaction were pooled and
 analyzed by GLC as outlined above.

Rat studies

Sprague-Dawley rats weighing 200 to 300 g were confined two to
a metabolic cage and maintained on Purina rat chow except for the
24 hours once weekly during which urines were collected. On the
day of collection the rats were injected intraperitoneally with
5.0 ml of experimental solution and 24-hour urine samples from 4
rats were pooled for analysis. Control animals were not injected.
Ten to 15 ml of urine were analyzed using the two-way system
consisting of anion exchange chromatography followed by GLC as
described above.   Creatinine was measured by the Jaffe reaction.
The quantity of GSA excreted in 24 hrs. per mg of creatinine was
calculated and the results are reported as a ratio of study animals
to controls obtained on the same day.

RESULTS

Table 1 shows the results of binding studies. Neither GSA or
GAA appear to bind to the proteins of physiologic fluids on either
side of the cell membrane. The separation technique used (centricone C-25) discards proteins with a molecular weight above 25,000.

Table 1.   Binding and diffusion studies of GSA and GAA

	Bovine Albumin (4)*	Human Plasma (7)	Hemolyzed Red Cells (6)	Homogenized Red Cells (3)	Intact Red Cells (9)
GAA	100±4**	103±15	95±4	101±1	64±5
GSA	99±7	98± 5	96±8	98±2	109±9

* indicates number of individual analyses
** mean±S.D. percent recovery following incubation

Therefore, the most common binding proteins, do not bind the guanidino compounds, a finding supported by the observation that the
renal clearance of GSA approximates that of inulin[13]. GAA, but not
GSA, appears to be freely diffusible across the erythrocyte membrane
as shown by the studies on intact red cells. The mean hematocrit
in the series was 39 % which accounts for the 36% reduction in post-
incubation concentration of GAA.

Figure 2 compares the reactivity of GAA and GSA incubated
with guinea pig liver homogenates at 37 °C, pH 7.4 with methionine,
ATP and oxygen. GAA reacts, as shown by Borsook and Dubnoff[11], at
a rate proportional to the quantity of substrate present. GSA
shows no evidence of reactivity. This confirms findings of Takahara, Nakanishi, and Natelson[14] using several animal tissues.
In an earlier study[15], the same investigators injected 100 mg.
GSA intraperitoneally in rats and recovered 94 mg in urine in 5
days. In a similar study, we injected 2.5 mg intracardiac in
eight rats weighing 298±20 g. The serum GSA concentration obtained
by exsanguination at 24 hours was 53±8 µg/ml. Assuming an extracellular volume of 45 ml, this level is predictable only in the
absence of GSA metabolism.

Table 2 shows the hepatic concentration of GSA in liver obtained at post mortem from 4 subjects dying with uremia compared
with 4 non-uremic tissues which serve as controls. Note that a
value of zero simply implies a concentration below the limit of
sensitivity of the method and not the total absence of GSA in the
tissue. There is a fair correlation with hepatocellular urea concentration having a coefficient of 0.8308. Table 3 compares the
concentration of GSA in a variety of tissues from subjects dying
with uremia and suggests that the liver is the source of this
compound.

The plasma urea nitrogen concentration was measured in 9 rats

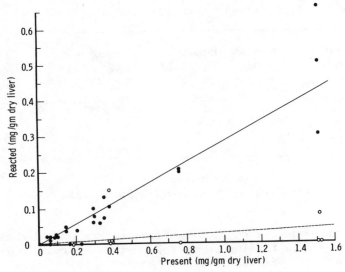

Fig. 2.   The figure presents results of the reactivity studies.
The closed circles are values obtained for GAA; the open
circles represent GSA. The reactivity of GAA is a linear
function of the concentration (r=0.8743) as shown by the
solid line. GSA, the broken line, is essentially non-
reactive.

following the intraperitoneal injection of a 5 ml. solution containing 10 mmoles of urea. The concentration rose promptly to 150
mg/dl persisting for 12-18 hours before returning to control
levels at 24 hours. Table 4 and 5 show the effect of similar injections on the 24 hour excretion of GSA which is reported per
mg of creatinine to correct for the osmotic diuresis induced by the
injected material.

DISCUSSION

The measurements shown in Table 4 confirm urea as a source of
GSA and, therefore, support the proposed guanidine cycle. Urea
increases the production of GSA in proportion to the amount injected and hydroxyurea appears to be an intermediate. Small quantities of hydroxyurea do not increase GSA synthesis while larger
amounts effect GSA production to the same extent as urea. This
suggests that the equilibrium between hydroxyurea and urea favors
urea formation as previously reported[5].

On the other hand, when urea levels are concurrently increased,
such as when both are injected (or in uremia), the oxidation

Table 2.   Serum and hepatic concentrations of GSA

| | Serum (mg/dl) | | | Hepatic (mg/dl)* | | |
	GSA**	Urea	Creatinine	GSA***	Urea	Creatinine
UREMICS						
E.T.	0.92	39	8.2	0.48	73	10.8
M.O.	0.48	63	7.6	0.84	64	7.4
H.R.	0.86	104	14.8	1.15	82	6.5
W.H.	0.29	61	10.1	0.23	45	7.4
CONTROLS						
G.C.	0.00	17	1.4	0.00	45	3.6
F.C.	0.00	21	1.3	0.00	27	2.7
D.R.	0.00	26	1.6	0.00	–	–
E.L.	–	13	0.8	0.00	18	0.9

*   Tissue water was measured by the weight change following lyoph-
      ylization. The computation of GSA was corrected for the con-
      centration in the mean cell water (77%).
**   By GLC the limit of sensitivity is 2 µg so that if 1 ml is
      processed, a reading of 0.00 indicates less than 0.20 mg/dl.
*** Estimated by Sakaguchi reaction and corrected for the mean
      tissue recovery of GSA (74%). The limit of sensitivity is
      25 µg so that if 29 g are processed, a reading of 0.00
      indicates less than 0.18 mg/dl.

Table 3.   Concentration of GSA in a variety of tissues in uremia
           GSA (mg/dl) in tissue water*

	Serum	Liver	Pancreas	Muscle	Brain
A.R.	–	2.60	0.27	0.22	trace
M.A.	0.95	1.40	0.09	0.14	trace
O.R.	0.46	0.58	0.11	0.05	–

* not corrected for recovery

reaction proceeds increasing the incorporation of exogenous
hydroxyurea into GSA. The oxidation reaction, therefore, appears
to have slower kinetics and accelerates when urea concentration
increases in the body fluids such as in uremia.

Table 4. Excretion of GSA in micrograms per milligram of crea-
          tinine in the 24-hour urine of control rats and rats
          injected with urea, hydroxyurea (OH-U) and amino acids

Injection	Control	N_C ***	Treated	N_T	$\dfrac{\text{Treated}}{\text{Control}}$	p°
Urea (10)*	8.12±2.50**	32	17.70± 8.50	36	2.2	<0.005
Urea (20)	7.25±2.60	32	32.27±15.30	40	4.4	<0.0025
OH-U (5)	8.22±2.08	20	8.58± 3.02	24	1.0	0.45
OH-U (10)	8.39±0.89	20	20.81± 7.89	20	2.5	0.005
Urea (10)+ OH-U (5)	8.72±1.36	20	26.94± 6.75	32	3.1	<0.0025
Urea (10)+ OH-U (10)	5.69±2.26	8	24.72±10.22	12	4.3	<0.0025
Urea (10) + Homoserine (1.25)	6.89±3.15	16	14.62± 4.92	32	2.1	>0.20°°
Urea (10) + Aspartate (1.25)	8.30±2.80	32	21.50± 7.20	26	2.6	0.40°°

*   The number in parenthesis refers to the quantity injected in
      millimoles per rat.
**   Mean ± S.D.
*** N is the number of animals
°   Significance was calculated using the Student's T-test
°°   Significance was calculated by comparing with results in rats
      given 10 mmole urea alone.

     Aspartate and homoserine, while participants in the guanidine
cycle, do not increase GSA output and, therefore, do not enter
into rate-limiting steps in the biosynthetic pathway yielding GSA
as a byproduct.

     The situation with regard to methionine is different as seen
in Table 5. Methionine, which is not an intermediate in the guanidine cycle, reduces GSA excretion. This observation, previously
reported [7,8] was once thought to result from the reaction of
methionine and GSA to form methyl GSA, and analogue of creatine.
As seen from the foregoing, GSA is non-reactive even in the presence
of added methionine (Fig's. 1 & 2). The mechanism, therefore, for
the methionine-induced reduction in GSA synthesis remains obscure.

     Apart from advancing the proposition that there exist circum-
stances in which urea is not biochemically inert, elucidation of

Table 5. Daily GSA excretion in rats following injections
          of urea and methionine

Injection		Treated Control
Urea 10 mmoles	(10)	1.9±0.6
Urea 10 mmoles + Methionine 1.25 mmoles	(12)	0.8±0.1
Urea 20 mmoles	(15)	6.1±2.2
Urea 20 mmoles + Methionine 1.25 mmoles	(9)	2.5±1.6

the factors controlling the guanidine cycle can have practical,
clinical consequences. Long considered as candidates for the role
of toxins in uremia, various guanidines have been shown to disrupt
the function of platelets, contribute to erythrohemolysis, inhibit
the activity of pyridoxal phosphate, depress the central nervous
system and denature proteins by substituting preferentially during
synthesis. Understanding the factors and conditions which lead to
urea oxidation could be of value in reducing toxicity in the presence of failing kidneys.

SUMMARY

     Retention of nitrogen is at the core of the uremic syndrome.
Recently, evidence was offered in support of the proposition that
urea is not metabolically inert but, under certain circumstances,
participates in a series of reactions leading to potentially toxic
or beneficial substances: creatinine and the guanidines. This
study presents data in support of this hypothesis and examines
some of the necessary conditions. Using guanidinosuccinic acid
(GSA), a demonstrably inert by-product of the reaction, as marker
we show that 1) GSA is both biochemically and physiologically
inert, 2) oxidation of urea is substrate dependent and 3) the reaction rate can be altered, notably by means of methionine.

ACKNOWLEDGEMENTS

     The authors wish to thank Roger Kornhauser for his technical
assistance and Ruth Geller for typing the manuscript.

REFERENCES

1. A. L. Sellers, J. Katz and J. Marmrston, Effect of bilateral
    nephrectomy on urea formation in rat liver slices,
    Am. J. Physiol., 191:345 (1957).
2. M. Walser, Determinants of ureagenesis, with particular ref-
    erence to renal failure, Kidney Int., 17:709 (1980).
3. S. Natelson and J. W. Sherwin, Proposed mechanism for urea
    nitrogen re-utilization: relationship between urea and pro-
    posed guanidine cycle, Clin. Chem., 25:1343 (1979).
4. B. D. Cohen, I. M. Stein and R. S. Kornhause, Guanidine
    retention and the urea cycle, in:"Proceedings of the Inter-
    national Congress of Nephrology," Karger, Basel (1970).
5. M. Colvin and F. H. Bono Jr., The enzymatic reduction of
    hydroxyurea to urea by mouse liver, Cancer Res., 30:1516
    (1970).
6. J. E. Bonas, B. D. Cohen and S. Natelson, Separation and
    estimation of certain guanidino compounds: Application to
    human urine, Microchem., J. 7:63 (1963).
7. I. M. Stein, B. D. Cohen and R. S. Kornhause, Guanidino-
    succinic acid in renal failure, experimental azotemia and
    inborn errors of the urea cycle, N. Engl. J. Med., 280:926
    (1969).
8. B. D. Cohen, H. Patel and R. S. Korhauser, Alternate reasons
    for atherogenesis in uremia, Proc. Dialysis Transplant
    Forum, 7:178 (1977).
9. H. Dobbelstein, H. H. Edel, M. Schmidt, G. Schubert and
    M. Weinzierl, Guanidinbernsteinsaure and Uremia, Klin.
    Wochenschr., 49:348 (1971).
10. H. Patel and B. D. Cohen, Gas-liquid chromatographic
    measurement of guanidino acids, Clin. Chem., 21:838 (1975).
11. H. Borsook and Dubnoff, J. W., On the role of oxidation in
    the methylation of guanidinoacetic acid, J. Biol. Chem.,
    171:363 (1947).
12. S. Ratner, Determination of Argininosuccinate in normal blood
    serum and liver, Analytical Biochemistry, 63:141 (1975).
13. B. D. Cohen, Guanidinosuccinic acid in uremia, Arch. Int.
    Med., 126:846 (1970).
14. K. Takahara, S. Nakanishi and S. Natelson, Studies on the
    reductive cleavage of canavanine and canavaninosuccinic
    acid, Arch. Biochem. Biophys., 145:85 (1971).
15. K. Takahara, S. Nakanishi and S. Natelson, Cleavage of
    canavaninosuccinic acid by human liver to form guanidino-
    succinic acid, a substance formed in the urine of uremic
    patients, Clin. Chem., 15:397 (1969).

ARGININE UTILIZATION IN MUSCLE AND KIDNEY

OF FROG DURING AMMONOTOXEMIA

P. Neeraja and K.S.Swami

Department of Zoology
Sri Venkateswara University
Tirupati-517 502, India

INTRODUCTION

The lethal effects of ammonia toxicity are well established. To resist the toxemia, the detoxification mechanism may occur in animals[1] and some of these involve arginase metabolism.

Arginine forms an important source for urea formation and guanidine synthesis[2,3]. Alterations in arginase activity levels were estimated in muscle and kidney of frog under induced ammonotoxemia.

MATERIAL AND METHODS

Medium sized frogs Rana hexadactyla were acclimated to laboratory conditions for a period of one week. As per standardisation described by Neeraja and Swami[4], a near lethal concentration of 550 µmol/animal of ammonium acetate was administered intraperitoneally to one set of frogs as this concentration will invariably induce ammonotoxemia. Another set of animals serving as controls received equivalent volume of physiological saline (0.75 % NaCl). The muscle and kidney tissues were quickly excised and chilled to 5 °C after double pithing the animals at 0, 5, 60, 120 and 180 min subsequent to the ammonium acetate administration. The tissues were analysed for arginase activity levels by the method of Campbell[5] and urea content by Natelson[6]. The values are represented as µmol/mg protein/hr and µmol/g wet wt. respectively.

Table 1.   Changings in urea and arginase activity levels at different
           time intervals in the muscle and kidney of Rana hexadactyla
           receiving 500 μmol/animal of ammonium acetate

Tissue & Component		Control	Time in minutes			
			5	60	120	180
Muscle	mean	8.15	3.10	4.39	7.87*	5.39
	S.D.	0.48	0.12	0.40	0.50	0.79
Urea			(−62.0)	(−46.0)	(−3.41)	(−33.8)
Muscle	mean	0.31	0.24	0.15	0.33*	0.16
	S.D.	0.04	0.03	0.01	0.02	0.01
Arginase			(−22.6)	(−51.6)	(+6.5)	(−48.4)
Kidney	mean	11.32	5.62	7.70	8.63	33.07
	S.D.	0.99	0.70	1.09	1.02	0.0]
Urea			(−50.4)	(−31.9)	(−23.8)	(+192.0)
Kidney	mean	16.75	12.04	7.67	7.87	32.99
	S.D.	1.60	1.10	0.50	0.41	3.90
Arginase			(−28.1)	(−54.2)	(−53.0)	(+96.9)

Values are mean and S.D. of eight observations.   % Difference is
represented in parenthesis.   Values represented by asterisk are
not significant.   All the other values are significant at p<0.001.

RESULTS AND DISCUSSION

     The arginase activity levels and urea content in the muscle
and kidney tissues showed variable trends (Table).

     The kidney urea content was higher than the muscle tissue in
the control animals.   Since kidney is known to be an active centre
for ureogenesis[7], the present findings are in conformity.

     The arginine activity levels and urea content in the control
animal was fairly constant during the period of investigation indicating a balanced metabolic synthesis.   But following the administration of ammonium acetate, the muscle and kidney tissues
showed alterations in the existing pattern.   The muscle urea content showed a progressive temporal decrease after ammonium acetate
administration upto 2 hrs as did the arginase activity levels.

The low arginase and urea content was indicative of low ornithine production and consequent metabolism. Hence, it is likely that the arginine is diverted into the synthesis of other guanidine compounds[8].

The kidney urea levels decreased initially upto 2 hrs and then increased at 3 hrs following ammonium acetate administration. The arginase activity levels showed a similar pattern. The low arginase activity is suggestive of low arginine substrate levels suggesting arginine depletion by other channels. Arginine depletion for synthesis of other compounds such as guanidines has been reported in similar circumstances[9-12]. In the present study a metabolic diversion of arginine could have resulted in low enzyme activity levels and consequently reduced urea production. But after longer periods of ammonium acetate stress, the kidney tissue restores the arginine-arginase system. Under stress such as ammonium loading, the kidney resorts to synthesizing guanidine compounds recovering urea synthesis sometime later. The observed shift in the kidney metabolism was not seen in muscle. Similar diverse responses to ammonia intoxication were reported in the brain and liver of frogs[13].

In the muscle and kidney tissues of the animals under ammonia stress, the low arginase activity levels and reduced urea is indicative of low ornithine production. While ornithine metabolism is the primary target of ammonotoxemia in these two tissues, synthesis of other guanidine compounds occurs. Since guanidine synthesis is also a mechanism to detoxify ammonia[14], this could be the propose of the shift in arginine metabolism observed.

REFERENCES

1. C. L. Prosser, "Comparative Animal Physiology,"
   W. B. Saunders company, (1973).
2. N. V. Thoai and J. Roche, Biological guanidine derivarive,
   Fortsch Chem. org Naturstoffe., 18:83 (1960).
3. F. J. R. Hird, D. Sivaprasad and R. M. McClean, Evolutionary
   relationships between arginine and creatine in muscle, in:
   "Urea cycle diseases," A. Lowenthal, A. Mori and
   B. Marescau, eds., Plenum Press, New York & London (1981).
4. P. Neeraja and K. S. Swami, Induced transient hyperammonemia
   in Rana hexadactyla with reference to the ammonia and urea
   excretion, Ind J. Physiol. Pharmacol., 27:123 (1983).
5. J. W. Campbell, Studies on tissue arginase and ureogenesis in
   the elasmobranch Mustelus canis, Arch. Biochem. Biophys.,
   93:448 (1961).
6. S. Natelson, "Techniquies of clinical chemistry," Thomas. CC,
   ed., Spring field, Illinois (1971).

7. H. A. Harper, Catabolism of amino acids, in: "Review of physiological chemistry," Lange Medical Book Publ. California (1977).

8. S. Natelson and J. E. Sherwin, Proposed mechanism for nitrogen reutilization: Relationship between urea and proposed guanidine cycles, Clin. Chem., 25:1343 (1979).

9. A. L. Triffilis, M. W. Kahng and B. F. Trump, Metabolic studies of mercuric chloride-induced acute renal failure in the rat, Exp. Mol. Pathol., 35(1):14 (1981).

10. Y. Yonetani and K. Iwake, Catecholamine induced hyperuricemia in eviscerated rats with functional hepatectomy, J. Pharmacol., 31(3):323 (1981).

11. B. Marescau, A. Lowenthal, E. Esman, Y. Leuten, W. Alder T. Frank and G. Heinz, Isolation and identification of some guanidine compounds in the urine of patients with hyperargininaemia by liquid chromatography, thin layer chromatography & gas chromatography-mass spectrometry, J. Chromatogr., 224(2):185 (1981).

12. Y. Katayama, S. Shindo, A. Sawaki, Y. Watanabe, C. Hiramatsu and A. Mori, Biosynthesis of taurocyamine by transamidination reaction of taurine, Ganryu Aminosan., 2:297 (1979).

13. P. Neeraja and K. S. Swami, Effect of induced ammonia toxicity of urea cycle enzymes of frog, Indian. J. Comp. Anim. Physiol., 1:19 (1983).

14. N. V. Thoai, Guanidine derivatives, their biological roles, Collog Intern. Centre. Natl. Recherch. Sci. (Paris)., 2:297 (1959).

ARGININE METABOLISM IN CITRULLINEMIC PATIENTS

Takeyori Saheki, Mariko Sase, Kyoko Nakano
and Yukio Yagi

Department of Biochemistry, School of Medicine
Kagoshima University, Kagoshima, Japan

INTRODUCTION

Arginine is synthesized from citrulline by the catalytic actions of argininosuccinate synthetase (ASS) and argininosuccinase in the kidney[1,2], and utilized as a precursor of creatine biosynthesis or supplied to various organs for protein synthesis. The liver performing ureogenesis also synthesizes arginine, which, however, is rapidly split to ornithine and urea by the potent activity of arginase.

Under physiological conditions, citrulline is supplied from the small intestine as an end product of glutamine nitrogen metabolism[3], but not from the liver where citrulline is produced as an intermediate of ureogenesis. The metabolic pathway is shown in Fig. 1a.

However, citrulline is released in large amounts from the liver of patients with citrullinemia, which is hereditary urea cycle disorder caused by a deficient activity of ASS, and accumulates in blood at even more than 100 times normal.

Walser et al.[4] and Morrow et al.[5] described that citrullinemia tends to be arginine-deficient in spite of elevated levels of citrulline in the blood, suggesting that ASSs in the liver and kidney originate from the same gene.

Observations conflicting with the above conclusion have been reported by several Japanese investigators who demonstrate that serum arginine is elevated in citrullinemic patients, particularly adults (Table 1). We tried to resolve the conflict by analyzing

149

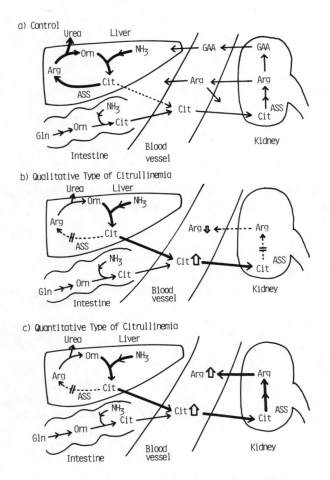

Fig. 1.   Citrulline and arginine metabolism in control subjects,
          and difference in the metabolism between qualitative and
          quantitative types of the enzyme abnormality in
          citrullinemia.

the abnormal enzymes in the liver and kidney of citrullinemic
patients in Japan.   Detailed case reports of several patients
mentioned in this paper have been published elsewhere[6-13].

Table 1. Concentrations of citrulline and arginine in the serum of citrullinemic patients and type of the enzyme abnormality

	Sex	Age of Hospitalization	Citrulline	Arginine	
			(nmol/ml serum)		
Control			20–40	80–130	
Patient (adult)					
No. 1	m	48Y	238	118	Quantitative
2	m	46Y	318	272	Quantitative
3	f	24Y	2000	43	Qualitative
4	m	24Y	373	358	Quantitative
5	m	18Y	370	381	Quantitative
6	f	26Y	88	69	Qualitative
7	m	16Y	130–280	162 57	Quantitative
8	m	20Y	667	–	Quantitative
9	m	21Y	173	–	Quantitative
10	m	20Y	751	213	Quantitative
11	m	28Y	684	269	Quantitative
12	m	29Y	290	255	Quantitative
13	m	22Y	266	219	Quantitative
14	m	42Y	194	247	Quantitative
15	f	20Y	105	17	Quantitative
16	f	16Y	240	139	Quantitative
17	f	40Y	208	147	Quantitative
18	m	33Y	1245	551	Quantitative
Patient (neonate)					
No. 1	f	24d	2950	101	Qualitative
2	f	21d	2485	35	Qualitative
3	f	7d	1140	31	Qualitative
4	f	3d	3925	79	Qualitative
5	f	10m	423	124	Quantitative

METHODS

Determination of enzyme activity

ASS activity was determined in a modified reaction medium of Schimke[14] which contained 5 mmol/l [^{14}C-ureido]citrulline (0.5 Ci), 5 mmol/l aspartate, 0.9 mmol/l ATP, 1.5 mmol/l phosphonolpyruvate, 2 units of pyruvate kinase, 4 units of adenylate kinase, 0.3 units of inorganic pyrophosphatase, 0.5 units of argininosuccinase, 80 units of arginase, 5 mmol/l $MgCl_2$, 20 mmol/l KCl and 50 mmol/l Tris-HCl, pH 7.5, in a final volume of 500 µl; the incubation was performed at 25 °C.

For the determination of the Km values, the enzyme sample was allowed to flow through a Sephadex G-25 column to remove low molecular weight substances. The concentration of each substrate was varied from 0.02 mmol/l to 20 mmol/l with the fixed concentration of the other substrates as described above. One unit was defined as the amount of enzyme producing 1 μmol of product per min.

Immunohistochemical staining was performed by the peroxidase anti-peroxidase (PAP) method of Sternberger et al.[15].

RESULTS AND DISCUSSION

Abnormality of argininosuccinate synthetase in the liver of citrullinemic patients

Incidence of citrullinemia is relatively higher in Japan than in any other country. More than 50 cases of citrullinemia have been reported in Japan. Most of them were characterized by a higher age of onset (20-50 years old) and moderately high level of serum citrulline (Table 1).

We succeeded in classifying the abnormality of ASS in the liver of citrullinemic patients into two groups by analyzing the kinetic properties and estimating the amount of enzyme protein in the liver[16], summarized in Table 2.

In one group, a decreased ASS activity in the liver represented a defectiveness of the enzyme protein, i.e. abnormal kinetic properties such as larger Km values (see Table 4). The amount of enzyme protein was compatible to or lower than that in the control liver. A lower amount of enzyme protein observed in some cases resulted from decreased stability of the enzyme judging from a heat stability test.

In a second group, a decreased ASS activity in the liver may be explained by the decrease in the amount of enzyme protein having the normal kinetic properties. We call the former and the latter qualitative and quantitative types, respectively. Most of analyzed neonatal citrullinemic patients were of the qualitative type, while the great majority of adult citrullinemic patients were of the quantitative type (Table 2).

ASS activity in the kidney of two types of citrullinemia

We analyzed ASS in the kidney of three adult and two neonatal citrullinemic patients whose abnormal hepatic enzymes were diagnosed as quantitative and qualitative, respectively (Table 3).

The kidney specimens of the two neonatal citrullinemic

Table 2. Comparison between qualitative and quantitative types of
the enzyme abnormality of citrullinemia

	Qualitative	Quantitative
Frequency		
# Adult	1-2/18	16/18
# Neonate	4/5	1/5
Liver ASS		
# Activity	Decreased	Decreased
# Kinetics (Km values)	Abnormal	Normal
# Enzyme amount determined by SRID	Normal or Decreased	Decreased
Kidney and fibroblast ASS		
# Activity	Decreased	Normal
# Kinetics	Abnormal (same as hepatic ASS)	Normal
Arginine in serum	Low	High
Pathogenesis		
# Abnormality in	Structural Gene	Organ-Specific Gene Expression

SRID   denotes single radial immunodiffusion

patients exhibited a decreased ASS activity with a very large Km
value for citrulline and aspartate quite similar to those of the
hepatic enzyme (Table 4). From the results, it may be concluded
that the ASS enzymes in the liver and kidney are not isozymes but
originated from the same gene.

On the contrary, all of the kidney specimens of the adult
patients tested showed ASS activities comparable to the controls
and no abnormal kinetic properties in the ASS reaction. These
results suggest that the quantitative abnormality of the hepatic
ASS of citrullinemic patients may be due to abnormal organ-specific
gene expression.

Serum-arginine levels of citrullinemic patients

The results listed above suggest that the qualitative type
of citrullinemia would show a decreased serum-arginine level
because of a decreased ASS activity in the kidney and that the
quantitative type would show an increased serum-arginine level
on the basis of an intact ASS activity in the kidney and a high
concentration of citrulline, i.e. an abundant supply of the
substrate.

Table 3. Argininosuccinate synthetase activity in the liver and
the kidney of control and citrullinemic patients, and its
relation to types of enzyme abnormality.

| | ASS Activity | | Type of Enzyme Abnormality of Hepatic ASS |
	Liver	Kidney	
	nmol/min/g tissue (nmol/min/mg protein)		
Adult			
Control	700±240(7.5±2.3)	160±40(3.3±1.5)	–
Patient			
No. 2	13(0.19)	220(6.7)	Quantitative
No. 9	96(1.6)	220(4.3)	Quantitative
No.10	43(0.80)	180(4.7)	Quantitative
Neonate (0-3 month)			
Control	180 ± 71(3.2±2.4)	42±23(0.92±0.51)	–
Patient			
No. 1	1.1(0.019)	1.8(0.036)	Qualitative
No. 3	2.9(0.056)	3.1(0.10)	Qualitative

Table 4. Km values (or substrate concentrations for half-maximum
activity) of argininosuccinate synthetase in the liver
and the kidney of control and citrullinemic patients
(neonatal type)

| | Km Value (mM) | | | |
| | Citrulline | | Aspartate | |
	Liver	Kidney	Liver	Kidney
Control	0.049	0.029	0.033	0.017
Patient(Neonate)				
No.1	20	15	37	22
No.3	7.0	8.0	12	20

Table 1 shows serum-arginine level and type of ASS abnormality
of adult and neonatal citrullinemic patients. Serum arginine of
the qualitative-type citrullinemia ranges from below normal to low
normal. On the contrary, serum arginine of the quantitative-type
citrullinemia is normal or increased.

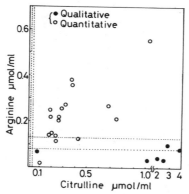

Fig. 2. Relation between serum citrulline and arginine of
         citrullinemic patients, and the difference between quali-
         tative and quantitative types of citrullinemia.
         Control ranges of serum citrulline and arginine are shown
         with dotted lines.

The results listed in Table 1 are illustrated in Fig. 2 which
clearly demonstrates the relation between serum citrulline and
arginine of citrullinemic patients and the difference between the
qualitative and quantitative types. The only exception is the
adult citrullinemic patient No 15, who shows a decreased serum
arginine in spite of a quantitative deficiency of hepatic ASS.

Localization of ASS in the kidney

It was reviewed by Ross and Guder[17] that since the kidney
contains various kinds of cells, there is a definite division of
various metabolic pathways in the kidney.

The localization of ASS in the kidney of rats was demonstrated by an immunohistochemical method[15] with anti-ASS antiserum.
As shown in Fig. 3, ASS was located almost exclusively in the cytoplasma of the proximal tubule, suggesting definite segregation of
arginine metabolism.

CONCLUSION

Increased levels of serum arginine have been observed in
citrullinemic patients in Japan, which is in contrast with the
reports of Walser et al.[4] and Morrow et al.[5] demonstrating that
citrullinemia tends to be arginine-deficient.

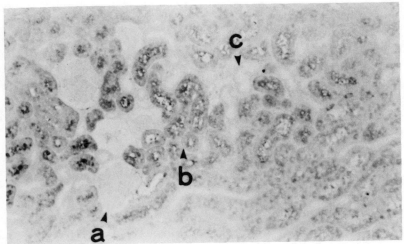

Fig. 3.   Immunohistochemical localization of argininosuccinate
          synthetase in rat kidney.
          a, glomerulus; b, proximal tubule; c, distal tubule.

This conflict was resolved by analyzing abnormal argininosuccinate synthetase in citrullinemic subjects in Japan. We found
two types of the enzyme deficiency in citrullinemia. One type
resulted in a qualitative deficiency of the enzyme where the
kidney exhibited a decreased activity with abnormal kinetic properties similar to those of the hepatic enzyme. The other type resulted in a quantitative deficiency of the enzyme in the liver,
but not in the kidney. The latter type is responsible for the
great majority of adult citrullinemia in Japan.

Argininosuccinate synthetase in the kidney plays a role in
the biosynthesis of arginine from citrulline as described by
Ratner[1] and Funahashi et al.[2]. The former type is defective in
synthesizing arginine from citrulline (Fig. 1b), while the latter
type has a great ability to synthesize arginine bacause of the
normal enzyme activity and an abundant supply of citrulline in the
kidney, resulting in increased levels of serum arginine (Fig. 1c).
We also showed by an immunohistochemical method that argininosuccinate synthetase is located mainly in the cytoplasma of the
proximal tubule, suggesting definite compartmentalization of
arginine metabolism.

ACKNOWLEDGEMENT

This work was done in the cooperation with the following
clinical investigators; Drs. T. Hirasawa, Y. Yajima, Y. Yamauchi,
K. Fujisawa, N. Yamada, H. Sibata, K. Akamatsu, Y. Ohta, K.

Kobayashi, K. Itahara, Y. Suzuki, S. Furuta, T. Kariya, M. Takamizawa, A. Watanabe, I. Akaboshi, I. Matsuda, K. Takahashi, Y. Hara, T. Tahira, T. Kiryu, H. Nara, A. Kondo, T. Kurihara, K. Araki, M. Saito, M. Matsuo, J. Ito, Y. Wada, Y. Maeda, T. Hosoyamada, Y. Itakura, and K. Oyanagi.

REFERENCES

1. S. Retner, Enzymes of arginine and urea synthesis, in: "Advances in Enzymology," A. Meister, ed., vol. 39, John Wiley and Sons, New York (1973).
2. M. Funahashi, H. Kato, S. Shiosaka and H. Nakagawa, Formation of arginine and guanidinoacetic acid in the kidney in vivo, J. Biochem., 89: 1347-1356 (1981).
3. H. G. Windmueller and A. E. Spaeth, Source and fate of circulating citrulline, Am. J. Physiol., 241: E473-E480 (1981).
4. M. Walser, M. Batshow, G. Sherwood, B. Robinson and S. Brusilow, Nitrogen metabolism in neonatal citrullinemia, Clin. Sci. Mol. Med., 53: 173-181 (1977).
5. G. Morrow, L. A. Brsness and M. L. Efron, Citrullinemia with defective urea production, Pediatr., 40: 565-574 (1967).
6. Y. Yajima, T. Hirasawa and T. Saheki, Treatment of adult-type citrullinemia with oral administration of citrate, Acta Hepatol. Jap., 21: 1682-1689 (1980).
7. M. Yamauchi, T. Kitahara, K. Fujisawa, H. Kameda, S. Takasaki, R. Komori, T. Saheki, T. Katsunuma and N. Katunuma, An Autopsied case of hypercitrullinemia in adult caused by partial deficiency of liver argininosuccinate synthetase, Acta Hepatol. Jap., 21: 326-334 (1980).
8. N. Yamada, M. Fukui, K. Ishii, H. Shibata, H. Okabe, H. Ohmiya, A. Matsunobu and M. Nishizima, A case of adult form hypertransaminasemia after delivery, Gastroenterologia Jap., 77: 1655-1660 (1980).
9. Y. Suzuki, N. Yamamura, K. Nozawa, Y. Akehane, K. Kiyosawa, A. Nagata, S. Furuta and K. Chiba, A case of adult-type congenital citrullinemia, Acta Hepatol. Jap., 21: 1215-1221 (1980).
10. M. Takamizawa, M. Toru, T. Kojima, A. Watanabe and K. Hirokawa, An autopsy case of juvenile hepato-cerebral degeneration with mental retardation, with special reference to anmonia and amino acid metabolism, Psychiatr. Neurolog. Jap., 75: 370-382 (1973).
11. T. Hara, T. Tabira, H. Shibasaki, Y. Kuroiwa and T. Saheki, Hepato-cerebral disease in a young adult with hyper-citrullinemia and partial deficiency of liver argininosuccinate synthetase, Clin. Neurol., 21: 879-884 (1981).
12. I. Akaboshi, F. Endo, I. Matsuda and T. Saheki, Kinetic analysis of argininosuccinate synthetase in a variant form of citrullinemia, J. Inher. Metab. Dis., 6: 36-39.

13. H. Nomura, Y. Ohara, S. Takase, H. Konno, S. Yamada and
    T. Sawai, A case of citrullinemia, _Igaku no Ayumi_,
    118: 871-882 (1981).
14. R. T. Schimke, Adaptive characteristics of urea cycle enzymes
    in the rat, _J. Biol. Chem._, 237: 459-468 (1962).
15. L. A. Sternberger, P. H. Hardy, J. J. Cuculis and H. G. Meyer,
    The unlabeled antibody enzyme method of immunohisto-
    chemistry: preparation and properties of soluble antigen-
    antibody complex (horseradish peroxidase-antihorseradish
    peroxidase) and its use in identification of Spirochetes,
    J. Histochem. Cytochem., 18: 315-333 (1970).
16. T. Saheki, A. Ueda, M. Hosoya, K. Kusumi, S. Takada, M. Tsuda,
    and T. Katsunuma, Qualitative and quantitative abnormalit-
    ies of argininosuccinate synthetase in citrullinemia,
    Clin. Chim. Acta, 109: 325-335 (1982).
17. B. D. Ross and W. G. Guder, Heterogeneity and compartmentation
    in the kidney, _in_:"Metabolic Compartmentation," H. Sies,
    ed., Academic Press, London (1982).

STUDIES ON THE OCCURRENCE OF GUANIDINO COMPOUNDS IN SERUM AND URINE OF PATIENTS WITH EPILEPSY, NEURONAL CEROID-LIPOFUSCINOSIS AND UREMIC PATIENTS

Claus Munk Plum

Aldersrogade 43 F DK 2200 Kobenhavn N Denmark

INTRODUCTION

Studies of guanidino compounds (GCs) in the serum and urine of patients with renal diseases are numerouns[1-14].

Some of these studies form the basis for the assumption that gastrointestinal, cardiovascular and neuromuscular problems in association with the presence of uremia could be the result of changes in GCs in the serum of these patients.

Natelson[15] found -as the first- significant elevation of guanidinosuccinic acid (GSA) in the urine of patients with chronic renal failure. Baker & Marshall[16] reported a significant increase of methylguanidine (MG) level in the serum of uremic patients compaired with normal controls.

Some GCs, such as γ-guanidinobutyric acid (GBA)[17], taurocy-amine (TC)[7], glycocyamine(guanidinoacetic acid)(GAA)[18], MG[19] and other compound[20] have convulsive effects on rabbits and cats when administered intracisternally.

Several studies on the convulsive effect of GCs are on record in the literature in which also examinations on the content of these compounds in brain tissue and cerebrospinal fluid are included[7,17,18,20,25].

All these examinations and theories were the stimulus for the studies on GCs in serum and urine which are presented in this paper.

The examinations were made on a control group, in patients

with epilepsy, with neuronal-lipofuscinosis (= Spielmeyer-Vogt-
Batten-Stengel's syndrome) and a few patients with renal failure.
The reason why the group with Spielmeyer-Vogt-Batten-Stengel's
syndrome (SVBS's) is included is that epileptic seizures are often
observed in these patients. Also, the renal handling of free
amino acids in epilepsy and in patients with SVBS's seems to be
more like that found in cases with renal failure[26] than in normal
subjects.

MATERIAL

The control group consisted of 30 normal individuals, 15 women
and 15 men, aged 20-50 years. The group with epilepsy consisted of
15 women and 15 men, aged 20-45 years, all admitted for antiepileptic treatment. The samples were taken from patients, who had
been seizure-free for at least one week. The group with SVBS's
consisted of 9 women and 5 men aged 12-23 years. These patients
normally die at age 18-25.

The blood samples were taken in the morning, with the patients
having fasted. The urine samples were collected within the same 24
hours period as the blood samples.

Serum

Four hundred mg of sulphosalicylic acid were added to 5 ml
fresh serum. The samples were carefully mixed, and then left for
at least 10 min at room temperature. After this the sample was
centrifuged at 3000 rpm for 10 min. One ml of the clear protein-
free supernatant was used for the analysis.

Urine

To 10 ml of the 24 hour sample of urine 300 mg sulphosalicylic
acid was added. (The amount of sulphosalicylic acid was varied
according to the amount of protein present in the urine). After
adding the sulphosalicylic acid the sample was carefully mixed,
and left at room temperature for at least 10 min. After this, the
sample was centrifuged at 3000 rpm for 10 min. Five ml of the
clear protein-free supernatant was used for the analysis.

Analytical methods

In all cases the amount of free amino acids and GCs were de-
terminated using the Technicon automatic analyser. The resin used
was Chromobeads Type B (T-15-0335-42), column 750 mm in lenght,
diameter 6 mm. Elution temperature 60°C. The determination of
the free amino acids and the GCs were made on two different columns.
The system was built up in such a way that the two columns closely

followed each other, i.e. starting time was the same, and the time
for the arginine (Arg) peak was the same. Both columns were supplied with elution fluid from the same "auto-grade". This complicated system was used in order to faciliate the identification of
the individual GCs in relation to the peaks of the individual
amino acids.

The colorimetric determination of the GCs was performed according to the method previously described[27], developed from the
method given by Carles & Abravamel[28]. The colorimetric determinations of the amino acids was carried out according to the method
previously described[29].

RESULTS AND DISCUSSION

Serum

The serum concentration of the different GCs in control subjects and in the three different groups of patients: Epilepsy,
neuronal ceroid-lipofuscinosis (SVBS's) are summarized in table 1.
The serum concentration of GSA of the group with epilepsy shows an
average which is higher than that observed in the controls; patients
with SVBS's shows a reduced average, but both values were within
the range observed in the controls[27].

In patients with renal failure there is a marked and statistically significant increase in the values of GSA. These results
conform to the results found in the literature
(Table 1a,1b,1c)[6,10,11,13,27,30,31].

The serum concentration of GSA were increased in renal failure.
The highest value was 211 nmol/ml. A level which has be shown to inhibit the activation of ADP-induced platelet factor 3 in vitro[32,33].

GSA is derived from Arg and is absent in those inborn errors
of metabolism involving interference with the syntheses or utilization of Arg in the urea cycle. It may be derived from argininosuccinic acid or from the combination of Arg with aspartic acid[2,3].

The endogenous clearance of creatinine (CRN) and GSA were
closely correlated.

The values of GAA, TC, creatine (CR), CRN and Arg shows a
slight increase over average values, but here again without statistical significance in the case of epilepsy and SVBS's.

In the case of renal failure a decrease in GAA was found.
These findings confirm results given in the literature, but here it
should be mentioned that the values given in the literature differ

Table 1   Concentration of guanidino compounds in serum (nmol/ml)

	Controls	Epilepsy	SVBS's	Renal failure
Guanidino-acetic acid	5.9±3.4 (2.6–9.9)	7.2±3.5 (2.7–13.5)	7.4±2.6 (3.3–12.3)	(0–4.5)
Creatine	30±6 (10–53)	35±15 (12–52)	33±7 (12–56)	(10–186)
Creatinine	8.1±3.0 (3.7–19.4)	9.4±4.2 (5.6–23.2)	9.1±3.3 (5.5–19.9)	* (72–123)
Guanidino-succinic acid	8.2±3.1 (3.6–15.6)	9.0±2.7 (3.4–16.2)	7.6±3.2 (2.0–11.2)	* (62–211)
γ–Guanidino-butyric acid	ND	T	T	* (3.2–13.3)
Methylguanidine	T	1.4±0.6* (1.0–2.2)	T	* (1.2–2.8)
γ–Guanidino-β-OH butyric acid	T	T	T	T
β–Guanidino-propionic acid	ND	T	T	* (2.8–10.9)
Taurocyamine	5.2±3.1 (1.6–12.6)	7.4±3.2* (2.0–14.0)	7.0±2.9 (2.8–12.3)	* (23–53)
Arginine	80±26 (56–123)	85±19 (63–144)	76±31 (59–183)	(69–223)
α–Acetylarginine	T	T	T	T
N	30	30	14	6

T: Trace   ND: not detected
SVBS: Spielmeyer-Vogt-Batten-Stengel's syndrome
*: p<0.05   v.s. Control

markedly from one another for which are offered a multiplicity of explanations.

Guanidines are normal metabolic products of protein catabolism. GAA is an intermediate in CR synthesis.

It is also possible that the disordered fat metabolism of renal failure gives rise to the fatty moieties of GAA and GBA [11].

In table 1, an increase is found in the levels of GBA and MG. Here the values show a great disparity from individual to individual (Table 1a,1b,1c) [34,35]. MG is thought to arise from the oxidation of CRN[14]. The long-term administration of MG to dogs can simulate some of the symptoms of the uremic syndrome [5,34].

Table 1a. Concentration of guanidino compounds in serum (nmol/ml)

	Cohen, 1968		Sawynok, 1975	
	Control	Renal failure	Control	Renal failure
Guanidino-acetic acid			(2.6–6.8)	(2.6–23.1)
Creatinine	7.0±2.0	67±14	9.7±4.4	97±24
			(3.5–20.4)	(52–128)
Guanidino-succinic acid	2.0±0.9	5.8±3.6	<2.5	42±51
				(3.9–273)
Methylguanidine			0.3±0.1	0.3±0.2
			(0.1–0.5)	(0.1–1.1)
Arginine	9.4±3.4	10.3±3.7		
N	90	100	20	32

Table 1b. Concentration of guanidino compounds in serum (nmol/ml)

	Shainkin, 1975		Perez, 1976	
	Control	Renal failure	Control	Renal failure
Guanidino-acetic acid	2.5 (0–6.3)	1.4 (0–5.3)	2.9±0.6 (1.5–5.0)	6.4±2.0 (1.7–12.3)
Creatine	31.2 (0–55.6)	45.0 (1.0–218)	29.5±3.1 (16.2–37.0)	276±163 (237–845)
Creatinine	8.5 (7–11)	63.2 (9–182)	8.8±2.3 (5.3–12.2)	91.3±8.8 (61–115)
Guanidino-succinic acid	1.14 (0–3.2)	15.1 (0–56.7)	2.0	27.9±4.7 (12.2–44.9)
γ-Guanidino-butyric acid	12.2 (0–57)	29.5 (0–81)	7.0±2.2 (0.8–13.3)	12.1±2.8 (3.3–17.9)
Methylguanidine	0.2 (0–1.0)	0.3 (0–1.4)		
β-Guanidino-propionic acid	0.2 (0–1.0)	18.4 (0–62.3)		
Arginine	114.2 (38–256)	131.9 (0–266)	108±9 (88–144)	85±11 (61–121)
N	9	10	6	5

Table 1c.   Concentration of guanidino compounds in serum (nmol/ml)

| | Matsumoto, 1976 | | Kikuchi, 1981 | |
	Control	Renal failure	Control	Renal failure
Guanidino-acetic acid	7±3	T	2.8±0.9	2.4±0.9
Creatinine			84.6±26.2	440.2±44.2
Guanidino-succinic acid	7±3	79±69	0.4±0.2	20.1±1.6
Methylguanidine	ND	T		
γ-Guanidino-β-OH butyric acid	T	T		
β-Guanidino-propionic acid	T	T		
Taurocyamine	4±1	27±12		
Arginine	139±31	162±47	99±23	107±39
N	13	14	6	4

T: Trace   ND: not detected

Urine

The excretion of the different GCs are given in table 2.

GSA tends to increase in cases of renal failure, whereas the average of GSA excreted by patients suffering from SVBS's and epilepsy was within the range of normal. These results confirm those found in the literature (Table 2a,2b)[1-3,9,10,27,36].

The reason why uremics have an increased excretion of GSA is not yet clear since the origin of this substance is still an open question.

The amount of GAA excreted was nearly normal in the cases with epilepsy and SVBS's, but here a marked decrease in the amount excreted was to be found in cases with renal failure, perhaps due to a decreased ability of the kidney to clear GAA from the blood.

These findings confirm those from the literature[1,9,26,27,36] .

A decreased urinary excretion of GAA suggests impaired kidney function. This concept is consistent with the observation by Bonas et al.[36].

Table 2.   Excretion of guanidino compounds in urine (mg/24 hours)

	Controls	Epilepsy	SVBS's	Renal failure
Guanidino- acetic acid	43.3±17.9	49.3±20.6	45.1±19.9	4.3±2.9*
Creatine	76.2±48.3	72.1±51.2	66.2±39.6	79.9±25.2
Creatinine	1172±512	1422±533	1514±514*	1010±378
Guanidino- succinic acid	9.0±7.4	12.3±8.9	10.1±4.8	33.6±21.7
γ-Guanidino- butyric acid	T	T	T	T
Methylguanidine	1.9±1.3	2.1±2.2	1.7±1.9	6.7±5.9
γ-Guanidino-β- OH butyric acid	T	1.8±1.6*	T	2.0±2.3*
β-Guanidino- propionic acid	T	2.6±4.9*	T	T
Taurocyamine	3.9±2.9	6.8±5.9*	5.3±4.2	7.2±6.8
Arginine	30±12	40±14*	32±16	33±19
N	30	30	14	6

T: Trace   * p<0.05 v.s. controls
SVBS's: Spielmeyer-Vogt-Batten-Stengel's syndrome

        In contrast to GAA, production of Arg occurs mainly in the
liver[16]. Only in liver diseases Arg production seriously
decreased.

        Sasaki[9] has calculated the ratio of GSA/GAA and found this
ratio in his control group to be nearly 1/5 and in his patients
with renal failure it was 5/1. The results given in this paper
show similar differences in the ratio of GSA/GAA. Elevation of
GSA/GAA in the urine is evidently the most sensitive indicator of
impaired kidney function. The retention of GAA alone could not
account for the marked decrease in urinary excretion of GAA. It is
more likely caused by a decreased rate of GAA formation, which would
be expected because a substantial portion of this substans is syn-
thesised in the kidney by transamidination from Arg and glycine[37].

        MG tends to increase in renal failure. This increased excre-
tion suggests increased production in chronic renal insufficiency.
This finding confirms those reported in the literature[10-13].

        In some cases of renal failure increased excretion of γ-
guanidino-β-hydroxybutyric acid is to be found. In cases of

Table 2a. Excretion of guanidino compounds in urine (mg/24 hours)

| | Cohen, 1968 | | Sawynok,1975 | |
	Control	Renal failure	Control	Renal failure
Guanidino-	27.5±17.2	2.5±2.5	25.0±12.9	4.3±3.4
acetic acid			(5.4–41.5)	(0.5–13.8)
Creatinine	940±822	695±480	1650±900	750±370
			(500–2100)	
Guanidino-			3.7±1.7	29.5±22.1
succinic acid			(1.6–6.0)	(4.3–79.5)
Methylguanidine			1.6±1.1	8.6±5.7
			(0.3–3.9)	(1.1–24.7)
Arginine	8.6±4.3	8.7±5.4		
N	15	30	20	32

Table 2b. Excretion of guanidino compounds in urine (mg/24 hours)

| | Sasaki, 1973 | | | |
| | (Male) | | (Female) | |
	Control	Renal failure	Control	Renal failure
Guanidino-	46.7±14.4	10.3±8.6	60.6±13.9	8.5±3.7
acetic acid	(30.5–84.0)		(48.3–68.9)	
Creatine	54.7±51.6	65.2±159.4	98.8±38.9	79.6±108.8
	(15.1–171)		(42.2–154.5)	
Creatinine	1631±498	1090±380	1143±368	910±257
	(681–2182)		(773–1876)	
Guanidino-	11.9±5.8	51.7±32.7	12.4±6.4	70.1±33.5
succinic acid	(3.5–22.7)		(6.8–28.0)	
N	17	16	10	13

epilepsy increased excretion of β-guanidinopropionic acid as well
as TC is noted.

Although the exact role of the GCs in the production of uremic
symptoms is difficult to define, it is likely that several GCs
contribute to the toxicity of uremia.

SUMMARY

Some guanidino compounds (GCs) have proved to have a convulsive effect when injected intracisternally into animals. In patients with uremia, some GCs are often present in increased amounts, and such patients often experience epileptic seizures. Epileptic seizures are often encountered in cases with neuronal ceroid-lipofuscinosis (SVBS's), and based on previous observation of the convulsive effect of GCs, a number of studies were performed on the occurrence of GCs in the serum and urine of patients with epilepsy and with SVBS's and some cases of renal failure. In these studies, minor shifts in the concentration of GCs were demonstrated both in patients with epilepsy and SVBS's. In cases of uremia some values of GCs were found elevated, some decreased, but in these cases no epileptic seizures were observed.

REFERENCES

1. B. D. Cohen, I. M. Stein and J. E. Bonas, Guanidinosuccinic aciduria in uremia, Am. J. Med., 45: 63 (1968).
2. B. D. Cohen, Guanidinosuccinic acid in uremia, Arch. Intern. Med., 126: 846 (1970).
3. B. D. Cohen, The origin and effect of guanidinosuccinic acidemia in uremia, 1st Annual Report of National Institute of Health Bethesda, Maryland, U.S.A. (1970).
4. H. Dobbelstein, H. H. Edel, M. Schmidt, G. Schubert und M. Weinzierl, Guanidinbernsteinsaure and Uramie, Kline. Wochenschr., 49: 348 (1971).
5. S. Giovanetti, M. Biagnini and L. Cioni, Evidence that methylguanidine is retained in chronic renal failure, Experientia, 24: 341 (1962).
6. T. Kikuchi, Y. Orita, A. Ando, H. Mikami, M. Fuji, A. Okada and H. Abe, Liquid-chromatographic determination of guanidino compounds in plasma and erythrocytes of normal persons and uremic patients, Clin. Chem., 27: 1899 (1981).
7. A. Mizuno, J. Mukawa, K. Kobayashi and A. Mori, Convulsive activity of taurocyamine in cats and rabbits, IRCS Med. Sci., 3: 35 (1975).
8. J. F. van Pilsum, R. P. Martin, E. Kiot and J. Hess, Determination of creatine, creatinine, arginine, guanidinoacetic acid, guanidine and methylguanidine in biological fluids, J. Biol. Chem., 222: 225 (1956).
9. M. Sasaki, M. Takahara and S. Natelson, Urinary guanidinoacetate guanidinosuccinate ratio: An indicator of kidney dysfunction, Clin. Chem., 19: 315 (1973).
10. J. Sawynok and J. K. Dawborn, Plasma concentration and urinary excretion of guanidine derivates in normal subjects and patients with renal failure, Clin. Exp. Pharm. Physiol., 2: 1 (1975).

11. R. Shainkin, Y. Berkenstadt, Y. Giatt and G. M. Berlyne, An
    automated technique for the analysis of plasma guanidino
    acids, and some findings in chronic renal disease, Clin.
    Chem., 60: 45 (1975).
12. I. M. Stein, G. Perez, R. Johnson and N. N. B. Cummings, Serum
    levels and urinary excretion of methylguanidine in chronic
    renal failure, J. Lab. Clin. Med., 77: 1020 (1971).
13. I. M. Stein, B. D. Cohen and R. S. Kornhauser, Guanidino-
    succinic acid in renal failure, experimental azotemia and
    inborn errors of the urea cycle, New Engl. J. Med.,
    280: 926 (1973).
14. I. M. Stein and M. J. Micklus, Concentration in serum and
    urinary excretion of guanidine, 1-methylguanidine, 1,1-di-
    methyl-guanidine in chronic real failure, Clin. Chem.,
    19: 583 (1973).
15. S. Natelson, I. Stein and J. E. Bonas, Improvements of the
    method of separation of guanidino organic acids by column
    chromatography, Microchem. J., 8: 371 (1964).
16. L. R. Baker and R. D. Marshall, A reinvestigation of methyl-
    guanidine concentration in sera from normal and uraemic
    subjects, Clin. Sci., 41: 563 (1973).
17. D. Jinnai, A. Sawai and A. Mori, γ-Guanidinobutyric acid as
    convulsive substance, Nature, 212: 617 (1966).
18. D. Jinnai, A. Mori, J. Mukawa, H. Ohkusu, M. Hosotani,
    A. Mizuno and L. C. Tye, Biochemical and physiological
    studies on guanidino compound induced convulsion,
    Jpn. Brain Physiol., 106: 27 (1969).
19. M. Matsumoto, K. Kobayashi, H. Kishikawa and A. Mori, Convul-
    sive activity of methylguanidine in cats and rabbits, IRCS
    Med. Sci., 4: 65 (1976).
20. A. Mori, M. Akagi, Y. Katayama and Y. Watanabe, α-Guanidino-
    glutaric acid in cobalt-induced epileptogenic cerebral
    cortex of cats, J. Neurochem., 32: 603 (1980).
21. M. Matsumoto, H. Kishikawa and A. Mori, Guanidino compounds
    in the sera of uremic patients and in sera and brain of
    experimental uremic rabbits, Biochem. Med., 16: 1 (1976).
22. M. Matsumoto and A. Mori, Effects of guanidino compounds on
    rabbits brain microsomal Na^+ -K^+ ATPase activity,
    J. Neurochem., 27: 635 (1976).
23. A. Mori, M. Hosotani and L. C. Tye, Studies on brain guanidino
    compounds by automatic liquid chromatography, Biochem. Med.,
    10: 8 (1974).
24. A. Mori, M. Hiramatsu, T. Takahashi and M. Kohsaka, Guanidino
    compounds in rats organs, Comp. Biochem. Phisol., 51B: 143
    (1975).
25. K. A. Szilagy, F. Lavinha and Y. Mardens, Study of free amino
    acidpattern in human cerebraospinal fluid along the cerebro-
    spinal axis, Acta Neurol. Belg., 74: 329 (1974).
26. C. M. Plum, Renal handling of free amino acids in normal
    adults, in patients with epilepsy and in patients with

Spielmeyer-Vogt-Batten's disease, Dan. Med. Bull., 22: 194 (1975).

27. C. M. Plum, Studies on the occurrence of guanidino compounds in serum and urine of patients with epilepsy and patients with Spielmeyer-Vogt-Batten's syndrome, Dan. Med. Bull., 27: 207 (1980).

28. J. Carles and G. Abravamel, Dosage automatic et simultane des acides amines et des guanidines, Bull. Soc. Chim. Biol., 52: 453 (1970).

29. C. M. Plum, Free amino acid levels in the cerebrospinal fluid of the normal humans and their variation in cases of epilepsy and Spielmeyer-Vogt-Batten disease. J. Neurochem., 23: 595 (1974).

30. A. Mori and Y. Watanabe, Recovery rates for guanidino compound analyses and variance of measured values by different laboratories, Rinshoukagaku, 9: 183 (1980).

31. G. Perez, A. Rey, M. Micklus and I. Stein, Cation-exchange chromatography of guanidine derivates in plasma of patients with chronic renal failure, Clin. Chem., 22: 240 (1976).

32. H. I. Horowitz and B. D. Cohen, Defectice ADP-indued platelets factors-3-activation in uremia, Blood, 30: 331 (1967).

33. H. I. Horowitz, I. M. Stein, B. D. Cohen and J. G. White, Further studies on the platelet-inhibitory effect of guanidinosuccinic acid and its role in uremic bleeding, Am. J. Med., 49: 336 (1970).

34. P. L. Balestri, M. Biagini, P. Rindi and S. Giovanetti, Uremic toxins, Arch. Intern. Med., 126: 843 (1970).

35. S. Giovanetti, L. Cioni, L. Balestri and M. Biagini, Evidence that guanidines and some related compounds cause haemolysis in chronic uraemia, Clin. Sci., 34: 141 (1968).

36. J. E. Bonas, B. D. Cohen and S. Natelson, Separation and estimation of certain guanidino compounds. Application to human urine, Microchem. J., 7: 63 (1963).

37. H. Borsook and J. W. Dubnoff, The formation of glycocyamine in animal tissue, J. Biol. Chem., 138:389 (1941).

DETERMINATION OF GUANIDINO COMPOUNDS IN PLASMA AND URINE OF

PATIENTS WITH ARGININEMIA BEFORE AND DURING THERAPY

B. Marescau, I. A. Qureshi*, P. De Deyn, J. Letarte*
M. Yoshino** and A. Lowenthal

Laboratory of Neurochemistry, Born-Bunge Foundation
U.I.A., 2610 Wilrijk, Belgium
*Centre de Recherche pédiatrique, Hôpital Ste.Justine
Montreal, Qué. H3T 1C5, Canada
**Department of Pediatrics, Kurume University School of
Medicine, Kurume 830, Japan

INTRODUCTION

The first clinical and biochemical description of two sisters affected with argininemia, last of the five primary disorders of the urea cycle, was published in 1969[1-3]. A third sister homozygote was described shortly after birth five years later[4]. Five other families including eight cases have been reported in the literature[5-10]. The first clinical symptoms seen in patients with argininemia are irritability, coma and epilepsy. The children show also pyramidal spasticity and mental retardation. All the patients described are still alive. The patient's biochemistry is characterized by an arginase deficiency in liver shown after biopsy as well as in erythrocytes and leucocytes. As a consequence to this arginase deficiency, the patients accumulate arginine in their cells and biological fluids. The arginine accumulation leads to an increase of its catabolites: the guanidino compounds. Already in 1972 it was reported that guanidinoacetic acid, N-α-acetylarginine, argininic acid, γ-guanidinobutyric acid, arginine and an unknown guanidino compound (later identified as being γ-keto-δ-guanidinovaleric acid[11]) were elevated in urine of these patients[12]. These determinations were done applying the colorimetric Sakaguchi detection method.

To obtain information about the occurrence of guanidino compounds in the biological samples of these patients with a higher alkalinity than arginine we changed our chromatographic conditions. For detection of the guanidino compounds in serum and cerebrospinal fluid, a fluorometric detection method was chosen. With these changes we are able to report the first determinations of guanidino compounds in plasma of patients with argininemia. Aware of the neurological symptoms of these patients and the epileptogenic character of some guanidino compounds the concentration of the guanidino compounds in plasma and urine was measured before and during therapy.

PATIENTS

A.Y. is a six year old girl. Clinical and biochemical data of this patient were described by Yoshino et al. at the age of four years[10]. The patient was treated with a low protein diet: 0.56 g/kg/24 hours. To this diet an essential amino acid mixture (0.42 g/kg/24 hours) and 800 calories were added. The body weight was 12.5 kg. Determinations of guanidino compounds in serum and urine were done only after this therapy.

F.F.L. is a five year old boy. Before therapy the protein intake was 2.5 g/kg/24 hours. At three years and six months he was set on a low protein diet: 1.64 g/kg/24 hours. The body weight was 16.5 kg.

L.C. is an eighteen year old girl. At the age of fifteen, the patient was admitted to the hospital with anorexia, being in a catabolic state. There was a marked hyperammonemia. She was started on a low protein diet: 0.53 g/kg/24 hours. The diet represented about 1650 calories. Her body weight at that time was 30 kg. At seventeen years the therapy was changed to a low arginine diet along with sodium benzoate therapy: 0.5 g protein /kg/24 hours. Arginine in the diet was restricted to 12.5 mg/kg/24 hours (normally 50 mg/kg/24 hours). At the age of seventeen her body weight was 40 kg. The diet represented about 1800 calories.

Clinical and biochemical data of F.F.L. and L.C. were described by Qureshi et al.[8,9]. Of these patients we report the follow up of guanidino compounds in serum and urine before and during therapy. For F.F.L. only we compared the values of guanidino compounds with those of his heterozygous parents.

METHODS

The concentration of guanidino compounds was determined with a Biotronic LC 6001 amino acid analyser adapted for guanidino determination. The method of Hiraga et al.[13] was modified: column size 140 x 4 mm; cation exchange resin BTC 2710 (Biotronik); column temperature 50 °C. 0.2 M Sodium citrate buffers were applied successively: (a) pH 3 (5 minutes), (b) pH 3.3 (25 minutes), (c) pH 4.9 (25 minutes), (d) pH 10.5 (50 minutes), and finally 0.6 N NaOH for 25 minutes. To the column eluent 0.75 N NaOH was pumped and passed through a mixing coil. To this mixture 0.6 % ninhydrin dissolved in 50 % water and 50 % ethyleneglycolmonoethylether was pumped into a reaction coil. The flow rate for the buffers was 20 ml/hour, for the sodium hydroxide 17.14 ml/hour and for the nin-hydrin 11.43 ml/hour. Fluorophor was formed at 50 °C ina PTFE-tubing reaction coil (20 m x 0.3 mm I.D.). Fluorescence was meas-ured using a fluorometer Biotronik BT 6630 at Ex = 380 and Em = 460 nm.

RESULTS

   Table 1 gives the results of the determinations of guanidino compounds in serum and urine of patient A.Y. during a low protein diet.

   Table 2 compares the plasma values for guanidino compounds of patient F.F.L. with those of his heterozygous parents and with controls, before therapy and during treatment with a low protein diet.

   Table 3 gives a follow up of the guanidino compounds in the plasma of patient L.C. before therapy, during treatment with a low protein diet and finally during a period on a low arginine diet along with sodium benzoate therapy.

   Table 1, 2 and 3 show that a low protein diet has little influence on the arginine concentration. This leads to similar concentration values for guanidino compounds before and during the period of low protein diet. After institution of a low arginine diet along with benzoate therapy, the arginine values in plasma return almost to normal. Together with the decrease of arginine values there is a significant decrease for guanidinoacetic acid, N-α-acetylarginine and δ-guanidinobutyric acid. May be there is a small decrease in argininic acid and creatine. Low arginine diet does not influence the serum concentration of homoarginine which remains high.

   Table 4 and 5 give the results in urine for two patients. The results obtained for urine are not as sharp as those for plasma but the valuses run parallel.

_____

Abbreviations: α-k-δ-GVA = α-keto-δ-guanidinovaleric acid; GSA = guanidinosuccinic acid; CT = creatine; GAA = guanidinoacetic acid; N-α-AA = N-α-acetylarginine; ARG.A = argininic acid; CTN = creati-nine; γ-GBA = γ-guanidinobutyric acid; ARG = arginine; H.ARG = homoarginine; G = guanidine; MG = methylguanidine.

Table 1. Levels of guanidino compounds in the serum and urine of
          A.Y. during a low protein diet.

	SERUM nmol/100ml		URINE μmol/g creatinine	
	A.Y.	controls(n=20)	A.Y.	controls(n=10)
α-k-δ-GVA	350	<5	867	10-60
GSA	<2.5	36±18	1.4	65-130
CT	16129	5778±2450	15638	2500-25000
GAA	262	190±68	987	245-1155
N-α-AA	65	16±8	195	40-130
ARG.A	237	8±4	571	20-65
CTN	5147	7385±2500		
γ-GBA	57	<2.5	272	10-35
ARG	52268	11500±2500	201	35-190
H.ARG	557	186±75	17	<2-30
G	20	<20-40	12	15-30
MG	<10	<10	13	2-30

Table 2. Levels of guanidino compounds in the plasma (nmol/100 ml)
          of patient F.F.L. and his heterozygous parents before
          and during therapy with a low protein diet

	* F.F.L.(1) 21.04.81	F.F.L.(2) 23.04.82	Mother 21.04.81	Father 25.04.81	controls n=20
**	X	L.P.D.	X	X	X
α-K-δ-GVA	828	414	<5	<5	<5
GSA	11	<2.5	24	23	36±18
CT	18336	14780	7302	8845	5778±2450
GAA	298	219	106	277.7	190±68
N-α-AA	98	98	18	36	16±8
ARG.A	788	381	13	15	8±4
CTN	3990	3247	5531	9124	7385±2500
γ-GBA	71	45	2.5	11.7	<2.5
ARG	83673	53982	5164	13897	11500±2500
H.ARG	800	1139	457	391	186±75
G	20	20	20	<20	<20-40
MG	<10	<10	<10	<10	<10

*   sample number and date of collection.
** therapy: L.P.D. = low protein diet; X = no diet restriction

Table 3. Levels of guanidino compounds in plasma (nmol/100 ml) of patient L.C. before and during therapy

*	L.C.(1) 28.10.80	L.C.(2) 30.10.80	L.C.(3) 3.11.80	L.C.(4) 6.10.82	L.C.(5) 7.10.82	controls n=20
**	X	X	L.P.D.	L.A.D. + Bz.	L.A.D. + Bz.	X
α-k-δ-GVA	<5	125	146	229	271	<5
GSA	2.5	2.5	2.5	2.5	<2.5	36±18
CT	25157	13830	14173	12950	14322	5778±2450
GAA	883	921	1568	396	238	190±68
N-α-AA	474	408	246	35	39	16±8
ARG.A	227	103	111	91	95	8±4
CTN	6292	6294	4978	5886	5693	7385±2500
γ-GBA	330	825	301	5	2.5	<2.5
ARG	61953	62414	52483	16350	14704	11500±2500
H.ARG	845	899	719	976	1026	186±75
G	<20	<20	<20	<20	<20	<20-40
MG	<10	<10	<10	<10	<10	<10

* sample number and date of collection
** therapy: L.P.D. = low protein diet, L.A.D. + Bz = low arginine diet + benzoate therapy, X = no diet restriction

Table 4. Excretion values of guanidino compounds in urine (µmol/g creatinine) of patient F.F.L. and his heterozygous parents, before and during therapy with a low protein diet

*	F.F.L.(1) 21.04.81	F.F.L.(2) 23.04.81	Mother 21.04.81	Father 25.04.81	Controls n=10
**	X	L.P.D.	X	X	X
α-k-δ-GVA	2458	2758	30	17	10-60
GSA	4	2	55	55	65-130
CT	31620	9315	5190	2574	2500-25000
GAA	1175	485	1183	601	245-1155
N-α-AA	436	183	29	57	40-130
ARG.A	1865	563	15	23	20-65
γ-GBA	328	120	21	42	10-35
ARG	204	73	21	27	35-190
H.ARG	30	21	10	17	<2-30
G	14	6	13	18	15-50
MG	13	4	19	95	2-100

* sample number and date of collection
** therapy: L.P.D. = low protein diet, X = no diet restriction

Table 5. Excretion values of guanidino compounds in urine (μmol/
g creatinine) of patient L.C. before and during therapy

*	L.C.(1) 28.10.80	L.C.(2) 30.10.80	L.C.(3) 3.11.80	L.C.(4) 6.10.82	L.C.(5) 7.10.82	controls n=10
**	X	X	L.P.D.	L.A.D. + Bz.	L.A.D. + Bz.	X
α-k-δ-GVA	traces	140	traces	124	702	10-60
GSA	10	4	2	0.5	2	65-130
CT	58425	23184	9110	2961	13227	2500-25000
GAA	8570	4205	3502	490	2067	245-1155
N-α-AA	973	470	286	15	137	40-130
ARG.A	670	294	210	47	215	20-65
γ-GBA	1512	375	351	38	161	10-35
ARG	152	74	60	11	57	35-190
H.ARG	10	1.5	1.4	1.3	3.9	<2-30
G	48	18	13	2	12	15-50
MG	86	27	6	1	8	2-30

* sample number and date of collection
** therapy: L.P.D. = low protein diet; L.A.D. + Bz. = low arginine
diet + benzoate therapy; X = no diet restriction

DISCUSSION

Epilepsy is one of the first clinical symptoms observed in
patients with argininemia. Furthermore it has been shown that
different guanidino compounds like taurocyamine[14], guanidinoacetic
acid[15], γ-guanidinobutyric acid[16], N-α-acetylarginine[17], α-guanidinoglutaric acid[18], methylguanidine[19] and α-keto-δ-guanidinovaleric acid[20] are epileptogenic. This knowledge has led us to
make an analytical study of the guanidino compounds in the biological fluids of these patients. Previous results demonstrated
that guanidinoacetic acid, α-keto-δ-guanidinovaleric acid, N-α-
acetylarginine, argininic acid, γ-guanidinobutyric acid and
arginine are excreted at higher levels in urine[12]. In this work,
we report the first determinations of guanidino compounds in serum
of patients with argininemia. Determinations of the guanidino
compounds in serum of the patients before therapy demonstrate that
the compounds known to be increased in urine are also increased in
serum. Important to notice is the decrease of guanidinosuccinic
acid in serum, also observed in urine. Homoarginine is also
increased in the serum of patients with argininemia. Guanidine
and methylguanidine however are normal.

The major catabolic pathway of arginine in man is the hydrolysis to ornithine and urea. Next to this major pathway, there are

at least four secondary catabolic routes. As a consequence of the arginase deficiency the other pathways are activated in patients with argininemia. This results in increased levels of some guanidino compounds: α-keto-δ-guanidinovaleric acid, creatine, guanidinoacetic acid, N-α-acetylarginine, γ-guanidinobutyric acid and homoarginine. These compounds are formed by transamination and dehydrogenation, by transamidination or by acetylation. The increase of homoarginine can be explained by the arginase deficiency. Indeed, Scott-Emuakpor has shown that homoarginine can be hydrolyzed by arginase to urea and lysine[21]. Homoarginine is not a catabolite of arginine but can be catabolized by arginase. This explains why increased homoarginine concentration in the serum of the patients does not decrease during a low arginine diet.

Why is guanidinosuccinic acid decreased in serum as well as in urine? There are two hypotheses concerning the biosynthesis of guanidinosuccinic acid. According to Perez et al.[22], guanidino-succinic acid is a catabolite of arginine and should be formed through a transamidination. Natelson[23] however suggests that guanidinosuccinic acid is synthesized along the guanidine cycle starting from urea. This correlates with the biochemical findings in uremia where the high urea levels correspond with high levels of guanidinosuccinic acid in serum. Patients with argininemia have lower urea values in the blood and low concentration values of guanidinosuccinic acid in serum and urine. This phenomenon is seen in the urine of all patients with the urea cycle disorders[24].

Patients with argininemia have normal serum values of guanidine and methylguanidine. These two compounds are also elevated in the serum of patients with uremia. Uremic patients characteristically have increased serum values of guanidinosuccinic acid, creatinine, guanidine and methylguanidine. Remarkably these compounds are normal or even decreased (guanidinosuccinic acid) in argininemic serum.

The serum values of the guanidino compounds of the heterozygous parents are almost normal. This is probably due to the normal serum values of arginine.

Aware of the epileptogenic character of different guanidino compounds the follow up of these compounds before and during therapy was very important. A low protein diet does not decrease the arginine concentration markedly. However a low arginine diet in conjunction with benzoate therapy brings the arginine values of serum almost to normal levels. Next to the serum arginine, the plasma ammonia and glutamine and the orotate excretion become also normal. The concentrations of guanidinoacetic acid, N-α-acetyl-arginine and γ-guanidinobutyric acid are also decreased. Homoarginine and creatine do not decrease during arginine restriction.

For the analytical study of α-keto-δ-guanidinovaleric acid and α-guanidinobutyric acid in biological samples it is very important to limit the time between collection and analysis. Indeed, α-keto-δ-guanidinovaleric acid is not stable and degrades, among others, to γ-guanidinobutyric acid by chemical oxidation[25]. This is clearly seen by comparing sample 1 and 2 of patient L.C. (Table 3 and 5) with sample 1 of patient F.F.L. (Table 2 and 4). Most of the γ-guanidinobutyric acid in sample 1 and 2 of patient L.C. is a consequence of the chemical degradation of α-keto-δ-guanidinobutyric acid. Only a small portion is due to increased enzymatic transamidination. If the time between collection and analysis is limited, the concentration values of α-keto-δ-guanidinovaleric acid are always several times higher than those of γ-guanidinobutyric acid. We have observed the same phenomenon for cerebrospinal fluid, stored for many years. No α-keto-δ-guanidinovaleric acid was observed, as in sample 1 of patient L.C. (Table 3 and 5), however increased amounts of γ-guanidinobutyric acid suggest that increased amounts of α-keto-δ-guanidinovaleric acid are present in cerebrospinal fluid of patients with argininemia.

ACKNOWLEDGEMENTS

This work was supported by the "Fonds voor Geneeskundig Wetenschappelijk Onderzoek" (Grant nr. 3.0004.81) and the "Ministerie van Nationale Opvoeding en Nederlandse Cultuur".

REFERENCES

1. H. G. Terheggen, A. Schwenk, A. Lowenthal, M. Van Sande and J. P. Colombo, Argininemia with arginase deficiency, Lancet, 2:748-749 (1969).
2. H. G. Terheggen, A. Schwenk, A. Lowenthal, M. Van Sande and J. P. Colombo, Hyperargininämie mit Arginäsedefekt. Eine neue familiäre Stoffwechselstörung, I. Klinische Befunde, Z. Kinderheilk., 107:298-312 (1970).
3. H. G. Terheggen, A. Schwenk, A. Lowenthal, M. Van Sande and J. P. Colombo, Hyperargininämie mit Arginäsedefekt. Eine neue familiäre Stoffwechselstörung. II. Biochemische Untersuchungen, Z. Kinderheilk., 107:313-323 (1970).
4. H. G. Terheggen, A. Lowenthal, F. Lavinha and J. P. Colombo, Familial hyperargininemia, Arch. Dis. Child., 50:57-62 (1975).
5. S. D. Cederbaum, K. N. F. Shaw and M. Valente, Hyperargininemia, J. Pediatr., 90:569-573 (1977).
6. V. V. Michels and A. L. Beaudet, Arginase deficiency in multiple tissues in argininemia, Clin. Genet., 13:61-67 (1978).
7. S. E. Snyderman, C. Sansaricq, W. F. Chen, P. M. Norton and S. V. Phansalkar, Argininemia, J. Pediatr., 90:563 (1977).

8. I. A. Qureshi, J. Letarte, R. Quellet, M. Lelièvre and C. Laberge, Ammonia metabolism in a family affected by hyperargininemia, Diabète et Métabol., 7:5 (1981).

9. I. A. Qureshi, J. Letarte, R. Quellet and J. Larochelle, A new French-Canadian family affected by hyperargininemia, Pediatr. Res., 16:194A (1982).

10. M. Yoshino, K. Kubota, I. Yoshida, T. Murakami and F. Yamashita, Argininemia: report of a new case and mechanisms of orotic aciduria and hyperammonemia, in:"Urea Cycle Diseases," A. Lowenthal, A. Mori and B. Marescau, eds., Advances in experimental medicine and biology, vol. 153, Plenum Press, New York, pp. 121-125 (1982).

11. B. Marescau, J. Pintens, A. Lowenthal, E. Esmans, Y. Luyten, G. Lemiere, R. Domisse, F. Alderweireldt and H. G. Terheggen, Isolation and identification of α-keto-δ-guanidinovaleric acid in urine of patients with hyperargininemia by chromatography and gas chromatography-mass spectrometry, J. Clin. Chem. Biochem., 19:61-65 (1981).

12. H. G. Terheggen, F. Lavinha, J. P. Colombo, M. Van Sande and A. Lowenthal, Familial hyperargininemia, J. Genet. Hum., 20:69-84 (1972).

13. Y. Hiraga and T. Kinoshita, Post-colomn derivatization of guanidino compounds in high-performance liquid chromatography using ninhydrin, J. Chromat., 226:43-51 (1981).

14. A. Mizuno, J. Mukawa, K. Kobayashi and A. Mori, Convulsive activity of taurocyamine in cats and rabbits, IRCS Med. Sci., 3:385 (1975).

15. D. Junnai, A. Mori, J. Mukawa, H. Ohkusu, M. Hosotani, A. Mizuno and L.C. Tye, Biochemical and physiological studies on guanidino compounds induced convulsions, Jpn. J. Brain Physiol., 160:3668-3673 (1969).

16. D. Jinnai, A. Sawai and A. Mori, γ-Guanidinobutyric acid as a convulsive substance, Nature, 212:617 (1966).

17. H. Ohkusu, Osaka-Igakkai-Zashi, 21:49-55 (1970).

18. A. Mori, Y. Watanabe, S. Shindo, M. Akagi and M. Hiramatsu, γ-Guanidinoglutaric acid and epilepsy, in:"Urea Cycle diseases," A. Lowenthal, A. Mori and B. Marescau, eds., Advances in experimental medicine and biology, vol. 153, Plenum Press, New York, pp. 419-426 (1982).

19. M. Matsumoto, K. Kobayashi, H. Kishikawa and A. Mori, Convulsive activity of methylguanidine in cats and rabbits, IRCS Med. Sci., 4:65 (1976).

20. B. Marescau, M. Hiramatsu and A. Mori, α-keto-δ-guanidinovaleric acid induced electroencephalographic, epileptiform discharges in rabbits, Neurochem. Pathol., in press.

21. A. Scott-Emuakpor, On the hydrolysis of homoarginine by arginase, J. West Afr. Sci. Ass., 17:161-170 (1972).

22. G. Perez, A. Rey and E. Schiff, The biosynthesis of guanidinosuccinic acid by perfused rat liver, J. Clin. Invest., 57:807-809 (1976).

23. S. Natelson and J. E. Sherwin, Proposed mechanism for urea
        nitrogen re-utilization. Relationship between urea and pro-
        posed guanidine cycles, Clin. Chem., 25:1343-1344 (1979).
24. A. Lowenthal and B. Marescau, Urinary excretion fo monosub-
        stituted guanidines in patients affected with urea cycle dis-
        eases, in:"6th International congress of neurogenetics and
        neuro-ophthalmology," A. Huber and D. Klein, eds.,
        Elsevier / North Holland Biomedical Press, Amsterdam,
        pp. 347-350 (1981).
25. N. V. Thoai and J. Roche, Biochemie du groupement guanidique,
        Exp. Ann. Biochim. Med., 18:165-185 (1956).

HYPERARGININEMIA: TREATMENT WITH SODIUM BENZOATE AND PHENYLACETIC ACID

Naoki Mizutani, Mitsuo Maehara, Chiemi Hayakawa,
Tomoaki Kato, Kazuyoshi Watanabe and Sakae Suzuki

Department of Pediatrics, Nagoya University School
of Medicine, Nagoya, Japan

INTRODUCTION

Hyperargininemia is a rare hereditary disorder of the urea cycle due to arginase deficiency. This disorder is the most uncommon among five inborn errors resulting from enzyme deficiencies of the urea cycle. Hyperargininemia was first described by Terheggen et al. in 1969[1], and to our knowledge, there have been at least fifteen case reports in the literature[1-12]. Patients with this disorder all have the same clinical symptoms, which include vomiting, irritability, lethargy, convulsions and coma. Severe mental retardation and a marked degree of spasticity are also observed.

There have been a few attempts to treat patients with this disorder. Restriction of protein intake[2,4], oral administration of an essential amino acid mixture[5,7,9] or lysine[6,13], and exchange transfusion or gene replacement therapy[3,6,11,12] have been tried, but most of these trials are unsatisfactory. Recently, Brusilow and Batshaw have reported that administration of sodium benzoate and phenylacetic acid reduces the plasma ammonia levels in patients with urea cycle enzymopathies[14,15]. There are also reports showing that sodium benzoate is effective in increasing urinary excretion of non-urea nitrogen metabolites by activating alternative pathways in patients with other enzyme defects of the urea cycle[14-16]. But there are only a few papers concerning the effect of sodium benzoate or phenylacetic acid in patients with hyperargininemia[8]. In this report, we studied the effects of sodium benzoate and phenylacetic acid on hyperammonemia in a male patient with hyperargininemia, and the changes of urinary excretion of orotic acid and hippuric acid during sodium benzoate administration.

181

Furthermore, the effects of these agents on the plasma and CSF concentrations and urinary excretion of the amino acids concerned with the urea cycle and of various guanidino compounds were studied.

CASE REPORT

The patient is a 4-year and 7-month-old boy, who is the first of two children of healthy consanguineous parents. The patient is the product of an uncomplicated term pregnancy and normal spontaneous deliverly. Mental and physical development for the first three years was almost normal. There was no history of convulsion or coma. Since 3 years and 6 months of age, he had suffered from unsteady gait. At 4 years and 7 months of age, he was found to have hyperammonemia (239-402 µg/dl).

On admission, neurological examination revealed severe spasticity of the lower extremities, hyperactive deep tendon reflexes, positive ankle and patellar clonus, and Babinski reflexes. He could walk alone, but had a tendency to walk on tiptoe. His IQ was almost normal. There were no other abnormalities on physical examination.

Laboratory examinations revealed abnormal liver function tests (GOT: 562 IU/1, GPT: 930 IU/1, LDH: 734 IU/1). Plasma urea levels were low (2-8 mg/dl), and plasma ammonia levels were markedly elevated (208-475 µg/dl). The urinalysis was almost normal, but the urinary nitrogen and uric acid excretion were extremely low and the urinary excretion of orotic acid was markedly increased (577.9 mg/day) (Table 1). The EEG during hyperammonemia showed slow wave dysrhythmia. The cranial CT scan on admission showed slightly decreased density of white matter. There were no other abnormalities in laboratory findings.

Plasma and CSF concentrations of arginine were 5-10 times normal values, and urinary excretion of arginine was also increased. Plasma and CSF concentrations and urinary excretion of other amino acids concerned with the urea cycle, ornithine, citrulline, lysine and cystine were almost normal (Table 2).

MATERIALS AND METHODS

Blood samples were obtained early in the morning after an overnight fast, and plasma was deproteinized by adding 5% sulphosalicylic acid. Amino acids in plasma, CSF, and urine were determined with an automatic amino acid analyzer. Plasma ammonia was measured using a commercial kit "Amitest"[17] or enzymatic method[18]. Erythrocyte arginase activity was determined by a modification of the method of Shih et al.[19]. Urinary hippuric acid

Table 1. Labolatory Findings

CBC			T.P. (g/dl)		5.8
RBC (x 10^4/cmm)	477		Albumin (g/dl)		4.1
Hb. (g/dl)	12.5		α-1 globulin (g/dl)		0.2
Ht. (%)	36.5		α-2 globulin (g/dl)		0.5
WBC (/cmm)	7700		β-globulin (g/dl)		0.5
Pl. (x 10 /cmm)	20.4		γ-globulin (g/dl)		0.5
GOT (IU/1)	562		Immunoglobulin		
GPT (IU/1)	930		IgA (mg/dl)		100
LDH (IU/1)	734		IgM (mg/dl)		96
ACP (IU/1)	37.2		IgG (mg/dl)		1006
CPK (IU/1)	96		CSF		
Cho-E. (ΔpH)	0.92		pH		7.8
BUN (mg/dl)	7		Pandy		(-)
Creatinine (mg/dl)	0.5		C.C.		1/3
Uric acid (mg/dl)	1.4		Protein (g/1)		0.16
Ammonia (μg/dl)	250		Glucose (g/1)		0.65
Serum Electrolytes			Urinalysis		
Na (mEg/1)	141		Protein		(-)
K (mEg/1)	4.9		glucose		(-)
Cl (mEg/1)	114		urobilinogen		(-)
Ca (mEg/1)	4.4		bilirubin		(-)
P (mg/1)	5.2		Sediment		normal
T-Cholesterol (mg/dl)	198		Urinary UN (g/day)		1.3
F-Cholesterol (mg/dl)	58		Urinary Creat. (g/day)		0.25
Phospholipid (mg/dl)	272		Urinary Uric acid (g/day)		0.29
Triglyceride (mg/dl)	87				
NEFA (mEg/1)	0.26				
β-lipo. (mg/dl)	404				

was determined by a high pressure liquid chromatography. Urinary orotic acid was measured by the method of Adachi et al.[20]. Plasma and CSF concentrations and urinary excretion of monosubstituted guanidino compounds were measured by Professor A. Mori, Okayama University.

RESULTS

Figure 1 indicates the pedigree of the family affected by hyperargininemia and the values of erythrocyte arginase activities. The erythrocyte arginase acitivity of the patient was completely absent. Liver arginase activity of the patient was not determined. The levels of erythrocyte arginase activities of his parents, younger brother, paternal grandfather and uncle were less than a half of those of normal controls. They were presumed to be hetero-

Table 2. Plasma and CSF concentrations and urinary excretion of amino acids

Sample Amino acid	Plasma (nmol/ml) patient	Plasma (nmol/ml) normal range	CSF (nmol/ml) patient	CSF (nmol/ml) normal range	Urine (μmol/mg creatinine) patient	Urine (μmol/mg creatinine) normal range
Threonine	47.7	122.5-181.1	23.1	14.0- 38.0	0.11	0.10-0.48
Glutamic acid	30.7	22.8- 45.4	Tr	Tr	0.27	0.10-0.51
Glutamine	567.4	478.3-658.5	1615.1	65.6-111.8	0.52	0.06-0.75
Proline	105.4	165.6-243.6	ND	Tr	-	0
Glycine	98.4	181.1-268.9	Tr	2.0- 8.0	0.70	0.72-2.74
Alanine	205.4	321.9-479.9	15.4	14.0- 54.0	0.50	0.15-1.10
Citrulline	34.6	28.2- 40.8	ND	1.0- 3.0	0.04	0.02-0.18
Valine	93.0	224.1-276.3	ND	3.0- 33.0	0.03	0.04-0.16
Cystine	21.0	29.8- 49.0	ND	Tr	0.09	0.03-0.11
Methionine	33.2	26.4- 36.0	11.1	3.0- 9.0	0.05	0.05-0.12
Isoleucine	27.3	63.4- 88.0	Tr	5.0- 7.0	0.05	0.05-0.18
Leucine	52.1	107.3-144.1	3.8	9.0- 17.0	0.11	0.06-0.13
Tyrosine	66.3	54.6- 75.2	26.9	7.0- 15.0	0.15	0.06-0.41
Phenylalanine	51.1	57.1- 68.7	14.4	6.0- 16.0	0.05	0.07-0.26
Histidine	70.5	67.4- 98.6	39.7	5.0- 15.0	1.06	0.42-3.06
Ornithine	40.6	47.1- 72.5	1.8	1.0- 5.0	0.17	0.02-0.08
Lysine	139.6	142.5-208.3	8.2	9.0- 27.0	0.28	0.18-.097
Arginine	604.5	64.6- 97.8	51.7	11.0- 25.0	0.36	0.03-0.22

Tr: trace   ND: not detected

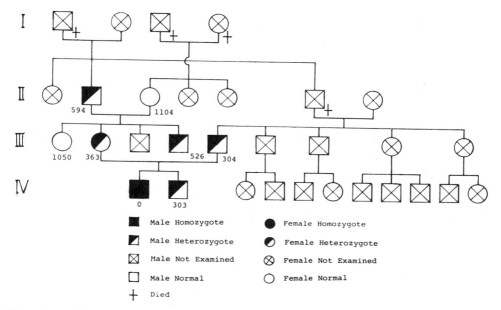

Fig. 1. The pedigree of the family affected by hyperargininemia.
Values under the symbols indicate erythrocyte arginase
activities (µmol Urea/hour/g hemoglobin). Controls (n=7):
971.1 ± 93.0 µmol Urea/hour/g hemoglobin.

zygotes of hyperargininemia. This disorder was thought to be in-
herited as an autosomal recessive trait. Plasma concentration and
urinary excretion of amino acids in these heterozygotes were normal.

With restriction of protein intake, the plasma ammonia levels
were initially reduced. However, within 1.5 months of this
treatment, plasma ammonia levels rose again and he became unable
to walk alone because of the increased spasticity of his lower extremities. Moreover, slight spasticity of the upper extremities
appeared. The abnormal liver function tests were unrelated to the
plasma ammonia levels. Changes in the plasma ammonia levels and
other laboratory findings during treatment are summarized in
Fig. 2.

Then, 250 mg/kg/day of sodium benzoate was administered orally
to the patient resulting in a prompt reduction in hyperammonemia.
Urinary excretion of total non-protein nitrogen and urea were almost
unchanged. urinary excretion of orotic acid initially diminished,
but gradually increased to the pretreatment levels or more. On the
other hand, urinary excretion of hippuric acid markedly increased
to 50-250 times pretreatment levels (Fig. 3).

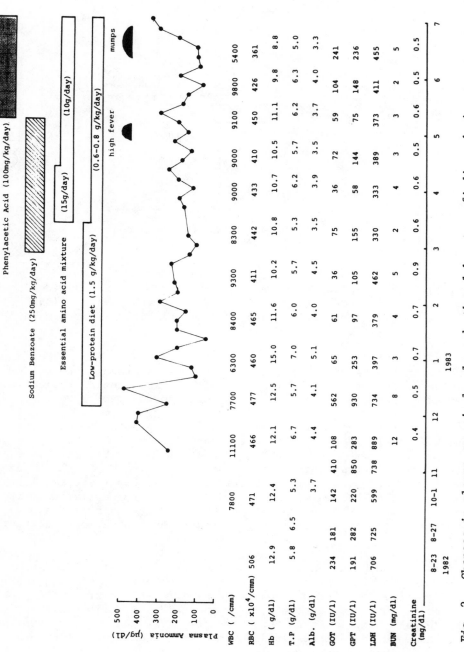

Fig. 2. Changes in plasma ammonia levels and other laboratory findings during treatment.

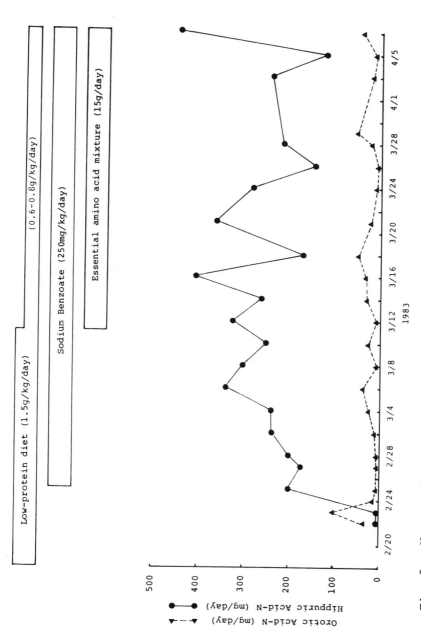

Fig. 3. Changes in urinary excretion of orotic acid-N and hippuric acid-N during the oral administration of sodium benzoate (250 mg/kg/day).

Two weeks later, we added an essential amino acid mixture according to Snyderman et al.[7] in order to control plasma and CSF concentrations of arginine. Protein intake was further restricted to 0.6-0.8 g/kg/day. One month later, we tried 100 mg/kg/day of phenylacetic acid because plasma ammonia levels rose again. This agent also decreased plasma ammonia levels. By these treatments, plasma and CSF concentrations of arginine showed a slight decrease, but was still far above the normal range. Clinically, spasticity of the lower and upper extremities gradually increased, and mild mental deterioration appeared (Fig. 4).

The data showing guanidino compound concentrations in plasma, CSF, and urine are summarized in Table 3. One of these, α-keto-δ-guanidinovaleric acid (KGVA), which is not usually found in normal subjects, was detected in the plasma and urine of the patient. A large amount of γ-guanidinobutyric acid (GBA), which is also normally undetectable, was excreted in his urine. On the other hand, guanidinosuccinic acid (GSA), which is usually present in normal urine, was lacking in all samples. Furthermore, plasma concentration of homoarginine markedly increased compared with that of normal adults. These findings seem to be typical of patients with hyperargininemia but there seems to be no correlation between the treatments or clinical findings and the concentrations of guanidino compounds in plasma, CSF and urine.

DISCUSSION

Hyperargininemia is a rare hereditary disorder of urea cycle metabolism, and can be diagnosed by demonstrating a deficiency of erythrocyte arginase activity. The clinical features of the reported patients with this disorder are almost the same, but different from those of the most patients with other urea cycle enzymopathies or other amino acid disorders.

Generally, hyperammonemia is considered a cause of central nervous system damage in patients with enzyme defects of the urea cycle. However, in many patients with hyperargininemia, hyperammonemia is observed only intermittently whereas this is always observed in other enzyme defects of the urea cycle. The pathogenesis of this hyperammonemia in hyperargininemia is not fully understood. The accumulation of ammonia in the body fluid may not be the sole etiology of the brain damage. The accumulation of arginine and its metabolites may produce central nervous system damage as well. Thus, the treatment of hyperargininemia should aim not only at the prevention of hyperammonemia but also at the reduction of plasma and CSF arginine concentrations.

Recently, Snyderman et al.[5,7], and Cederbaum et al.[9] have reported that an essential amino acid mixture is effective in

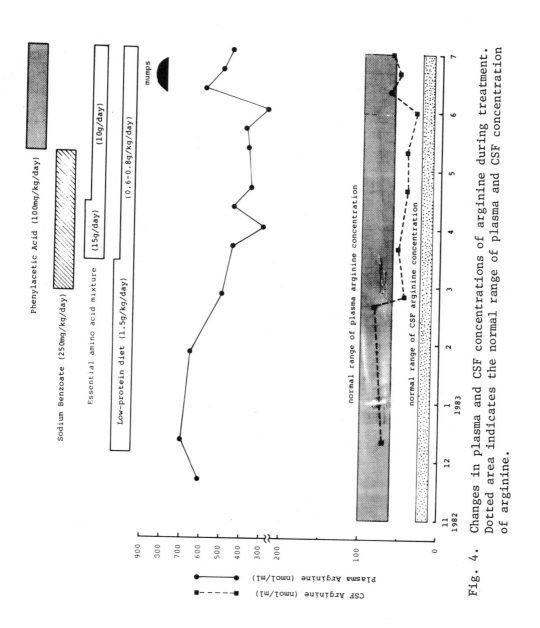

Fig. 4. Changes in plasma and CSF concentrations of arginine during treatment. Dotted area indicates the normal range of plasma and CSF concentration of arginine.

Table 3. Plasma and CSF concentrations and urinary excretion of guanidino compounds in the patient with hyperargininemia during low protein diet, sodium benzoate, and phenylacetic acid therapy.

	Plasma (nmol/ml)			CSF (nmol/ml)			Urine (μmol/g creatinine)		
	a	b	c	a	b	c	a	b	c
α-keto-δ-GVA	1.47	2.74	1.16	ND	ND	ND	190.81	608.25	43.79
GSA	ND	ND	ND	ND	ND	ND	ND	ND	ND
GAA	2.43	1.21	3.00	0.03	0.03	0.08	801.88	991.55	1334.71
NAA	5.34	2.79	2.13	0.50	0.29	0.08	293.01	418.97	144.43
GBA	ND	ND	ND	ND	ND	ND	217.13	118.35	50.48
Arg	456.18	465.64	354.55	58.78	45.77	38.06	104.46	9.48	47.78
HArg	9.46	20.09	8.75	1.16	1.25	1.16	3.83	17.94	3.50
G	ND	ND	ND	ND	ND	ND	10.93	8.04	15.60
MG	ND	ND	0.03	ND	ND	ND	3.06	1.03	1.91

ND: not detected

a: low protein diet only, b: low protein diet plus sodium benzoate,
c: low protein diet plus phenylacetic acid,

α-keto-δ-GVA: α-keto-δ-guanidinovaleric acid,   GSA: guanidinosuccinic acid,
GAA: guanidinoacetic acid,   NAA: N-α-acetylarginine,   GBA: γ-guanidinobutyric acid,
Arg: arginine,   HArg: homoarginine,   G: guanidine,   MG: methylguanidine.

reducing plasma arginine concentration and preventing the episodic hyperammonemia, resulting in a significant improvement in mental activity.

In our patient, it was difficult to control plasma ammonia levels with a low protein diet alone. When dietary protein is restricted, it is important to supply adequate amounts of essential amino acids, calories, vitamins and minerals. An insufficient caloric intake may result in decomposition of body protein leading to an increase in plasma ammonia levels. Also, a low protein diet is unsavory, which may result in inadequate caloric intake. Therefore, it may be difficult to maintain normal ammonia levels. The oral administration of sodium benzoate and phenylacetic acid also reduced the plasma ammonia levels in our case, and the effect of sodium benzoate was confirmed by the increase in urinary excretion of hippuric acid. Unfortunately, we could not estimate the urinary excretion of phenylacetylglutamine, because the reagent was not available. There were also no adverse effects from the oral administration of these agents on either the laboratory or clinical findings. Furthermore, plasma and CSF concentrations of glycine and glutamine also remained unchanged.

At present, there is no rational treatment for patients with enzyme defects of the urea cycle. Oral administration of sodium benzoate, phenylacetic acid, or an essential amino acid mixture, and exchange transfusion are all temporary. Together they represent a good way to increase urinary excretion of non-urea nitrogen metabolites by activating alternative pathways.

In Japan, there have been two other cases with hyperargininemia. These patients were also treated with low protein diet and the oral administration of an essential amino acid mixture without clinical improvement[10,11]. The second case was treated with blood and erythrocyte transfusions, which normalized the biochemical abnormalities together with clinical improvement[11]. We also perfoemed transfusion of 200 ml/kg of fresh blood, which normalized the biochemical abnormalities without clinical improvement.

In our case, plasma and CSF concentrations and urinary excretion of some monosubstituted guanidino compounds were considered to be typical of patients with hyperargininemia, as Marescau et al.[21] reported. The role of guanidino compounds in patients with hyperargininemia is still unclear. Some of the guanidino compounds are considered to be the catabolites of arginine, and the accumulation of arginine may produce an increased urinary excretion of guanidino compounds with the exception of guanidinosuccinic acid. In patients with hyperargininemia, arginine is catabolized via other pathways and it is suggested that nitrogen is excreted partially in the urine in the form of guanidino compounds.

CONCLUSION

In a patient with hyperargininemia, we attempted the oral administration of sodium benzoate or phenylacetic acid together with an essential amino acid mixture to decrease the plasma levels of ammonia and arginine. Sodium benzoate reduced the plasma ammonia levels, which was confirmed by an increase in the urinary excetion of hippuric acid. Phenylacetic acid also controlled hyperammonemia. By these treatments, plasma and CSF concentrations of arginine showed a slight decrease, but was far above the normal range. There was no clinical improvement, and spasticity of the lower and upper extremities was progressive with mild mental deterioration.

The data concerning guanidino compound concentrations in plasma, CSF and urine were considered to be typical of patients with hyperargininemia. There seemed to be no correlation between these treatments or clinical findings and guanidino compound con-contrations in plasma, CSF or urine.

ACKNOWLEDGEMENTS

The authors are grateful to Professor A. Mori for the mea-surement of guanidino compounds in the samples and for his invalu-able advice.

REFERENCES

1. H. G. Terheggen, A. Schwenk, A. Lowenthal, M. Van Sande and
    J. P. Colombo, Argininemia with arginase deficiency,
    Lancet, 2:748 (1969).
2. H. G. Terheggen, A. Lowenthal, F. Lavinha and J. P. Colombo,
    Familial hyperargininemia, Arch. Dis. Child., 50:57 (1975).
3. H. G. Terheggen, A. Lowenthal, F. Lavinha, J. P. Colombo and
    S. Rogers, Unsuccessful trial of gene replacement in
    arginase deficiency, Z. Kinderheilk., 119:1 (1975).
4. S. D. Cederbaum, K. N. F. Shaw and M. Valente,
    Hyperargininemia, J. Pediatr., 90:569 (1977).
5. S. E. Snyderman, C. Sansaricq, W. J. Chen, P. M. Norton and
    S. V. Phansalkar, Argininemia, J. Pediatr., 90:563 (1977).
6. V. V. Michels and A. L. Beaudet, Arginase deficiency in multi-
    ple tissues in argininemia, Clin. Genet., 13:61 (1978).
7. S. E. Snyderman, C. Sansaricq, P. M. Norton and F. Goldstein,
    Argininemia treated from birth, J. Pediatr., 95:61 (1979).
8. I. A. Qureshi, J. Letarte, R. Ouellet, M. Lelievre and
    C. Laberge, Ammonia metabolism in a family affected by
    hyperargininemia, Diabet. Metabol., 7:5 (1981).
9. S. D. Cederbaum, S. J. Moedjono, K. N. F. Shaw, M. Carter,
    E. Naylor and M. Walzer, Treatment of hyperargininaemia

due to arginase deficiency with chemically defined diet,
J. Inher. Metab. Dis., 5:95 (1982).

10. M. Yoshino, K. Kubota, I. Yoshida, T. Murakami and
F. Yamashita, Argininemia: Report of a new case and mechanisms of orotic aciduria and hyperammonemia, in:"Urea
Cycle Diseases," A. Lowenthal, A. Mori and B. Marescau,
eds., pp 121, Plenum Press, New York (1983).

11. T. Sakiyama, H. Nakabayashi, Y. Kondo, H. Shimizu, S. Kodama
and T. Kitagawa, Argininemia: Clinical course and trial of
enzyme replacement therapy, Biomedicine & Therapeutics (in
Japanese), 8:907 (1982).

12. K. Adriaenssens, D. Karcher, A. Lowenthal and H. G. Terheggen,
Use of enzyme-loaded erythrocytes in in-vitro correction of
arginase-deficient erythrocytes in familial hyperargininemia, Clin. Chem., 22:323 (1976).

13. W. M. Pardridge, Lysine supplementation in hyperargininemia,
J. Pediatr., 91:1032 (1977).

14. M. L. Batshaw, G. H. Thomas and S. W. Brusilow, New approaches to the diagnosis and treatment of inborn errors
of urea synthesis, Pediatrics, 68:290 (1981).

15. M. L. Batshaw, S. Brusilow, L. Waber, W. Blom, A. M. Brubakk,
B. K. Burton, H. M. Cann, D. Kerr, P. Mamunes, R. Matalon,
D. Myerberg and I. A. Schafer, Treatment of inborn errors
of urea synthesis: Activation of alternative pathways of
waste nitrogen synthesis and excretion, N. Engl. J. Med.,
306:1387 (1982).

16. E. Takeda, Y. Kuroda, K. Toshima, T. Watanabe, E. Naito and
M. Miyao, Effect of long-term administration of sodium
benzoate to a patient with partial ornithine carbamoyl
transferase deficiency, Clin. Pediatr., 22:206 (1983).

17. K. Tada, K. Okuda, K. Watanabe, Y. Iimura and S. Yamada,
A new method for screening for hyperammonemia,
Eur. J. Pediatr., 130:105 (1979).

18. A. Mondzac, G. E. Ehrlich and J. E. Seegmiller, An enzymatic
determination of ammonia in biological fluids, J. Lab. &
Clin. Med., 66:526 (1965).

19. V. E. Shih, T. C. Jones, H. L. Levy and P. M. Madigan,
Arginase deficiency in Macaca fascicularis. I. Arginase
activity and arginine concentration in erythrocytes and in
liver, Pediatr. Res., 6:548 (1972).

20. T. Adachi, A. Tanimura and M. Asahina, A colorimetric determination of orotic acid, J. Vitaminol., 9:217 (1963).

21. B. Marescau, A. Lowenthal, H. G. Terheggen, E. Esmans and
F. Alderweireldt, Guanidino compounds in hyperargininemia,
in:"Urea Cycle Diseases," A. Lowenthal, A. Mori and
B. Marescau eds., pp 427, Plenum Press, New York (1983).

III: PHYSIOLOGICAL, PHARMACOLOGICAL AND TOXICOLOGICAL

ASPECTS OF GUANIDINO COMPOUND

EFFECT OF METHYLGUANIDINE ON MUSCLE PROTEIN SYNTHESIS

Masamitsu Fujii*, Akio Ando*, Hiroshi Mikami*,
Akira Okada*, Enyu Imai*, Yukifumi Kokuba*,
Yoshimasa Orita*, Hiroshi Abe*, Yaeta Endo**,
Kazuo Chiku** and Yasuo Natori**

* First Department of Medicine, Osaka University
Medical School, Osaka 553, Japan
** Department of Nutritional Chemistry, Tokushima
University Medical School, Tokushima 770, Japan

INTRODUCTION

Methylguanidine (MG) is increased in the serum of uremic
patients[1]. Giovannetti et al. considered it one of the most important uremic toxins, since the administration of MG to dogs
resulted in weight loss, gastrointestinal disturbance, anemia and
so on[2].

In uremic patients, weight loss and loss of muscular mass are
frequently observed. This pathological state is known as the
"wasting syndorome". Uremic toxins have been suggested as a
possible cause of this syndrome[3].

The present investigation attempts to clarify the mechanism
of the "wasting syndrome" which we propose is an effect of MG on
the protein synthesizing activity in muscle. The influence of MG
on the RNA/DNA ratio, protein/DNA ratio, polysome size distribution and cell-free protein synthesizing activity of polysomes in
muscle from MG-infused rats was investigated. Further, the effect
of MG on protein synthesizing activity of muscle polysomes in
vitro was also studied.

Table 1. Composition of Solution

Glucose	208.3	g/l
Na	41.7	mEq/l
K	25.0	mEq/l
Mg	5.0	mEq/l
Cl	19.5	mEq/l
Ca	13.3	mEq/l
P	13.3	mM/l
Mixed amino acids	35.5	g/l
Vitamin concentrate	10.0	ml/l
Trace elements	10.0	ml/l

Non-protein energy	42	kcal/rat/day
Total nitrogen	253	mg/rat/day
Total volume	50	ml/rat/day

MATERIALS AND METHODS

Treatment of animals

Healthy male Sprague-Dawley rats, weighing approximately 180 g, were anesthetized with ketamine hydrochloride injected intra-peritoneally at a dose of 70 mg/kg body weight. Following anes-thesia, a surgical catheterization of the superior vena cana via the right external jugular vein was performed. Animals were pre-pared for continuous infusion by connecting them to freely rotat-ing swivels.

Experimental infusions

Rats were divided into two groups; control rats and the MG-infused rats. All were sustained on total parenteral nutrition. MG was continuously administered to the MG-infused rats at a dose of 5 mg/rat/day for 7 days. The basal composition of the infusate is shown in Table 1.

After 7 days, the rats were sacrificed by exsanguination, and the gastrocnemius muscle, the liver and epididymal fat pads were excised. The muscle was immediately placed in liquid nitrogen and stored at -70 °C until the determination of MG, nucleic acid and protein were undertaken and the preparation of polysomes was begun.

Determination of MG, nucleic acid and protein

Analysis of MG in serum and in the muscle was carried out with a Guanidine Analyzer G-520 (Japan Spectroscopic Co., Tokyo, Japan) as previously described[4]. Muscle RNA and DNA were extracted by the Schmidt-Thannhauser-Schneider procedure except that RNA was extracted in alkali for 1 hour[5,6]. RNA was estimated colorimetrically with orcinol[7]. DNA was measured colorimetrically by the diphenylamine method as modified by Giles and Myers[8]. Protein was measured by the method of Lowry[9].

Preparation of polysomes

For the preparation of muscle polysomes, all operations were performed in a cold room. 3 g of the muscle were minced with scissors and homogenized in 5 volumes of medium M (0.25 M KCl, 10 mM $MgCl_2$, 10 mM Tris-HCl, 6 mM 2-mercaptoethanol, pH 7.6)[10] within 60 seconds with a POLYTRON® homogenizer (Kinematica Co., Switzerland) at a control setting of 7.2. The post-mitochondrial supernatant was prepared by centrifuging at 10,000 g for 10 minutes. The supernatant was layered over a discontinuous gradient comprised of 0.5 and 2.0 M sucrose containing medium M and centrifuged for 24 hours at 105,000 x g to pellet the ribosomes.

Measurement of polysome size distribution

The polysome pellet was suspended in distilled water and applied to a linear 0.5 to 1.5 M sucrose gradient and centrifuged at 105,000 x g for 70 minutes. After centrifugation the ultraviolet absorption of the gradient at 254 nm was continuously recorded with ISCO flow cell. To measure the size distribution of polysomes, the recorded absorbance profile was divided into two segments; (1) the area encompassed by subunits to disomes (monosome-disomes area), (2) the other area. The value of the monosome-disomes ratio per total polysomes was computed from the area measurement.

Cell-free protein synthesizing activity of muscle polysomes

In the control and the MG-infused rats, the time course of ^{3}H-leucine incorporation into protein in the muscle polysome system was determined by the method of Schackelford and Lebherz[11].

Furthermore, to study the effect of MG on protein synthesizing activity in vitro, 1, 10 and 100 µM (final concentration) of MG with ^{3}H-leucine were each added to a cell-free system of muscle polysomes from control rats.

Table 2.   Concentrations of methylguanidine in serum
           and gastrocnemius muscle

(n)	Control (8)	MG-infused (7)
Serum (μg/dl)	ND	35.2 ±3.0
Gastrocnemius (μg/g)	0.02±0.02	2.85±0.68*

Each value represents the mean±S.D. of control or MG-
infused rats.   ND: not detected, *: p<0.01

Table 3.   Effect of methylguanidine on body weight and organ weight

(n)		Control (8)	MG-infused (7)
Body weight initial	(g)	183 ± 9	181 ± 8
final	(g)	193 ±13	190 ±13
Liver	(g)	6.60± 0.40	6.94± 0.86
Gastrocnemius	(g)	2.54± 0.30	2.50± 0.16
Epididymal fat pads	(g)	1.12± 0.27	1.09± 0.19

Each value represents the mean±S.D. of control or MG-infused rats.

RESULTS

The MG concentrations in serum and muscle of the MG-infused
rats greatly exceeded those of control rats, and the serum levels
of the MG-infused rats were comparable to those of severely uremic
patients (Table 2).

Body weight and the weight of the gastrocnemius muscle, liver
and epididymal fat pad of the MG-infused rats were nearly equal to
those of the control rats (Table 3).

No significant difference could be observed in the RNA/DNA
ratios in muscle between the two groups.   Protein/DNA ratios
generally followed the pattern of RNA/DNA ratios.   The values of
monosome-disomes per total polysomes were almost the same for the
two groups (Table 4).

Table 4. Effect of methylguanidine on RNA/DNA ratios, protein/DNA ratios and polysome profiles

(n)	Control (8)	MG-infused (7)
RNA/DNA	2.85± 0.50	2.80± 0.26
Protein/DNA	265 ±74	33 ±63
Monosome-disomes per total polysomes (%)	12.4 ± 2.6	10.4 ± 3.8

Each value represents the mean±S.D. of control or MG-infused rats.

Cell-free incorporation of ^{3}H-leucine into protein by muscle polysomes of control and MG-infused rats was examined in vitro. Polysomes prepared from the MG-infused rats showed increased protein synthesizing activity as compared with controls at 10 and 20 minutes of elapsed time (p<0.05) (Fig. 1).

When MG was added to the cell-free system, no effect was observed on the cell-free protein synthesizing activity of muscle polysomes from control rats (Fig. 2).

DISCUSSION

Giovannetti et al. reported that upon administration of MG to dogs subcutaneously every eight hours at a dose of 10 mg/kg body weight for twenty days, body weight loss was observed within five to ten days, and therefore MG seemed to be the cause of "wasting syndrome" in chronically uremic patients[2]. In their study, however, animals were fed ad libitum. Therefore, it was unknown whether the cause of body weight loss was decreased dietary intake, malabsorption in the gastrointestinal tract, the direct inhibitory effect of MG on protein synthesis or other possible factors. Furthermore, the serum concentrations of MG in their experimental animals were not constant during the period between injections and MG levels ranged several times higher than those of uremic patients.

In the present study, we utilized total parenteral nutrition to eliminate the effect of decreased dietary intake and malabsorption in the MG-infused rats. Moreover, the continuous intravenous administration of MG at a dose of 5 mg/rat/day for 7 days maintained the serum MG concentration constant and equal to those seen in chronically uremic patients. Under these experimental conditions, decreases in body weights or organ weights were not observed in the MG-infused rats.

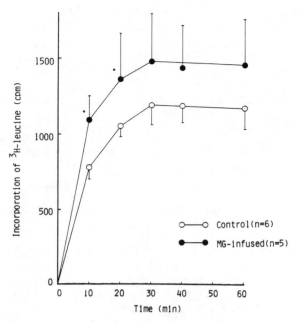

Fig. 1.   Time-course of ³H-leucine incorporation into protein by
          muscle polysomes from control and MG-infused rats
          Each value represents the mean±S.D..   *: p<0.05

The RNA content of tissues represents mostly the quantity of
ribosomes and the quantity of ribosomes is closely related to the
protein synthesizing capacity of the tissue.   RNA content is,
however, affected by the water, lipid and glycogen content.
Therefore, the RNA/DNA ratio is considered a better index of the
capacity for protein synthesis than is the RNA content.   The state
of protein synthesis in vivo is generally represented by the state
of ribosomal aggregation in the form of polysomes, and polysome
size distribution reflects the protein synthesizing activity of
the tissue.   The rate of incorporation of ³H-leucine into protein
in a cell-free system is a measure of the protein synthesizing
activity of polysomes in vitro.

Estimated by the RNA/DNA ratios, protein/DNA ratios, polysome
profiles and cell-free protein synthesizing systems, the muscle
protein synthesizing activity of the MG-infused rats was not decreased.

In the regulation of protein synthesis, two main steps are
considered: transcription and translation.   Regarding transla-

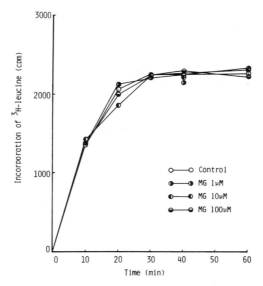

Fig. 2. Effect of the addition of MG to a cell-free system on the
protein synthesizing activity of muscle polysomes

tional control, two main components are proposed: ribosomal activity
and ribosomal content. Translation can be divided into three processes: initiation, elongation, and termination. Ribosomal activity
depends on the rate of these three processes[12].

In our study, increased cell-free protein synthesizing activity of muscle polysomes from MG-infused rats was observed. The
RNA content per cell was not affected by MG administration in vivo.
Judging from the results of the addition of MG to the cell-free
system, it could be inferred that MG did not affect the process of
elongation or termination during translation in protein synthesis
by polysomes. Consequently, MG administration might increase the
rate of the transcriptional process or initiational process of
translation.

In conclusion, MG does not decrease but increases protein
synthesis in muscle, and there is a possibility that MG might be a
stimulator rather than an inhibitor of muscle protein synthesis.
With respect to the role of MG in the "wasting syndrome" of uremic
patients, metabolic and endocrinological abnormalities other than
direct inhibition of protein synthesis might be involved.

REFERENCE

1. Y. Orita, A. Ando, Y. Tsubakihara, H. Mikami, T. Kikuchi,
    K. Nakata and H. Abe, Tissue and blood cell concentration
    of methylguanidine in rats and patients with chronic renal
    failure, Nephron., 27:35 (1981).
2. S. Giovannetti and G. Barsotti, Methylguanidine in uremia,
    Arch Intern Med., 131:709 (1973).
3. J. U. Koppel, Nutritional management of chronic renal failure,
    Postgrad. Med., 4:135 (1978).
4. T. Kikuchi, Y. Orita, A. Ando, H. Mikami, M. Fujii, A. Okada
    and H. Abe, Liquid-chromatographic determination of
    guanidino compounds in plasma and erythrocyte of normal
    persons and uremic patients, Clin. Chem., 27:1899 (1981).
5. G. Schmidt and J. Thannhauser, A method for the determination
    of deoxyribonucleic acid, ribonucleic acid, and phospho-
    proteins in animal tissues, J. Biol. Chem., 161:83 (1945).
6. W. C. Schneider, Phosphorus compounds in animal tissues,
    J. Biol. Chem., 164:747 (1946).
7. W. C. Schneider, Determination of nucleic acids in tissues
    by pentose analysis, in:"Method in enzymology, vol 3,"
    Academic Press, New York, pp. 680 (1957).
8. K. W. Giles and A. Myers, An improved diphenylamine method
    for the estimation of deoxyribonuleic acid, Nature,
    4979:93(1965).
9. O. H. Lowry, N. J. Rosebrough, A. L. Farr and R. L. Randall,
    Protein measurement with Folin phenol reagent,
    J. Biol. Chem., 193:265 (1951).
10. A. Yoshikawa and T. Masaki, Increase in protein synthetic
    activity in chicken muscular dystrophy, J. Biochem.,
    90:1775 (1981).
11. J. E. Shackelford and H. G. Lebherz, Cell-free synthesis of
    fructose diphosphate aldolases A, B, and C, J. Biol. Chem.,
    254:4220 (1979).
12. J. C. Waterlow, P. J. Garlick and D. J. Willward, Protein
    synthesis and its regulation, in:"Protein turnover in
    mammalian tissues and in the whole body," North-Holland,
    Amsterdam, pp.24 (1978).

EFFECT OF METHYLGUANIDINE ON ERYTHROCYTE MEMBRANES

Hiroshi Mikami, Akio Ando, Masamitsu Fujii, Akira Okada,
Enyu Imai, Yukifumi Kokuba, Yoshimasa Orita, and
Hiroshi Abe

First Department of Medicine, Osaka University Medical
School, Osaka Japan

INTRODUCTION

Hemolysis is an important contributing cause of anemia in
uremic patients. Uremia causes several erythrocyte abnormalities
such as increased mechanical fragility[1], increased autohemolysis[2],
and shortened erythrocyte survival[1,2]. These abnormalities may be
due to a decrease in erythrocyte deformability, which is measured
by filtrability as an index of cell rigidity. Several workers
demonstrated that erythrocyte deformability is decreased in uremic
patients[3,4]. Erythrocyte deformability is reduced by a rise in
intracellular Na content, which may be due in part to a decline in
the ATPase activity of the erythrocyte membrane. However, the relationship of erythrocyte deformability to uremic toxins has not
been clarified. Giovannetti reported that guanidino compounds,
especially methylguanidine (MG), caused hemolysis both in vivo and
in vitro[5,6]. Depressed Na-K ATPase activity of uremic erythrocytes was demonstrated by Cole[7].

This study aims to determine the effect of MG both on erythrocyte deformability and the ATPase activity of erythrocyte membranes.

MATERIALS AND METHODS

Erythrocytes and ghosts

Heparinized blood was collected from healthy volunteers and
uremic patients maintained on chronic hemodialysis. Blood from

uremic patients was collected just before and after a hemodialysis.
Erythrocyte were isolated by centrifugation at 100 g for 20 min.
The supernatant and buffy coat were discarded by aspiration and
erythrocytes were washed three times with the first supernatant
volume of isotonic saline solution containing 4 mM $CaCl_2$ and
100 mg/dl glucose. Ghosts were prepared by osmotic hemolysis of
washed erythrocytes for 30 min at 0 °C in 10 volumes of 0.1 mM EDTA-
Tris (pH 7.4) followed by centrifugation at 10,000 g for 10 min
and washing seven times with distilled water.

Measurement of erythrocyte deformability

Erythrocyte deformability was determined by the micro-filter
method. One ml of packed erythrocytes was layered on a Millipore
filter (FSLW, 3 μm) and a filtration volume through the filter
under a negative pressure of 20 cm H_2O was measured with a gradu-
ated syringe.

Assay of ATPase activity

ATPase activity was assayed by measuring the inorganic phosphate released from ATP during incubation for 60 min at 37 °C in a
standard medium containing 0.2-0.3 mg protein of erythrocyte
ghosts according to the method of Ichida et al.[8]. The standard
medium for measuring Na-K ATPase activity containing 40 mM Tris-
HCl buffer (pH 7.4), 3 mM ATP-Tris (pH 7.4), 2 mM $MgCl_2$, 5 mM KCl,
100 mM NaCl in a final volume of 2 ml. Na-K ATPase activity was
calculated by subtracting the ATPase activity assayed in the
medium except NaCl and KCl. The standard medium for measuring
Ca-Mg ATPase activity contained 40 mM Tris (pH 7.4), 3 mM ATP-Tris
(pH 7.4), 2 mM $MgCl_2$, 100 mM KCl, 10^{-4}M ouabain, 0.1 mM EGTA,
8×10^{-5} M $CaCl_2$. Ca-Mg ATPase activity was calculated by subtracting
Mg-ATPase activity in the medium to which was added all the constituents of the standard medium except $CaCl_2$. The reaction was
stopped by adding 0.5 ml of 50 % trichloroacetic acid. Precipitation was removed by the method of Fiske-Subbarow. The protein
content of erythrocyte ghosts was assayed by Lowry's method[9].

Addition and assay of methylguanidine

Packed erythrocytes from normal subjects were washed and incubated for 60 min at room temperature in the saline-glucose solution with MG added at concentrations of 0, 100, 500, and 1000 μg/dl
to determine the effect of MG on erythrocyte deformability. After
incubation, erythrocytes were equilibrated with each solution.

MG concentration of MG-treated erythrocytes and uremic erythrocytes were determined by Automated Guanidine Analyzer JASCO G-520
as previously described[10].

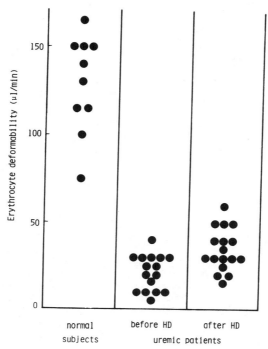

Fig. 1.   Erythrocyte deformability in normal subjects and uremic
patients before and after hemodialysis (HD).
p<0.05 : normal subjects vs uremic patients before HD,
p<0.05 : before HD vs after HD

ATPase activity of the erythrocyte membranes from normal subjects was assayed through the use of the reaction mixture containing MG (0, 50, 100, 500 µg/dl) to determine the effect of MG on
ATPase activity in vitro.

RESULTS

Erythrocyte deformability

Erythrocyte deformability of uremic patients before hemodialysis [20.6±10.1(S.D.) µl/min] was significantly lower than
that of normal subjects [129±27.7 µl/min] (p<0.05), but significantly improved to 34.7±11.8 µl/min after hemodialysis (p<0.05)
(Fig. 1).

Fig. 2 shows the relationship between erythrocyte deformability and the MG concentration of erythrocytes in uremic patients.
There was a significant negative correlation between erythrocyte

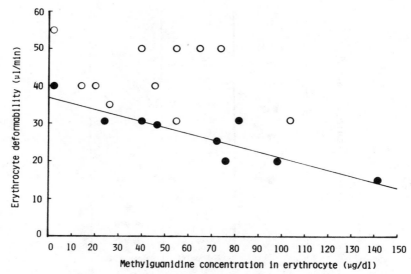

Fig. 2. Relationship between erythrocyte deformability (y) and
erythrocyte concentration of methylguanidine (x) in
uremic patients before hemodialysis(●) and after hemodialysis (o).
●: y = -0.159 x + 36.75,   r = -0.902, p<0.001.

deformability and the MG concentration before hemodialysis (r=
-0.902, p<0.001). After hemodialysis, erythrocyte deformability
was more improved than that expected from the fall in erythrocyte
MG concentration.

    Fig. 3 shows the effect of MG on erythrocyte deformability in
vitro. The erythrocyte deformability was decreased from 129±27.7
(S.D.) μl/min to 105.5±23.9, 100.5±33.7 and 77.7±23.1 (p<0.01) at
concentrations of MG made up to 100, 500 and 1000 μg/dl, respectively.

ATPase activity

    The Na-K ATPase activity of erythrocyte membrane of uremic
patients was significantly lower than that of normal subjects
(p<0.05), however, no significant difference of Ca-Mg ATPase activity was observed between the two groups (Table 1). Table 2 shows
the effect of MG on Na-K and Ca-Mg ATPase activity in vitro. The
Na-K ATPase activity was significantly decreased from 0.200±0.022
(S.E.) μmoles Pi/mg protein/h to 0.097±0.032 with MG concentrations
of 100 and 500 μg/dl, respectively (p<0.05). But the Ca-Mg ATPase
activity was not reduced by treatment with MG.

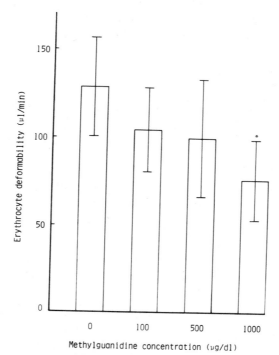

Fig. 3.   Effect upon erythrocyte deformability of methylguanidine
added to normal blood samples. * : p<0.01 compared with
the control.

DISCUSSION

It is proposed that uremic anemia is partly due to an acceleration of hemolysis, which may relate to guanidino compounds,
such as MG[5, 6]. The decrease in erythrocyte deformability, which
is an important factor in hemolysis, is believed to be caused by
an impairment of ATPase, a rise in intracellular Na content, a fall
in intracellular K, and lower pH. Depressed Na-K ATPase activity
and increased Na content have been demonstrated in uremic erythrocytes[7, 14]. This is accompanied by an increased ATP content and
elevated glycolitic activity in erythrocytes in uremia[11, 12]. The
present results confirm that erythrocyte deformability and the
Na-K ATPase activity of erythrocytes of uremic patients are lower
than those of normal subjects.

With regard to the effects of MG on hemolysis and ATPase activity, it was reported that autohemolysis[5, 6] and depressed Na-K
ATPase activity[13] were found at concentrations of MG above 2
mg/dl, a level not actually observed in the erythrocytes of uremic

Table 1. ATPase activity of erythrocyte membranes in normal
          subjects and uremic patients

	Normal subjects (n=7)	uremic patients (n=4)
Na-K ATPase	0.200±0.022	0.145±0.010*
Ca-Mg ATPase	0.235±0.027	0.211±0.007

Data given are mean values±S.E.. (μmole Pi/mg protein/h)
Blood samples of uremic patients were collected before
hemodialysis.
* : p<0.05 compared with normal subjects.

Table 2. Effect of methylguanidine in vitro on the Na-K and Ca-Mg
          ATPase activity of erythrocyte membranes.

MG concentration (μg/dl)	(n)	Na-K ATPase activity (μmole Pi/mg protein/h)	Ca-Mg ATPase activity
0 (control)	7	0.200±0.022	0.235±0.027
50	5	0.172±0.018	0.213±0.034
100	7	0.097±0.032*	0.181±0.031
500	7	0.094±0.030*	0.176±0.023

Data given are mean value±S.E.. MG : methylguanidine
* : p<0.05 compared with the control. (calculated by paired
Student's t-test).

patients. The present study, demonstrated that both erythrocyte
deformability and the Na-K ATPase activity were decreased in proportion to the MG concentration in vitro, at levels comparable
to that seen in uremia. We also found that there was a significant negative correlation between the erythrocyte deformality and
erythrocyte MG concentration in uremic patients. These results
suggest that MG may be a cause of hemolysis in uremia.

SUMMARY

     The effects of methylguanidine (MG) on erythrocyte deforma-
bility and ATPase activity of erythrocyte membranes were investi-

gated in vivo and in vitro. Erythrocyte deformability and the Na-K ATPase activity of erythrocyte membranes were decreased in uremic patients and were also reduced by the exposure of normal erythrocytes to MG at concentrations compared to those reported in uremia. In uremic patients, there was a significant negative correlation between erythrocyte deformability and erythrocyte MG concentration before hemodialysis. After hemodialysis, erythrocyte deformability was more remarkably improved than that which might be expected from the fall in erythrocyte MG concentration observed. These results that MG may cause a decrease in erythrocyte deformability through inhibition of Na-K ATPase activity. It also suggest that homodialysis improves erythrocyte abnormalities through the removal of hemolytic factors such as MG.

REFERENCE

1. J. F. Desforges and J. P. Dawson, The anemia of renal failure, Arch. Intern. Med., 101:326 (1958).

2. S. Giovannetti, P. Giagnoni and ·P. L. Balestri, Red cell survival in chronic uremia: its relationship with the spontaneous in vitro autohemolysis and with the degree of anamia, Experientia, 22:739 (1966).

3. S. Forman, M. Bischel and P. Hochstein, Erythrocyte deformability in uremic hemodialyzed patients, Ann. Intern. Med., 79:841 (1973).

4. A. Rosenmund, U. Binswanger and P. W. Straub, Oxidative injury to erythrocytes, cell rigidity, and splenic hemolysis in hemodialyzed uremic patients, Ann. Intern. Med., 82:460 (1975).

5. S. Giovannetti, M. Biagini, P. L. Balestri, R. Navelesi, P. Giagnoni, A. de Mattetis, P. Ferro-Milone and C. Perfetti, Uremia like syndrome in dogs chronically intoxicated with methylguanidine and creatinine, Clin. Sci., 36:445 (1969).

6. S. Giovannetti, L. Cioni, P. L. Balestri and M. Biagini, Evidence that guanidine and some related compounds cause hemolysis in chronic uremia, Clin. Sci., 34:141 (1968).

7. C. H. Cole, Decreased ouabain-sensitive adenosine triphosphatase activity in the erythrocyte membrane of patients with chronic renal disease, Clin. Sci., 45:775 (1973).

8. S. Ichida, C. H. Kuo and T. Matsuda, Effects of La^{+++}, Mn^{++} and ruthenuim red on Mg·Ca-ATPase activity and ATP-dependent Ca-binding of synaptic plasma membrane, Japan. J. Pharmacol., 26:39 (1976).

9. O. H. Lowry, N. H. Rosebrough, A. L. Farr and R. J. Randall, Protein measurement with the folin phenol reagent, J. Biol. Chem., 193:265 (1951).

10. T. Kikuchi, Y. Orita, A. Ando, H. Mikami, M. Fujii, A. Okada, and H. Abe, Liquid-chromatographic determination of

guanidino compounds in plasma and erythrocyte of normal persons and uremic patients, Clin. Chem., 27:1899 (1981).

11. M. A. Lichtman, D. R. Miller and R. I. Weed, Energy metabolism in uremic red cells: relationship of red cell adenosine triphosphate concentration to extracellular phosphate, Trans. Ass. Amer. Phycns., 82:331 (1969).

12. C. H. Wallas, Metabolic studies on the erythrocyte from patients with chronic renal disease on hemolysis, Brit. J. Haematol., 27:145 (1974).

13. M. Matsumoto and A. Mori, Effects of guanidino compounds on rabbits brain microsomal Na-K ATPase activity, J. Neurochem., 27:635 (1976).

14. E. K. M. Smith and L. G. Welt, The red blood cell as a model for the study of uremic toxins, Arch. Intern. Med., 126:827 (1970).

TAURINE AND THE ACTIONS OF GUANIDINOETHANE SULFONATE

R. J. Huxtable, D. Bonhaus, K. Nakagawa*,
H. E. Laird and H. Pasantes-Morales**

University of Arizona, Department of Pharmacology
Health Sciences Center, Tucson, Arizona 85724
* On sabbatical leave from Kyoto's Women's College
Kyoto, Japan
** Centro de Investigaciones Fisiologia Celular
University of Mexico, Mexico

It has been recognized that major handicaps to the elucidation of the functional significance of taurine in mammals were the absence of simple methods for depleting tissues of taurine, and the absence of a pharmacological antagonist to taurine[1,2]. These handicaps have been lessened since the introduction of the structural analogue of taurine, guanidinoethane sulfonate (GES), as a tool for modifying taurine levels in certain species[3].

GUANIDINOETHANE SULFONATE AND TAURINE TRANSPORT

GES is transported into mammalian heart by a transport system specific for β-amino acids (Fig. 1). In muscle, and perhaps other tissues, it is also transported by the same carrier used by creatine[4,5]. As Fig. 2 illustrates for brain synaptosomes and isolated heart, GES is a competitive inhibitor of taurine transport, being, indeed, a better inhibitor than GABA. As a result of this observation, we tested the ability of GES, given in vivo, to deplete tissue taurine levels[3,6].

GUANIDINOETHANES SULFONATE AS A TAURINE-DEPLETER

GES was administered to mice, rats, cats, and guinea pigs as a 1 % solution in drinking water. This dose level was arbitrarily chosen. Animals were maintained on GES-containing drinking water

213

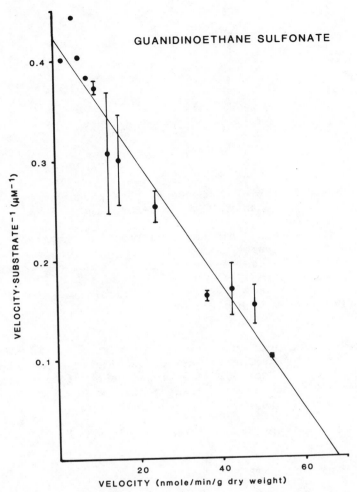

GUANIDINOETHANE SULFONATE

Fig. 1.   Eadie-Hofstee plot of transport of GES into isolated
          Langendorff-perfused rat heart.   Kinetic constants,
          derived by a non-linear fit to the Michaelis-Menten equa-
          tion (MLAB program) are (mean ± SEM): Km 142 ± 22 μM;
          Vmax 12.9 ± 0.8 nmole/min/g dry weight.

for periods of weeks or months. We found that in rats and mice
tissue taurine levels began to fall almost immediately. Some of
our findings are summarized on Fig. 3. Detailed findings are reported in Huxtable and Lippincott[6]. Peripheral tissues were de-
pleted to a greater extent than CNS tissues. Furthermore, in all
tissues there was a limit below which further depletion could not
be achieved. Taurine was the only amino acid to be affected by

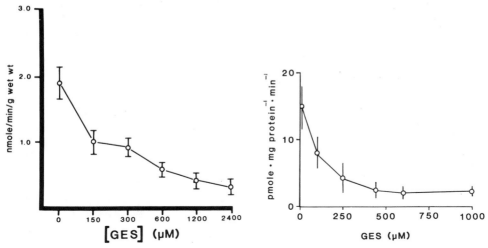

Fig. 2.   Inhibition of taurine transport
         (a) Left: Taurine transport in heart. The rate of up-
         take of 25 µM taurine (y-axis) in the Langendorff-per-
         fused rat heart is shown as a function of the co-perfused
         concentration of GES.
         (b) Right: Taurine transport in synaptosomes. The rate
         of uptake of 4 µM taurine into the P_2 fraction of 5-day
         old rat brain is shown as a function of the co-incubated
         concentration of GES.
         Four preparations per point; means ± SD.

the GES treatment. There was an accumulation of GES in all
tissues, and this accumulation was particularly marked in skeletal
muscle, presumably as a result of transportation via both the
β-amino acid and the creatine transport systems.

     This finding of taurine depletion following GES administra-
tion was significant in that it provided the first means whereby
taurine levels could be lowered in laboratory rodents. Since the
initial report, numerous studies have appeared from around the
world utilizing GES as a taurine-depleter. Some examples are ref-
erenced[7-10].

     GES is ineffective as a taurine depleter in the cat and the
guinea pig. The cat is unable to synthesize taurine, and requires a dietary source. Cats maintained on taurine-deficient
diets, therefore, suffer a gradual fall in tissue taurine levels[11].
Cats maintained on a taurine-free diet plus 1 % GES, however, do

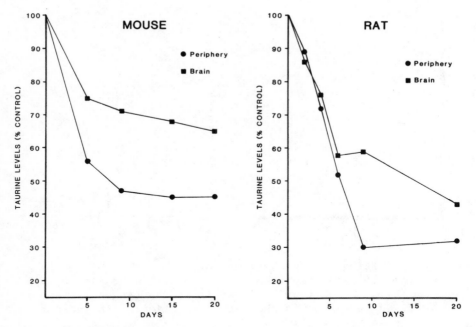

Fig. 3.   Taurine-depletion by GES in the mouse and rat.   Animals
          were maintained on 1 % GES solution as drinking water for
          the indicated number of days.   The percentage of initial
          taurine concentration remaining on average in peripheral
          and brain tissues is shown.   Tissues and organs separately
          analyzed were: mouse; heart, lung, liver, kidney, muscle,
          spleen, intestine, cerebellum, spinal cord, cerebral
          hemispheres.   For rat, in addition, hypothalamus, midbrain
          medulla pons, frontal cortex and inferior colliculus were
          analyzed.   The full data are reported in Huxtable and
          coworkers[3,6].

not exhibit changes in organ taurine concentrations.   This species
difference between the cat, and the rat and mouse, has been shown
to be due to the ability of the cat to utilize GES as an amidino
donor.   The GES is thereby converted to taurine.   The enzymology
of this conversion has not been studied, but it presumably involves an amidino transferase.   This has been discussed in
Huxtable and Lippincott[6].

     In its handling of GES, the guinea pig is intermediate
between the cat and the rodents studied.   Following administration
of GES, the guinea pig excretes both GES and an increased amount

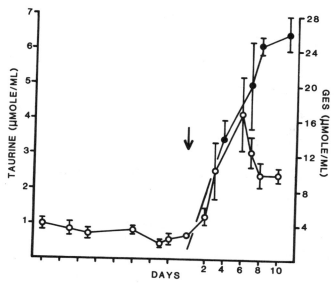

Fig. 4.   Urinary excretion of GES and taurine in the guinea pig.
Data are means ± SE for 3 animals. Solid symbols (right-
hand axis) indicate GES, and open symbols (left-hand axis)
indicate taurine levels. Animals were placed on a 1 %
solution of GES at the time indicated by the arrow. (From
Huxtable and Lippincott[6], by permission of the authors).

of taurine in the urine (Fig. 4). Over the first 4 days of exposure, there is an eightfold increase in taurine excretion. GES
concentrations increase steadily from an undetectable level to 26
μmole/ml urine after 10 days. Radiotracer studies confirmed that
the guinea pig has the ability to convert partially GES to taurine,
whereas the cat converts substantially all of administered GES to
taurine. Similar studies demonstrated the inability of the rat
and mouse to metabolize GES.

The comparative effects of GES on the 4 species studied are
summarized on Table 1.

Although GES was tested in vivo as a taurine depleter because
of its inhibitory action on taurine transport in vitro, recent
studies suggest that the mechanism of taurine depletion is more
complex than simple inhibition of transport[12]. In its taurine-
depleting action, GES has been shown to discriminate between
taurine derived from the diet and taurine derived from biosynthesis.
In rats receiving 5 mM ^{3}H taurine in the drinking water as sole
dietary source of taurine, GES, although it depletes the total
taurine content of a tissue, increases the percentage of the

Table 1. GES on taurine balance

	Effect on Taurine Levels	Do Tissues Accumulate GES?	Metabolism of GES
Guinea pig	No effect	Yes	Moderate(40%)
Rat	Depleted	Yes	None
Mouse	Depleted	Yes	None
Cat	No effect	Yes	High (80%+)

remaining taurine that is derived from the diet. If GES was
acting purely as an inhibitor of transport, it would be unable to
distinguish between biosynthetically derived and dietarily derived
taurine.

In the brain, for example, GES has little effect on the absolute amount of taurine obtained from the diet, but increases the
percentage contribution of the diet from an average of 60 % to an
average of 80 %. Furthermore, it has no effect on the brain to
serum distribution ratio of a tracer dose of ^{3}H taurine. It does
affect the distribution ratio in certain peripheral tissues,
however. The tissue most markedly affected is the liver, the
distribution ratio in GES-treated animals being only 36 % of that
in control animals. The liver is the major source of endogenous
taurine in the rats. These findings suggest that GES, in its
taurine-depleting effects, has a primary action on the liver.
What this action is must await further studies.

CHRONIC GES TREATMENT AND CHEMOSHOCK SEIZURE THRESHOLD

GES is strongly epileptogenic when injected into the inferior
colliculus, the hippocampus, or other areas of the brain[3,13].
Chronic oral administration of GES by our method leads to the accumulation of GES in the brain to a concentration of approximately
1 μmole/g. We examined, therefore, whether chronic treatment with
GES altered the threshold for pentylene tetrazole-induced seizures.
Adult rats were maintained on 1 % GES in drinking water for two
weeks. Chemoshock seizure thresholds were then determined by continuous intraperitoneal injection of pentylene tetrazole until
seizure occurred (Table 2). Under these conditions, there was no
alteration in seizure threshold between the control and GES-treated
groups.

Table 2. Chronic GES on pentylene tetrazole seizure threshold

Threshold (mg/kg Pentylene Tetrazole)

Control	GES
60.5 ± 5.2	54.5 ± 3.8

GES-treated rats were maintained on 1 % GES solution for 2 weeks. Pentylene tetrazole was infused intraperitoneally at a rate of 1.2 mg (0.12 ml) per minute until the animal convulsed. Thresholds are means ± SD for 5 rats per group (body weight 300 g).

CHRONIC EXPOSURE TO GES AND DEVELOPMENT

There is increasing evidence that taurine is an essential amino acid for humans. Children appear particularly vulnerable to taurine deprivation, because of its importance in development[14]. Babies normally obtain their taurine from milk, in which it is the most abundant free amino acid[15]. In the advanced nations, however, many babies are reared on cow-derived milk formulae which contain no taurine. There is increasing concern as to the developmental consequences of this. Children maintained on total parenteral nutrition regimens, which currently lack taurine, develop electro-retinogram abnormalities[16].

The developing rat has a high synthetic capacity for tau-rine[17,18]. However, as discussed above, chronic GES treatment leads to a marked depression of taurine level in rats. We have examined whether chronic exposure to GES leads to any adverse developmental consequences in this species.

Female rats were maintained on water containing 1% GES for two weeks prior to mating. The regimen was continued throughout pregnancy and lactation. The pups were weaned onto water containing 1 % GES.

At brain, rats exhibited depressed taurine concentration in all tissues examined, except for the lung (Table 3). By day 5, only the lung and carcass, composed primarily of musculature, showed significant decreases in taurine concentration. At ages of 25 and 40 days, the initial pattern had reestablished itself, and taurine levels were depressed in all tissues examined. Along with these changes in taurine levels were reciprocal changes in GES levels (Table 4). That is to say, all tissues contain GES at birth, indicating that placental transfer of GES had occurred. By day 5, GES levels had fallen in all tissues, but rose again subsequently.

Table 3. Taurine concentrations in control pups and pups exposed
          to GES (μmole/g)

| | At birth | | 5 Days of Age | |
	Control	GES	Control	GES
Brain	18.2 ± 1.6	11.8 ± 2.0*	11.4 ± 0.5	11.5 ± 1.3
Eye	9.0 ± 0.8	5.3 ± 0.4*	----------	----------
Heart	16.9 ± 1.7	10.1 ± 0.9*	10.4 ± 0.8	10.5 ± 0.2
Kidney	10.6 ± 0.6	6.6 ± 0.9*	6.9 ± 0.8	6.3 ± 0.3
Liver	7.2 ± 2.9	4.4 ± 0.6*	7.0 ± 0.5	6.1 ± 0.3
Lung	7.3 ± 2.2	6.1 ± 0.8	10.9 ± 0.3	7.0 ± 0.5*
Carcass	8.1 ± 0.6	5.3 ± 0.2*	5.6 ± 0.3	4.1 ± 0.4*

| | 25 Days of Age | | 40 Days of Age | |
	Control	GES	Control	GES
Brain	6.9 ± 0.4	5.6 ± 0.3	6.4 ± 0.3	3.7 ± 0.1*
Heart	18.9 ± 1.6	9.0 ± 0.8*	20.2 ± 0.4	4.4 ± 0.9*
Kidney	6.5 ± 0.3	3.0 ± 0.4*	7.9 ± 0.4	2.7 ± 0.3*
Liver	8.1 ± 2.3	2.5 ± 1.6*	2.5 ± 0.8	0.5 ± 0.1*
Lung	9.5 ± 0.6	4.4 ± 1.2*	9.6 ± 2.4	2.9 ± 0.6*
Carcass	7.5 ± 1.0	3.7 ± 0.3*	7.9 ± 0.4	1.9 ± 0.3*

Values are means ± SD for 4 animals per group.
*:$p < 0.05$ compared to corresponding control. Animals were exposed
to GES as described in the text.

These findings indicate that the milk, which is the only
source of GES following birth, is insufficient to maintain the
body burden of GES. The concentrations of GES and taurine in the
milk are reported in Table 5. The taurine value for the control
animals is in reasonable agreement with previously reported values
for rat milk[15].

Prenatal exposure to GES produced no detectable changes in
body weight or organ weights of 1 day old animals (Table 6).
Neither in 1 day old nor 45 day old animals were there detectable
changes in taurine transport into the brain P_2 synaptosomal fraction, or in taurine biosynthesis in brain and liver homogenates.

GES-exposed animals showed a slight but significant increase
in the CD_{50} for pentylene tetrazole-induced seizures, from 39.4 to
46.0 mg/kg (Fig. 5). It should be born in mind that this experi-

Table 4. GES concentrations in pups treated with GES (μmole/g)

Organ	Days of Age			
	0	5	25	40
Brain	2.1 ± 0.2	0.5 ± 0.2	1.0 ± 0.8	1.4 ± 0.1
Heart	6.4 ± 0.1	3.2 ± 1.9	15.5 ± 1.3	15.4 ± 1.2
Kidneys	3.2 ± 0.2	1.5 ± 0.8	7.1 ± 0.6	8.1 ± 0.4
Liver	4.6 ± 0.4	4.3 ± 2.0	17.3 ± 1.6	17.4 ± 1.6
Lung	2.5 ± 0.1	1.6 ± 1.0	6.0 ± 0.5	9.7 ± 0.5
Carcass	4.3 ± 0.3	2.3 ± 1.5	8.5 ± 0.3	9.6 ± 0.5

Values are the mean and standard deviation for 4 animals in each group.

ment was done using a different technique from the experiment reported on Table 2. The values obtained, therefore, are not directly comparable.

In contrast to its lack of effect on the other parameters examined, GES treatment produced a marked change in retinal structure. GES was given to pregnant rats as a 0.1 % solution in drinking water 1 to 2 days prior to delivery, and treatment continued to the pups postnatally for 3 or 6 weeks. Rats treated by such a means for 3 weeks show some degree of membrane disorganization[19]. Disc membranes were bent and separated. Loose and irregular associations of membranes were evident, mainly at the distral end of the outer segment. In some outer segments, irregularity of disc stacking was observed.

The plasma membrane of rod outer segments was clearly damaged. Only minimal lesions were observed at the inner segments, photoreceptor nuclei, or other retinal cells. After 6 weeks of treatment with GES, the photoreceptor layer was severely damaged and its size was decreased with respect to controls. Photoreceptors were severely disorganized; the outer segments being swollen, disoriented, and detached from the inner segments. At higher magnification, it was apparent that the outer segments had lost their characteristic cylindric shape and showed a marked shape distortion. This was accompanied by a disorganization of the disc membranes, all along the outer segments. At the proximal end, the outer segments were shrunk and showed irregular staining.At the distal end, profound membrane disorganization and profuse vesiculation was observed. At this stage, alteration at the inner segments, mainly swelling and vesiculation, was also apparent. A significant loss of photoreceptor cell nuclei could also be observed. The morpho-

Table 5. GES and taurine concentrations in rat milk (μmole/g Milk)

TREATMENT	TAURINE	GES
Control Dams	0.86 ± 0.03	0.06 ± 0.03
GES-treated Dams	0.39 ± 0.19*	0.49 ± 0.16*

Dams were exposed to GES as described in the text. Values are
means ± SD for 3 animals per group.
*:p<0.05 compared to control group.

Table 6. Body weight and organ to body weight ratios of one day
          old rat pups

	Ratio (x10^{-3})	
ORGAN	CONTROL	GES-TREATED
Brain	35.2 ± 5.7	36.9 ± 4.7
Eyes	5.6 ± 0.7	5.7 ± 0.6
Heart	5.2 ± 0.7	5.2 ± 0.6
Kidneys	11.5 ± 2.1	10.2 ± 0.7
Liver	29.5 ± 3.9	38.6 ± 6.2
Lung	23.9 ± 4.9	22.9 ± 3.8
Spleen	2.2 ± 0.6	1.8 ± 0.4
Body Weight(g)	6.8 ± 0.5	7.5 ± 0.8

GES-treated pups were exposed in utero to GES, as described in the
text. Data are means ± SD for 3 animals per group. None of the
intergroup differences were significant at the p<0.05 level.

logical alterations described were consistently observed in all
animals and sections examined.

    The morphological pattern of GES-induced degeneration was
compared to that of cats maintained on a taurine-free diet during
9 weeks. As observed in GES-treated rats, the retinas of taurine-
deficient cats showed swollen outer segments and considerable vesiculation of disc stacking. A deformation due to shrinkage in the
proximal region and swelling of the distal end of the outer segment
was also seen. The inner segments showed considerable intracellular
edema and a marked shape distortion. This pattern of damage was
the same in both GES-treated rats and in taurine-deficient cats.

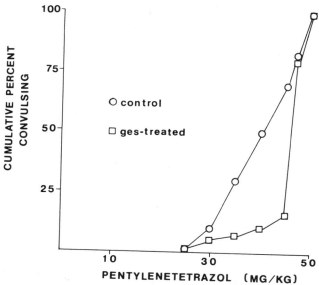

Fig. 5.   Chronic GES treatment on pentylene tetrazole chemoshock
seizure threshold.   Fifty day old rats were used.   GES-
treated rats had been born of GES-exposed mothers and
weaned to 1 % GES solutions as drinking water, as
described in the text.   Each animal received a single
intraperitoneal injection of pentylene tetrazole.   The
y-axis indicates the cumulative percentage of animals
convulsing at the given dose of pentylene tetrazole.
The CD_{50} of control and GES-treated rats was 39.4 ± 2.9
and 46.0 ± 1.3 mg/kg, respectively.   This difference was
statistically significant ($p < 0.05$) (Thakur and Fezio[22]).

These findings constituted the first report of pathological
consequence associated with decreased taurine levels in the rat[19].
GES promises to be, therefore, a useful tool in the study of the
effects of taurine depletion in mammals possessing the enzymatic
machinery for the synthesis of taurine.

The first evidence of a role for taurine in maintaining
photoreceptor structure was provided by the experiments of Hayes
and coworkers[11,20,21], who demonstrated that a critical level of
taurine is necessary to preserve the structural and functional
integrity of photoreceptors.   This was shown in the cat, a species
with a low ability for synthesizing taurine, and almost entirely
dependent on an exogenous source of taurine.   The present results
show that the requirement of taurine for the preservation of
photoreceptor structure is not restricted to the cats, and that
any experimental approach resulting in decreased taurine levels
will produce morphological alteration in photoreceptors.   The
similar pattern of degeneration observed in GES-treated rats and

in taurine-deficient cats suggests that taurine requirements in photoreceptors are the same in these species.

ACKNOWLEDGEMENTS

        Supported by Arizona Affiliate, American Heart Association, the Friedreich's Ataxia Association, Canada, and NIH grant EY-2540.

REFERENCES

1. R. J. Huxtable, Does taurine have a function ?, _Trends in Pharmacological Sciences_, 1:6-7 (1979).
2. R. J. Huxtable, Does taurine have a function ? Introduction, _Fed. Proceedings_, 39:2678-2679 (1980).
3. R. J. Huxtable, H. F. Laird and S. E. Lippincott, The transport of taurine in the heart and the rapid depletion of tissue taurine content by guanidinoethyl sulfonate, _J. Pharm. Exp. Therap._, 211:465-471 (1979).
4. R. P. Shields and C. K. Whitehair, Muscle creatine: in vivo depletion by feeding β-guanidinopropionic acid, _Can. J. Biochem._, 51:1046-1049 (1973).
5. C. D. Fitch, M. Jellinek and E. J. Mueller, Experimental depletion of creatine and phosphocreatine from skeletal muscle, _J. Biol. Chem._, 249:1060-1063 (1974).
6. R. Huxtable and S. E. Lippincott, Comparative metabolism and taurine depleting effects of guanidinoethane sulfonate in cats, mice and guinea pigs, _Arch. Biochem. Biophys._, 210:698-709 (1981).
7. J. B. Lombardini, Combined effects of guanidinoethane-sulfonate, a depletor of tissue taurine levels, and isoproterenol or methaxamine on rat tissues, _Biochem. Pharmacol._, 30:1698-1701 (1981).
8. N. Lake, Depletion of retinal taurine by treatment with guanidinoethyl sulfonate, _Life sciences_, 29:445-448 (1981).
9. N. Lake, Depletion of taurine in the adult rat retina, _Neurochem. Res._, 7:1385-1390 (1982).
10. M. C. Welty, J. D. Welty and M. J. McBroom, Effect of isoproterenol and taurine on heart calcium in normal and cardiomyopathic hamsters, _J. Mol. Cell. Cardiol._, 14:353-358 (1982).
11. S. Y, Schmidt, E. L. Berson and K. C. Hayes, Retinal degeneration in cats fed casein 1. Taurine deficiency, _Invest. Ophthalmol._, 15:47-52 (1976).
12. R. J. Huxtable, Guanidinoethane sulfonate and the disposition of dietary taurine in the rat, _J. Nutrition_, 12:2293-2300 (1982).
13. A. Mori, Y. Katayama, I. Yokoi and M. Matsumoto, Inhibition of taurocyamine (guanidinotaurine)-induced seizures by

taurine, in:"The Effects of Taurine on Excitable Tissue,"
S. W. Schaffer, S. I. Baskin and J. J. Kocsis, eds.,
Spectrum Press, 41–48 (1981).

14. J. A. Sturman, Taurine in nutrition research, in:"Sulfur
Amino Acids: Biochemical and Clinical Aspects," K. Kuriyama,
R. J. Huxtable and H. Iwata, eds., A. R. Liss, New York,
281–295 (1981).

15. D. K. Rassin, J. A. Sturman and G. E. Gaull, Taurine and
other free amino acids in milk of man and other mammals,
Early Human Develop., 2:1–13 (1978).

16. H. S. Geggel, M. E. Ament, J. R. Hackenlively and J. D. Kopple,
Evidence that taurine is an essential amino acid in
children receiving total parenteral nutrition,
Clin. Res., 30:486a (1982).

17. R. J. Huxtable, Sources and turnover rates of taurine in
nursing and weaned rat pups, J. Nutrition, 111:1275–1286
(1981).

18. R. J. Huxtable and S. E. Lippincott, Relative contribution
of the mother, the nurse and endogenous synthesis to the
taurine content of newborn and suckling rat, Nutrition and
Metabolism, 27:107–116 (1983).

19. H. Pasantes-Morales, O. Quesada, A. Carabez and R. J. Huxtable,
Effects of the taurine transport antagonists, guanidinoethane sulfonate and β-alanine, on the morphology of rat
retina, J. Neurosci. Res., 9:135–143 (1983).

20. K. C. Hayes, A. R. Rabin and E. L. Berson, An ultrastructural
study of nutritionally induced and reversed retinal
degeneration in cats, Am. J. Pathol., 78:505–524 (1975a).

21. K. C. Hayes, R. E. Carey and S. Y. Schmidt, Retinal degeneration associated with taurine dificiency in the cat,
Science, 188:949–951 (1975b).

22. A. K. Thakur and L. W. Fezio, A computer program for estimating LD$_{50}$ and its confidence limits using modified Behrens-
Reed-Muench cumuland method, Drug and Chemical Toxicology,
4:297–305 (1981).

EFFECTS OF GUANIDINOETHANE SULFONATE AND TAURINE ON ELECTROSHOCK SEIZURES IN MICE

Kanji Izumi, Chikara Kishita, Takeshi Koja,
Takao Shimizu, Takeo Fukuda and Ryan J. Huxtable*

Department of Pharmacology, Faculty of Medicine
Kagoshima University, Kagoshima 890, Japan and
*Department of Pharmacology, College of Medicine
Health Science Center, University of Arizona
Tucson, AZ 85724, U.S.A.

Guanidinoethane sulfonate is a compound which has been recently found to inhibit competitively taurine uptake and thereby decrease the concentration of this amino acid in various organs including the brain in rats[1] and mice[2,3]. Taurine, or 2-aminoethane sulfonate, is an inhibitory amino acid like γ-aminobutyric acid (GABA), glycine or β-alanine in the central nervous system (CNS) and is known to possess antiepileptic properties against various types of experimentally-induced seizures in mammals (for review see Barbeau and Huxtable[4], Huxtable[5], Durelli and Mutani[6]). Interference with GABA or glycine system in the CNS by specific receptor blockers such as bicuculline or strychnine causes convulsions[7]. Decreased GABA content of the brain caused by inhibiting glutamic acid decarboxylase activity also produces convulsions. In this study we investigate whether a selective decrease in the taurine concentration in the brain produced by guanidinoethane sulfonate can increase the susceptibility to electroshock seizures or modify the anticonvulsive action of phenobarbital or phenytoin against maximal electroshock seizures in mice.

Before discussing the effect of this guanidino compound on epilepsy, we would like to present data on the effect of taurine on maximal electroshock seizures in mice for the following reasons. In spite of a number of studies investigating the effect of taurine in experimental epilepsy, there are only two reports[8,9], to our knowledge, in which electroshock seizures were used. Thursby and Nevis[8] first described briefly the anticonvulsive action of taurine against electrically-induced seizures in mice. Because of the short

227

abstract, detailed information on experimental conditions such as
the electric current delivered or the dose of taurine or the route
and time of administration was not available in this report.   In
contrast, Boggan et al.[9] did not observe any beneficial effect of
this sulfur amino acid against electroshock seizures in mice.   In
their experiments, taurine was administered intraperitoneally (i.p.)
at a dose of 45 mg/kg 30 min before electroshock was given trans-
corneally at a current of 12 mA for 0.2 sec.   Since taurine is
thought to cross the blood-brain barrier with difficulty, the
negative data could be partially attributable to the barrier
problem.   Therefore, we decided to reassess first the effect of
intraventricularly (i.vt.) administered taurine on the maximal
electroshock seizures in mice.

EVALUATION OF MAXIMAL ELECTROSHOCK SEIZURES

Maximal electroshock seizures were elicited with a current
intensity of 50 mA (60 Hz) for 0.2 sec through corneal electrodes
in mice.   Convulsive movements were detected by an accelerometer,
amplified and recorded by a polygraph[10].   By this recording
method, the duration of tonic flexion (TF), tonic extension (TE)
and clonic convulsions were clearly distinguished.   Each duration
for TE and TF was measured from the recording paper, and the ex-
tensor flexor ratio (TE/TF ratio) was determined.   The TE/TF ratio
provides a reliable measure of seizure severity[11]; a ratio of the
test group larger than that of the control group indicates that
seizures in the test group are more severe; a ratio of the test
group smaller than that of the control group indicates that seizures in the test group are less severe.

EFFECTS OF TAURINE ON MAXIMAL ELECTROSHOCK SEIZURES

Male adult ddY mice (32–45 g) were used.   Test mice were injected i.vt. with 95 mM taurine in a volume of 2 µl,3 min prior to
an electroshock (50 mA for 0.2 sec).   The dose of taurine was
chosen from our previous data on the suppressive action of the
sulfur amino acid against ouabain-induced seizures in rats[12] and
pentylenetetrazol-induced seizures in mice[13,14].   Control animals
were injected with 5 mM phosphate buffered saline solution in an
identical volume and 3 min later received the electroshock.   All
solutions injected i.vt. were brought to neutral pH (7.2–7.4) and
isotonicity (280–290 mOsm/1).

As shown in Fig. 1, the administration of taurine prolonged
the duration of TF.   The duration of TE was slightly shortened.
The calculated TE/TF ratio in taurine-treated mice is significantly
smaller than that in control animals (p<0.05).   These results
suggest that i.vt. administered taurine has a suppressive effect

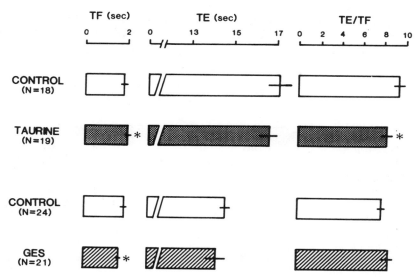

Fig. 1. The effect of taurine and guanidinoethane sulfonate (GES)
         on maximal electroshock seizures in mice. Taurine (95mM)
         was administered intraventricularly in a volume of 2 μl,
         3 min prior to an electroshock (50 mA for 0.2 sec). Con-
         trol animals were injected with 5 mM phosphate buffered
         saline solution without taurine in an identical volume
         and 3 min later received the electroshock. In a study
         investigating the effect of GES on maximal electroshock
         seizures, mice received drinking water containing 1 % GES
         for 9 days. Control animals received drinking water
         without GES. Significant differences between the taurine-
         or GES-treated and the control animals is shown by *
         (p<0.05). TF: duration of tonic flexion, TE: duration of
         tonic extension.

upon maximal electroshock seizures in mice, albeit mild.

EFFECTS OF GUANIDINOETHANE SULFONATE ON MAXIMAL ELECTROSHOCK
SEIZURES

     Male adult ddY mice (30 - 40 g) received drinking water con-
taining 1 % guanidinoethane sulfanate for 9 days. In mice receiving the guanidino compounds, neither excited nor depressed behavior
was observed. Control animals received drinking water without
guanidinoethane sulfonate. To see the selectivity and degree of
the reduction in taurine concentraion in the whole brain, taurine
and other amino acids were determined on an amino acid analyzer
(Hitachi Model 835). These results are summarized in Table 1.
Taurine concentration was selectively decreased to 76 % of control
value in the animals treated with guanidinoethane sulfonate.

Table 1. Alterations of amino acids concentrations in the whole
          brain following treatment with guanidinoethane sulfonate
          (GES) in mice

Amino acids	Control	GES
	μmol/g tissue	
Phosphoserine	0.45 ± 0.04	0.41 ± 0.07
Taurine	8.81 ± 0.19	6.71 ± 0.22*
Aspartate	2.77 ± 0.14	2.73 ± 0.11
Threonine	0.43 ± 0.02	0.37 ± 0.01
Serine	0.86 ± 0.04	0.87 ± 0.05
Glutamate	11.14 ± 0.19	11.19 ± 0.18
Glutamine	5.98 ± 0.12	6.47 ± 0.17
Glycine	1.26 ± 0.24	1.24 ± 0.24
Alanine	0.53 ± 0.04	0.64 ± 0.02
GABA	3.25 ± 0.04	3.18 ± 0.03

Test mice received drinking water containing 1 % GES for 9 days.
Control animals received drinking water without GES. Data are expressed by means±S.E. (N=6). *: $P<0.01$ compared to control value.

These results are in agreement with data reported by Huxtable and
Lippincott[2], who observed a decrease in taurine concentration in
the cerebral hemispheres in mice to 70 % of control value under
experimental conditions similar to ours.

    Under these conditions, maximal electroshock was delivered.
The duration of TF was slightly but significantly ($p<0.05$) shortened in the guanidinoethane sulfonate-treated mice as compared with
that in the control animals (Fig. 1). The duration of TE was not
altered. The calculated TE/TF ratio in mice with guanidinoethane
sulfonate was not significantly different from that in control
animals, although a tendency toward an increased ratio was observed
in the test mice. In separate experiments on determination of the
median convulsive current (CC_{50}), the CC_{50} for tonic convulsions
was found not to be varied by treatment with 1 % guanidinoethane
sulfonate in drinking water for 9 days in mice; CC_{50}s with 95 %
confidence limits are 9.4 (8.5-10.4) mA for the guanidinoethane
sulfonate group and 9.6 (8.9 - 10.3) mA for the control group.
Therefore, under the conditions employed in the present study,
guanidinoethane sulfonate does not seem to affect either the
severity of tonic convulsions of maximal electroshock seizures or
the threshold for tonic convulsions due to electroshock seizures
in mice. These results are not in accordance with the report by
Iwata et al.[15] who demonstrated that administration of 1 % taurocy-

amine (guanidinoethane sulfonate) in drinking water for 10 days
markedly decreased the threshold for EEG seizure activity induced
by cysteine sulfinic acid in rats. However, there are several
differences in experimental conditions between the two studies: in
their experiment, taurine concentrations in the brain was decreased
to 60 % of control levels, whereas in our studies the taurine
concentration in the whole brain was 76 % of the control value; in
their experiment, the seizure threshold was determined by the EEG
seizure pattern, while in our experiments, the threshold was determined by the number of animals with convulsions; they produced
seizures by cysteine sulfinic acid in rats, while we caused convulsions by electroshock in mice. Therefore, comparison of the two
reports is difficult.

EFFECTS OF GUANIDINOETHANE SULFONATE ON THE ANTICONVULSIVE ACTION OF PHENOBARBITAL AND PHENYTOIN AGAINST MAXIMAL ELECTROSHOCK SEIZURES

Baskin and Finney[16] demonstrated that phenytoin (25 mg/kg)
which had been injected i.p. in mice 1 hour prior to sacrifice
produced an increase in the taurine concentration in the brain.
Similarly, rats which had been fed a phenobarbital diet (2 mg/g
and 4 mg/g) for one day have been found to develop increased taurine levels in the cerebral cortex and hippocampus, accompanied by
a decrease in GABA concentration in the brain[17], although no change
in the taurine concentration in the rat brain was observed following the acute administration of barbital sodium (250 mg/kg i.p.)[18].
Baskin and Finney[16] suggested that phenytoin and possibly other
compounds such as barbiturates or carbamazepine might exert their
anticonvulsive effect by increasing cerebral taurine concentration.
On the basis of these findings and suggestions, we investigated
whether or not the anticonvulsive efficacy of phenobarbital and
phenytoin against maximal electroshock seizures in male adult ddY
mice (29 - 44 g) is modified under the conditions employed in the
present study, i.e. with the taurine concentration in the brain
decreased to 76 % of control value by means of treatment with 1%
guanidinoethane sulfonate in the drinking water for 9 days.
Phenobarbital (30 mg/kg) or phenytoin (15 mg/kg) was administered
i.p. 1 hour prior to maximal electroshock. Control mice received
a 0.9 % saline solution i.p. instead of the anticonvulsants.

As shown in Fig. 2, the duration of TF was greatly prolonged
and that of TE was shortened by phenobarbital or phenytoin. The
calculated TE/TF ratio was markedly reduced by both drugs. These
findings indicate that phenobarbital and phenytoin have powerful
suppressive action against tonic convulsions of maximal electroshock seizures as already known. Treatment with 1 % guanidinoethane sulfonate for 9 days lessened the action of these anticonvulsants: in mice with guanidinoethane sulfonate plus phenobarbital, the duration of TF was shortened (p<0.05) and that of TE

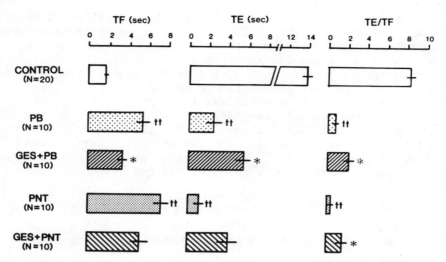

Fig. 2. The effect of guanidinoethane sulfonate (GES) on the
anticonvulsive action of phenobarbital (PB) and phenytoin
(PNT) against maximal electroshock seizures in mice. PB
(30 mg/kg) or PNT (15 mg/kg) was administered intraperitoneally 1 hour prior to an electroshock (50 mA for
0.2 sec). Control mice were injected with a 0.9 % saline
solution instead of the anticonvulsants. In a group with
GES+PB or GES+PNT, mice received drinking water containing
1 % GES for 9 days. In the other control group,
animals received drinking water without GES. Significant
difference between PB- or PNT-injected and control animals
is shown by ††(p<0.01). Significant difference between
GES+PB- and PB-treated animals or GES+PNT- and PNT-treated
animals is shown by *(p<0.05). TF: duration of tonic
flexion, TE: duration of tonic extension.

was prolonged (p<0.05) as compared with the duration of TF or TE
in animals with phenobarbital injection without guanidinoethane
sulfonate treatment; in the case of guanidinoethane sulfonate plus
phenytoin, similar effects were also observed; thus, the calculated
TE/TF ratio in mice with guanidinoethane sulfonate plus phenobarbital or phenytoin became significantly larger (p<0.05) than
the TE/TF ratio in animals not receiving guanidinoethane sulfonate.

As already mentioned, treatment with guanidinoethane sulfonate for 9 days does not increase the severity of tonic convulsions
of maximal electroshock seizures or decrease the threshold for
convulsions induced by electroshock. Nevertheless, the treatment
with guanidinoethane sulfonate diminished the antiepileptic efficacy
of phenobarbital and phenytoin. Selectively decreased taurine
concentration accompanied by accumulation of guanidinoethane

sulfonate in the CNS[1,2] or either one alone might be considered to be factors responsible for the reduced antipileptic potency of the drugs. Further studies on guanidinoethane sulfonate will be required to determine how much reduction of taurine concentration in the brain is critical to alter the threshold for electroshock convulsions. At the same time, it will be necessary to investigate how the accumulated guanidino compounds influence brain function.

SUMMARY

In the present study, we investigated whether or not taurine can suppress maximal electroshock seizures in mice. Furthermore, we examined whether guanidinoethane sulfonate, an inhibitor of taurine uptake, can worsen electroshock-induced convulsions or modify the antiepileptic action of phenobarbital and phenytoin against maximal electroshock seizures in mice. Intraventricular administration of taurine suppressed maximal electroshock seizures. Treatment with 1 % guanidinoethane sulfonate in drinking water for 9 days decreased taurine concentration in the brain to 76 % of control value. Under these conditions, neither the severity of tonic convulsions of maximal electroshock seizures nor the threshold for tonic convulsions caused by electroshock was altered. Nevertheless, anticonvulsive efficacy of phenobarbital or phenytoin was reduced. The decrease in taurine concentration or accumulation of guanidinoethane sulfonate in the CNS or both might be considered factors responsible for the reduced antiepileptic potency of these drugs.

ACKNOWLEDGEMENTS

We are grateful to Prof. H. Iwata, Osaka University, for his gift of taurocyamine for the initial study.

REFERENCES

1. R. J. Huxtable, E. Hugh, H. L. Laird, II and S. E. Lippincott, The transport of taurine in the heart and the rapid depletion of tissue taurine content by guanidinoethyl sulfonate, J. Pharmacol. Exp. Ther., 211:465 (1979).
2. R. J. Huxtable and S. E. Lippincott, Comparative metabolism and taurine-depleting effects of guanidinoethane sulfonate in cats, mice, and guinea pigs, Arch. Biochem. Biophys., 210:698 (1981).
3. M. Hiramatsu, H. Niiya and A. Mori, Effects of taurocyamine on taurine and other amino acid in brain, liver and muscle, Sulfur Amino Acids, 4:227 (1981).
4. A. Barbeau and R. J. Huxtable, "Taurine and Neurological

Disorders", Raven Press, New York (1978).

5. R. J. Huxtable, Insights on function: metabolism and pharma-
    cology of taurine in the brain, in:"The Role of Peptides
    and Amino Acids as Neurotransmitters," J. B. Lombardini and
    A. D. Kenny, eds., Alan R. Liss, Inc., New York (1981).
6. L. Durelli and R. Mutani, The current status of taurine in
    epilepsy, Clin. Neuropharmacol., 6:37 (1983).
7. D. M. Woodbury, Convulsant drugs: mechanisms of action, in:
    "Antiepileptic Drugs: Mechanisms of Action," G. H. Glaser,
    J. K. Penry and D. M. Woodbury, eds., Raven Press, New York
    (1980).
8. M. H. Thursby and A. H. Nevis, Anticonvulsant activity of
    taurine in electrically and osmotically induced seizures in
    mice and rats, Fed. Proc., 33:1494 (1974).
9. W. O. Boggan, C. Medberry and D. H. Hopkins, Effect of tau-
    rine on some pharmacological properties of ethanol,
    Pharmacol. Biochem. Behav., 9:469 (1978).
10. R. Oishi, N. Suenaga, T. Hidaka and T. Fukuda, Inhibitory
    effect of intraspinal injection of 6-hydroxydopamine on
    the clonic convulsion in maximal electroshock seizure,
    Brain Res., 169:189 (1979).
11. E. A. Swinyard, Electrically induced convulsions, in:"Experi-
    mental Models of Epilepsy," D.P. Purpura, J. K. Penry,
    D. Tower, D. M. Woodbury and R. Walter, eds., Raven Press,
    New York (1972).
12. K. Izumi, J. Donaldson, J. Minnich and A. Barbeau, Ouabain-
    induced seizures in rats: suppressive effects of taurine
    and GABA, Canad. J. Physiol. Pharmacol., 51:885 (1973).
13. K. Izumi, H. Igisu and T. Fukuda, Suppression of seizures by
    taurine-specific or non-specific ?, Brain Res.,
    76:171 (1974).
14. K. Izumi, H. Igisu and T. Fukuda, Effects of edetate on sei-
    zure suppressing actions of taurine and GABA, Brain Res.,
    88:576 (1975).
15. H. Iwata, T. Yamamoto, Y. Kumagae and A. Baba, Further study
    on cysteine sulfinate-induced EEG seizures in rats, in:
    "Sulfur Amino Acids: Biochemical and Clinical Aspects,"
    K. Kuriyama, R. J. Huxtable and H. Iwata, eds., Alan R.
    Liss, Inc., New York (1983).
16. S. I. Baskin and C. M. Finney, Factors that modify the tissue
    concentration or metabolism of taurine, in:"The Effects
    of Taurine on Excitable Tissues," S. W. Schaffer,
    S. I. Baskin and J. J. Kocsis, eds., MTP Press, Lancaster
    (1981).
17. K. Aoki and Y. Kuroiwa, Effect of acute and chronic pheno-
    barbital treatment on GABA and other amino acids contents in
    seven regions of the rat brain, J. Pharm. Dyn., 5:88 (1982).
18. H. Iwata, T. Matsuda, S. Yamagami, Y. Hirata and A. Baba,
    Changes of taurine content in the brain tissue of barbi-
    turate-dependent rats, Biochem. Pharmacol., 27:1955 (1978).

EFFECT OF TAURINE, TAUROCYAMINE AND ANTICONVULSANTS ON

DIBENZOYLGUANIDINE-INDUCED CONVULSIONS AND THEIR RELATION

TO BRAIN MONOAMINE LEVELS IN ddY AND EL MICE

Midori Hiramatsu, Hideaki Kabuto and Akitane Mori

Department of Neurochemistry
Institute for Neurobiology
Okayama University Medical School
2-5-1 Shikata-cho, Okayama 700, Japan

INTRODUCTION

It has been reported that taurocyamine (guanidinoethane-sulfonate)[1], guanidinoacetic acid[2], γ-guanidinobutyric acid[3], N-acetylarginine[4], methylguanidine[5] and α-guanidinoglutaric acid[6], are present in the mammalian brain and that these guanidino compounds induce violent convulsions after intracisternal injection into rabbits, dogs, cats and rats. N-amidinobenzamide[7] and di-benzoylguanidine[8], which do not occur naturally, have also been found to induce convulsions after intraperitoneal or intravenous injection into animals. Dibenzoylguanidine is thought to be a very suitable convulsant for the study of the convulsive mechanism, because it can easily pass the blood-brain-barrier and the latent time to induce convulsions is very long.

Matsumoto et al.[9] has reported that the taurocyamine level in the serum of uremic patients and in the brain of experimental uremic rabbits was high and that taurocyamine accumulating in the brain may be related to the convulsions. A high level of taurocyamine was also found in the CSF of epileptic patients[10]. In addition, the brain taurocyamine level in mice was higher in the preconvulsion period of convulsions induced with pentylenetetrazol and lower during convulsions due to ECS[11]. Such data suggest that taurocyamine is related to the seizure mechanism.

On the other hand, it has been reported that taurocyamine decreased the tissue taurine level[12, 13] and that abnormal behavior

235

induced by taurocyamine was depressed by the administration of taurine[14]. Recently, much effort has been focused on this reciprocal action of taurine and taurocyamine.

In the present study, the action of taurine and taurocyamine in nervous tissue and the effect of these substances along with anticonvulsants on dibenzoylguanidine (DBG)-induced convulsions were examined in ddY and El mice as well as the relation of these compounds to brain monoamine levels.

MATERIALS AND METHODS

Animals

Both ddY mice, obtained from Shizuoka Zikken Dobutsu (Shizuoka, Japan), and El mice, bred in our laboratory, were maintained under controlled lighting.

El mice are pure-strain mice, and convulsions are easily induced in them by throwing them into the air[15]. El mice were divided into a stimulated group and non-stimulated (control) group. Mice in the stimulated group were thrown 10 cm into the air once a week from 4 weeks of age. One stimulation consisted of 80 throws. Convulsions were induced in all El mice by this throwing procedure at about 7 weeks of age.

Drugs

DBG was dissolved in dimethylsulfoxide (DMSO), and 150 mg/kg of the solution was intraperitoneally injected into the ddY mice. A one percent solution of taurine and taurocyamine was administered orally to the ddY mice daily for 10 days, and 50 mg/kg of taurocyamine was intraperitoneally injected into the El mice for 10 days. Dipropylacetate (400 mg/kg), phenobarbital (100 mg/kg), diphenylhydantoin (20 mg/kg) and carbamazepine (50 mg/kg) were intraperitoneally injected into the ddY mice, followed by an injection of DBG 30 min later.

Preparation of tissue samples

The ddY mice were killed by microwave irradiation (JRC-NJE 2601) at 3 kW for 0.2 sec, 24 hours after the last administration of taurine and taurocyamine. The brains were removed and the cortex, striatum, hippocampus, hypothalamus, midbrain, medulla oblongata and cerebellum were rapidly separated on an ice-cooled plate by the method of Glowinski and Iversen[16] and stored at -80 °C until analysis.

Catecholamine (CA) analysis

Brain tissue was homogenized with 0.05 M $HClO_4$ in a Polytron homogenizer and centrifuged at 8,000rpm for 15 min. The supernatant layer was decanted, and 30 mg of aluminum oxide in Tris-HCl buffer(pH8.6) were added to the solution. After shaking for 15 min, the upper phase was discarded and the aluminum oxide solution was washed with water 3 times. The CA absorbed to aluminum oxide was eluted with 0.1N HCl, and 5-10 μl of the solution was analysed by high pressure liquid chromatography (Yanaco L-4000W, Kyoto) using an ODS-A column (4.6 x 250 mm) and 10 μM EDTANa /0.1M phosphate buffer (pH 3.1) as the solvent.

5-Hydroxytryptamine (5-HT) analysis

Brain 5-HT analysis was carried out by high pressure liquid chromatography. The procedure for 5-HT analysis has been reported elsewhere[17].

Amino acid analysis

The cerebrum and cerebellum were used for brain amino acid analysis. Five mice were pooled for one sample of cerebellum. The brain tissue was homogenized with 1% picric acid and centrifuged at 10,000rpm for 10min. The picric acid in the resultant supernatant was absorbed onto Dowex 2x8. The remaining solution was evaporated, and the dried sample was dissolved in 0.01 N HCl (pH 2.2). The amino acids were analysed with an amino acid auto analyzer (JLC-6US, Tokyo).

RESULTS

DBG induced convulsions

About 20 min after an intraperitoneal injection of 150 mg/kg of DBG into ddY mice, generalized myoclonus was induced, and 40-60 min after the injection clonic-tonic convulsions appeared. Mice were first stunned, the fore-limbs became tonic and the tail was raised vertically. The mice then rolled over on their sides with urinary incontinence and much salivation, jerked their heads backward followed immediately by tonic flexion. The entire seizure occurred within 20-30 secs. following which the mice died in tonic extension.

Effect of anticonvulsants on DBG induced convulsions in ddY mice

As shown in Table 1, phenobarbital completely inhibited the convulsions. Diphenylhydantoin and dipropylacetate slightly delayed the convulsions and decreased the incidence of both convul-

Table 1.   Effect of anticonvulsants on dibenzoylguanidine induced
           convulsions in ddY mice

	Incidence of convulsion	n	Latent time to convulsion (m±SD, min)	Incidence of death
Control(saline)	18/20 (90%)	18	64.6±19.6	18/18 (100%)
Phenobarbital (100 mg/kg)	0/10 (0%)			
Diphenylhydantoin (20 mg/kg)	5/ 9 (56%)	5	75.4±29.9	3/ 5 (60%)
Dipropylacetate (400 mg/kg)	6/ 9 (60%)	6	75.0±39.6	2/ 6 (33%)

The anticonvulsants were given 30 min before an injection of
dibenzoylguanidine.

sion and death.   Furthermore, in the case of diphenylhydantoin,
clonic convulsions appeared after tonic flexion and continued for
about 30-100 min, after which the mice died without tonic extension.

Effect of carbamazepine on DBG induced convulsions in ddY mice

        Carbamazepine reduced the latent time for convulsions to
begin.   However, the time to death was greatly delayed (Table 2),
and the incidence of death was reduced (Table 2).

Effect of taurocyamine on DBG-induced convulsions in ddY mice

        Taurocyamine markedly delayed the convulsions and reduced the
incidence of death (Table 3).   It did not affect the incidence of
convulsion.

Effect of taurine on DBG-induced convulsions in ddY mice

        Taurine affected neither the latent time of the convulsions
nor the incidence of convulsion or death (Table 4).

Effect of DBG on convulsions in El mice

        In the non-stimulated control group of El mice, the incidence
of convulsion by DBG was remarkably lower than in ddY mice.   When
taurocyamine (50 mg/kg) was administered i.p. for 10 days into
non-stimulated El mice, the incidence of convulsion by DBG was in-

Table 2. Effect of carbamazepine on dibenzoylguanidine induced convulsions in ddY mice

	Incidence of convulsion	n	Latent time to convulsion (m±SD, min)	Incidence of death	n	Latent time to death (m±SD, min)
Control (DMSO)	13/15 (87%)	13	39.5±18.2	13/13 (100%)	13	44.8±15.8
CBZ (50mg/kg)	15/15 (100%)	15	14.4± 8.2*	10/15 (67%)	15	98.2±44.8*

* p< 0.001, compared with the control group.
Carbamazepine was given 30 min before an injection of dibenzoylguanidine.

Table 3. Effect of taurocyamine on dibenzoylguanidine induced convulsions in ddY mice.

	Incidence of convulsion	n	Latent time to convulsion (m±SD, min)	Incidence of death
Control	16/19 (84%)	16	46.2±10.8	16/16 (100%)
Taurocyamine (1% orally for 10 days)	15/18 (83%)	15	71.7±27.9*	13/15 (87%)

* p<0.005, compared with the control group.

Table 4. Effect of taurine on dibenzoylguanidine induced convulsions in ddY mice.

	Incidence of convulsion	n	Latent time to convulsion (m±SD, min)	Incidence of death
Control	11/12 (92%)	11	59.6±22.9	7/ 7 (100%)
Taurine (1% orally for 10 days)	11/11 (100%)	11	45.4±15.6	12/12 (100%)

Table 5.   Effect of dibenzoylguanidine on convulsions in El mice

	Incidence of convulsion	n	Latent time to convulsion (m±SD, min)	Incidence of death
ddY	18/20(90%)	18	64.6±19.6	18/18(100%)
El Non-stimulated control				
Control	4/10(40%)	4	40.8±23.0	3/ 4 (75%)
Taurocyamine	7/ 9(78%)	7	32.6±13.3	7/ 7(100%)
(50mg/kg, i.p., 10 days)				
El Interictal period of stimulated group				
	7/10(70%)	7	21.4 6.9	7/ 7(100%)

creased. In the interictal period of stimulation, the latent time of the DBG-induced convulsions was markedly shorter than in ddY mice (Table 5).

Effect of taurocyamine on the brain taurine level in ddY mice

A one percent solution of taurocyamine was orally adminis-tered to ddY mice for 10 days, and the brain taurine level was studied. Taurocyamine decreased the cerebral taurine level and increased the cerebellar taurine level (Fig. 1).

Effect of taurine and taurocyamine on brain monoamine levels in ddY mice

One percent taurine and taurocyamine were administered orally to ddY mice for 10 days, and the brain levels of norepinephrine (NE), dopamine (DA) and 5-HT were studied. Taurine remarkably de-creased the NE levels in the striatum, hypothalamus and midbrain, and taurocyamine decreased the striatal NE level (Fig. 2). Taurine notably decreased the striatal DA level and taurocyamine decreased the DA level in the midbrain (Fig.3). Taurine greatly decreased 5-HT levels in the hippocampus and midbrain, while taurocyamine greatly increased the striatal 5-HT level (Fig. 4).

DISCUSSION

Taurocyamine is thought to be biosynthesized from taurine and arginine by transamidination. There are many reports on the recip-rocal action of taurocyamine and taurine in nervous tissue. For

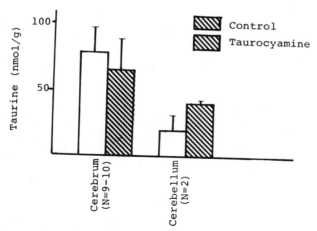

Fig. 1.   Effect of taurocyamine on the taurine level in mouse brain.

example, taurine inhibits the epileptic-like discharges induced by taurocyamine[1], and taurine receptor binding in rat heart muscle is inhibited by taurocyamine[17]. The brain taurine level is decreased by taurocyamine[12,13,18] and taurine transport is inhibited by taurocyamine in mammalian tissue. Taurine uptake by heart slices is also depressed by taurocyamine[20]. These results suggest a kind of opposing action. However, Okamoto and Sakai[21] have reported that taurocyamine is an agonist of taurine in the cerebellum.

Taurocyamine has been reported to induce seizures after intracisternal injection into animals[1], and accumulated taurocyamine in the brain is thought to be responsible for the convulsions in uremia[9]. In addition, a high level of taurocyamine has been found in the CSF of epileptic patients[10]. These facts suggest that taurocyamine is a excitatory substance of the central nervous system. On the other hand, taurine is considered an inhibitory neurotransmitter by many researchers. In our experiment, it was found that taurocyamine delayed the convulsions in ddY mice, contrary to these hypotheses. When a 1 % solution of taurocyamine was orally administered to ddY mice for 10 days, the average taurocyamine intake by the mice was about 0.6 mg/day. On the other hand, unlike the ddY mice the El mice rejected the 1 % solution of taurocyamine, so 50 mg/kg of taurocyamine was intraperitoneally injected into the non-stimulated El mice, and an intake of 1 mg/day was obtained. Taurocyamine tended to decrease the threshold for convulsions induced with DBG. The effects of taurocyamine on ddY mice and El mice were completely different. This difference may be dependent of the strain of mice and/or the dose of taurocyamine.

Oral administration of taurocyamine to ddY mice for 10 days produced an increase in the taurine level in the cerebrum and a

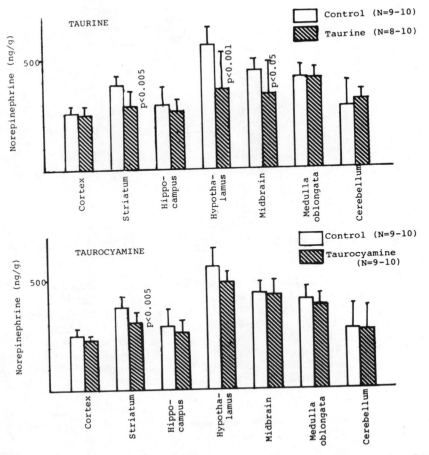

Fig. 2. Effect of taurine and taurocyamine on the norepinephrine
         level in mouse brain.

decrease in the cerebellar level. The decrease in the taurine
level due to taurocyamine agrees with previous reports[12, 13, 17].
Recently, after intracisternal injections of ^{35}S-taurocyamine, we
found high radioactivities in the hippocampus and cerebellum. In
addition, we found high redioactivity in the cerebellum by regional
analyses of radioactivity. Considering these facts, the increased
taurine level induced by taurocyamine in the cerebellum is thought
to be due to the transamidination of taurocyamine with arginine to
taurine, but more detailed study of taurine-taurocyamine metabolism
in nervous tissue is needed.

Iwata[23] has reported that serum taurine levels in non-stimulated El mice were higher than in ddY mice, which are the mother

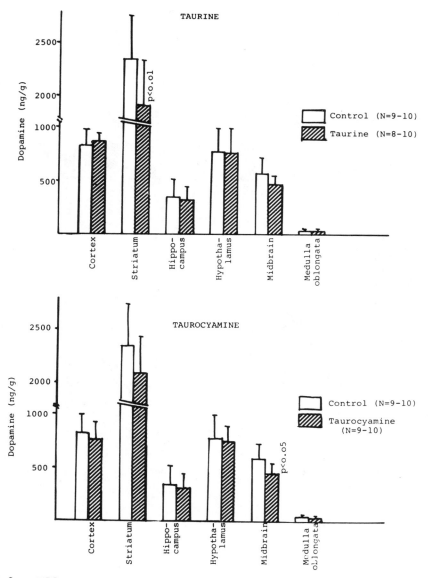

Fig. 3.   Effect of taurine and taurocyamine on the dopamine level
in mouse brain.

strain of El mice but do not have a convulsive disposition.   The
brain taurine levels in the interictal period, and even during the
pre-convulsion period, were also higher in El mice.   These facts
and our results suggest that the metabolism of taurine and
taurocyamine in El mice is abnormal.

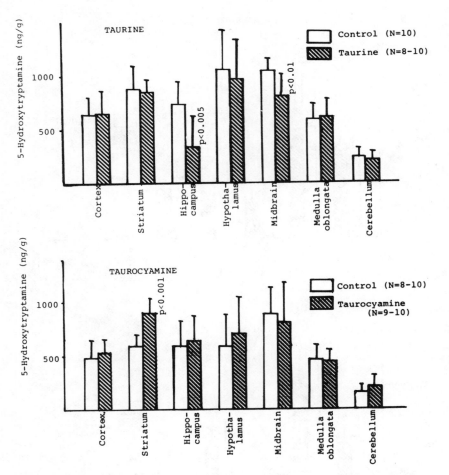

Fig. 4. Effect of taurine and taurocyamine on the 5-hydroxy-
         triptamine level in mouse brain.

In our experiments, phenobarbital completely inhibited DBG-
induced convulsions in the ddY mice, and other anticonvulsants
slightly depressed the convulsions. Taurine, however, did not
have any effect of them, a fact which probably depends on the dose
of taurine administered.

Recently, many authors have reported on brain monoamine levels
and their relationship to the susceptibility to seizures[24-27], and
have shown that the threshold for convulsions was decreased when
the monoamine levels were low and, conversely, that the threshold

for convulsions was elevated when the brain monoamine levels were high. In our experiment it was found that taurine decreased NE levels in the striatum, hypothalamus and midbrain; DA levels in the striatum and 5-HT levels in the hippocampus and midbrain. These regions are thought to be responsible for seizure development. Taurocyamine decreased the NE level in the striatum and the DA level in the midbrain, but increased the 5-HT level in the striatum. The accumulated 5-HT may increase the threshold for DBG-induced convulsions. Nakae[8] has reported that the cerebral 5-HT level was higher during the myoclonic period and during tonic flexion induced by DBG in ddY mice, but CA levels were not changed during those periods. These facts and our results suggest that the seizure mechanism of DBG is greatly dependent on the 5-HT neurone in the striatum. Furthermore, it was found that the actions of taurine and taurocyamine on DBG-induced convulsions may appear through CA nergic and 5-HTnergic neurones in specific brain regions.

SUMMARY

     The effect of taurocyamine, taurine and anticonvulsants on dibenzoylguanidine (DBG)-induced convulsions and their relation-ships to brain monoamine levels in ddY and El mice were studied.

     When 1 % solutions of taurine and taurocyamine were orally administered to ddY mice for 10 days, taurocyamine markedly delayed the convulsions, but taurine did not affect the DBG-induced con-vulsions. Phenobarbital completely inhibited the convulsions when administered i.p. 30 min before the injection of DBG, and diphenyl-hydantoin (20 mg/kg) and dipropylacetate (400 mg/kg) slightly delayed the convulsions and decreased the incidence of convulsion and death. Carbamazepine (50 mg/kg) shortened the latent time to convulsion, but markedly delayed death.

     The incidence of convulsion in the non-stimulated El mice was very low, but injection of taurocyamine (50 mg/kg) into non-stimu-lated El mice for 10 days increased the incidence of convulsion. The latent time to convulsion was very short in the interictal period of stimulated El mice.

     When 1 % solutions of taurine and taurocyamine were adminis-tered to ddY mice for 10 days, taurine decreased NE levels in the striatum, hypothalamus and midbrain; DA levels in the striatum, and 5-HT levels in the hippocampus and midbrain. In addition, taurocyamine decreased the striatal NE levels, the midbrain DA levels and the cerebral taurine levels. However, it increased the striatal 5-HT levels and the cerebellar taurine levels. These results suggest that the actions of taurine and taurocyamine on DBG-induced convulsions are mediated by monoaminergic neurones in special regions of the brain.

REFERENCES

1. A. Mizuno, J. Mukawa, K. Kobayashi and A. Mori, Convulsive
    activity of taurocyamine in cats and rabbits,
    IRCS Med. Sci., 3:385 (1975).
2. D. Jinnai, A. Mori, J. Mukawa, H. Ohkusu, M. Hosotani,
    A. Mizuno and L. C. Tye, Biological and physiological
    studies on guanidino compounds induced convulsion,
    Jpn. J. Brain Physiol., 106:3668 (1969).
3. D. Jinnai, A. Sawai and A. Mori, γ-Guanidinobutyric acid as a
    convulsive substance, Nature, 212:617 (1966).
4. H. Okusu and A. Mori, Isolation of α-N-acetyl-L-arginine from
    cattle brain, J. Neurochem., 16:1485 (1969).
5. M. Matsumoto, K. Kobayashi, H. Kishikawa and A. Mori, Convul-
    sive activity of methylguanidine in cats and rabbits,
    IRCS Med. Sci., 4:65 (1976).
6. A. Mori, Y. Watanabe, S. Shindo, M. Akagi and M. Hiramatsu,
    α-Guanidinoglutaric acid and epilepsy, in:"Urea Cycle
    Diseases," A. Lowenthal, A. Mori and B. Marescau, eds.,
    Plenum Publishing Corporation, New York, (1983).
7. E. Arrigoni-Martelli, A. Garzia and L. Vargin, Attivata'
    farmacologiche della benzoilguanidina, Boll. Soc. Ital.
    Biol. Sper., 38:1421 (1962).
8. I. Nakae, Synthesis of N'N-dibenzoylguanidine and its convul-
    sive action, Neurosciences, 7:205 (1981).
9. M. Matsumoto, H. Kishikawa and A. Mori, Guanidino compounds
    in the sera uremic patients and in the sera and brain of
    experimental uremic rabbits, Biochem. Med., 16:1 (1976).
10. A. Mori, Y. Watanabe and M. Akagi, Guanidino compound
    anomalies in epilepsy, in:"Advances in Epileptology,"
    H. Akimoto, H. Kazamatsuri, M. Seino and A. Ward, eds.,
    Raven Press, New York, (1982).
11. C. Hiramatsu, Guanidino compounds in mouse brain II. Guani-
    dino compound levels in brain in relation to convulsions,
    Okayama-Igakkai-Zasshi, 92:427 (1980).
12. M. Hiramatsu, H. Niiya-Nishihara and A. Mori, Effect of
    taurocyamine on taurine and other amino acids in the brain,
    liver and muscle of mice, Neurosciences, 8:289 (1982).
13. R. J. Huxtable and S. E. Lippincott, Comparative metabolism
    and taurine-depleting effects of guanidinoethanesulfonate
    in cats, mice and guinea pigs, Arch. Biochem. Biophys.,
    210:698 (1981).
14. M. Hiramatsu, S. Ohara, C. Hiramatsu, K. Nanba and A. Mori,
    Effects of taurocyamine on motor activity and brain mono-
    amine level of mouse, Sulfur-containing Amino Acids,
    2:79 (1979).
15. J. Glowinski and L. L. Iversen, Regional studies of catechol-
    amines in the rat brain, J. Neurochem., 13:655 (1966).
16. M. Hiramatsu, Brain monoamine levels and El mouse
    convulsions. Folia Psychiat. Neurol. Jpn., 35:261 (1981).

17. S. W. Schaffer, J. Chovan, J. Kramer and E. Kulakowski, The role of taurine receptors in the heart, in:"The Effects of Taurine on Excitable Tissues," S. W. Schaffer, S. I. Baskin and J. J. Kacsis, eds., Spectrum Publications, New York, (1981)

18. R. J. Huxtable, H. E. Laird and S. Lippincott, Rapid depletion of tissue taurine content by guanidinoethylsulfonate, in: "The Effects of Taurine on Excitable Tissues," S. W. Schaffer, S. I. Baskin and J. J. Kacsis, eds., Spectrum Publications, New York, (1981).

19. H. I. Yamamura, R. C. Speth, R. E. Hruska, N. Bresolin, B. A. Meiners and R. J. Huxtable, Effects of kainic acid lesions of taurine transport into rat brain synaptosomes, in:"The Effects of Taurine on Excitable Tissues," S. W. Schaffer, S. I. Baskin and J. J. Kicsis, eds., Spectrum Publications, New York, (1981).

20. J. Bahl, C. J. Frangakis, B. Larsen, S. Chang, D. Grosso and R. Bressler, Accumulation of taurine by isolated rat heart cells and rat heart slices, in:"The Effects of Taurine on Excitable Tissues," S. W. Schaffer, S. I. Baskin and J. H. Kocsis, eds., Spectrum Publications, New York, (1981).

21. K. Okamoto and Y. Sakai, Inhibitory actions of taurocyamine, hypotaurine, homotaurine, taurine and GABA on spike discharges of purkinje cells, and localization of sensitive sites, in guinea-pig cerebellar slices, Brain Res., 206:371 (1981).

22. S. Shindo, M. Hiramatsu, Y. Katayama, S. Ohara, S. Miyamoto and A. Mori, Distribution and metabolism of ^{35}S-taurocyamine administered to mouse in vivo, Sulfur Amino Acids, 5:197 (1982).

23. H. Iwata, S. Yamagami, E. Lee, T. Matsuda and A. Baba, Increase of brain taurine contents of El mice by physiological stimulation, Jpn. J. Pharmacol., 29:503 (1979).

24. K. Kobayashi and A. Mori, Brain monoamines in seizure mechanism (Review), Follia Psychiat. Neurol. Jpn., 31:483 31:483 (1977).

25. A. Mori, Clinical biochemistry of epilepsy-Specially regarding to neurotransmitters, No-shinkei, 34:1129 (1982) (in Japanese).

26. M. Hiramatsu, Brain 5-hydroxytryptamine level, metabolism and binding in El mice, Neurochem. Res., 8:1163 (1983).

27. E. W. Mynert, T. J. Marczynski and R. A. Browing, The role of the neurotransmitters in the epilepsies, Adv. Neurol., 131:79 (1975).

EFFECT OF CHRONIC ALCOHOL ADMINISTRATION ON THE CONCENTRATIONS OF GUANIDINO COMPOUNDS IN RAT ORGANS

Isao Yokoi, Junji Toma and Akitane Mori

Department of Neurochemistry
Institute for Neurobiology
Okayama University Medical School
2-5-1, Shikata-cho, Okayama City 700, Japan

INTRODUCTION

It is well known that chronic ethanol (EtOH) administration causes convulsions during the withdrawal period. It is also known that some guanidino compounds influence brain function; for example, methylguanidine[1], α-guanidinoglutaric acid[2] and taurocyamine[3] induce epileptic EEG discharges and/or convulsions after intracerebral administration, and the α-guanidinoglutaric acid concentration increases in the cobalt-induced epileptic focus of the cat cerebral cortex[4]. There has been no report about the effects of chronic EtOH administration on the metabolism of guanidino compounds, nor on the effects of metabolic changes of guanidino compounds on the observed convulsions during the withdrawal period from EtOH.

We report here the changes which occur in the concentrations of guanidino compounds in the brain, liver, kidney and blood, and the occurrence of homoarginine (HArg) in the kidney, as well as the effect of lysine on the guanidino compound levels in the liver and kidney of rats administered EtOH chronically.

MATERIALS AND METHODS

Preparation of animals

Sprague-Dawley rats initially weighing 300-350 g were anesthetized by inhalation of ether and mounted on a stereotaxic frame. Through a mild-scalp incision, 4 burr holes were made at the

249

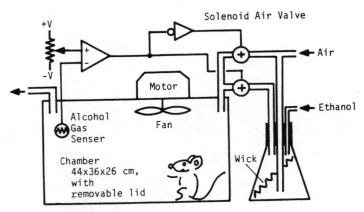

Fig. 1.   Apparatus for delivering alcohol vapor to rats. The con-
          centration of alcohol vapor in the chamber was adjusted
          by the potentiometer.

following sites: 2 mm anterior to the coronal suture, 2 mm lateral
to the sagital suture, 6 mm posterior to the coronal suture and
3 mm lateral to the sagital suture. Four small steinless steel
screw electrodes were placed epidurally. A reference electrode
was placed in the frontal sinus. The five electrodes were wired
to a connector that was fixed to the skull using dental cement.

Groups of 6 rats (EtOH-rats) were housed in 44 x 36 x 26 cm
chambers for 3 days for recovery from the operation and for domes-
tication. At 7:00 AM on the day following recovery, the EEGs of
the rats were recorded for 15 min to check the rat's condition,
without any anesthesia or restriction. After being given 2.5 g/kg
of EtOH (2.5 g EtOH in 7.5 ml of physiological saline solution)
intraperitoneally as the priming dose, the rats were housed in
chambers with 20 mg/1 of EtOH vapor for 5 days. The concentration
of EtOH vapor in the chamber was controlled by a feedback control
system using an alcohol gas senser (Figaro TGS #812). The air flow
was 6 l/min. Food and water were supplied liberally (Fig. 1).
At 7:00 A.M. on the sixth day, the blood of 6 rats was collected
for the determinations of EtOH, GOT and GPT contents, while another
group of rats was placed in another chamber where EEGs were recorded
for 12 h without any EtOH, anesthesia or restriction. At 7:00 P.M.
each animal was decapitated, and the brain, liver, kidney and serum
were promptly removed and cooled in liquidnitrogen.

Another group of 5 rats (control rats) underwent the above
procedure, except that the rats were given physiological saline
intraperitoneally (10 ml/kg) and housed in a chamber without EtOH
vapor.

For the lysine loading experiment, 5 rats (Lys-rats) were given a priming dose of 2.5 g/kg of EtOH along with 400 mg/kg of lysine (Lys) intraperitoneally, then housed in a chamber with 20 mg/1 of EtOH vapor for 5 days. They were given 400 mg/kg of Lys every morning during EtOH inhalation. At 7:00 A.M. on the sixth day, the rats were given Lys and housed in another chamber without EtOH vapor. At 7:00 P.M. each rat was decapitated, and the liver and kidney were promptly removed and cooled in liquid nitrogen.

Preparation of materials

The brain, liver and kidney were homogenized and the serum blended with 10 volumes of ice-cold 1 % picric acid and centrifuged at 3000 rpm for 20 min. The supernatant was passed through a column of Dowex 1x8 (Cl form), 1.0 x 2.0 cm, and the colorless eluate was dried in vacuo.

Guanidino compounds in these samples were fluorometrically analysed using a JASCO G-520 guanidino compound analyzer[5].

Determinations of the concentrations of alcohol, GOT and GPT in blood

The blood alcohol concentration was measured using a kit, Alcohol UV Test (C. F. Boehringer & Soehne GmbH, Mannheim), and the blood GOT and GPT concentration were measured using another kit, Transnase "Nissui" (Nissui Pharmaceutical Co. Ltd., Tokyo).

Identification of homoarginine

Gas chromatography-mass spectrometry (CG/MS) of HArg was performed after converting them to dimethylpyrimidyl butyl ester derivative by reaction with acetylacetone and esterification with butyl alcohol[6]. Standard HArg was purchased from Calbiochem (San Diego). Analysis was done with a Hitachi M-80 type mass spectrometer under the following conditions: for GC- column length, 100 cm; column packing, OV-1 3 %; column temperature, 290 °C, and carrier gas, He, and for MS- ion source, EI; accelerating voltage, 3.0 kV; ionizing voltage, 20 eV, and emission current, 100 μA.

RESULTS

The EEG findings

About 3 to 5 h after the withdrawal from EtOH, the EEGs of 95 % of the EtOH-rats (15/16 rats) showed the initiation of spike discharges (Sp-D) during the spindle-hill phase. The Sp-D from the 4 electrodes were either synchronous or dissynchronous and continued to the end of the recording at 7:00 P.M. (Fig. 2). The

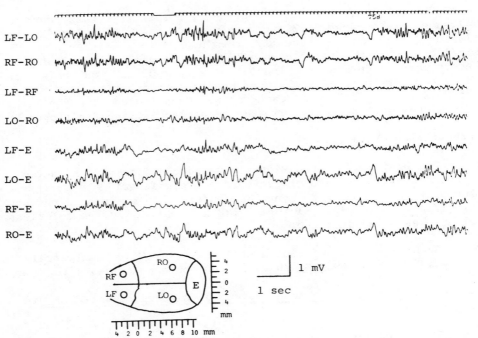

Fig. 2.   Effect of chronic alcohol administration on the EEGs of
          rats. The recordings were carried out from 4 epidural
          electrodes 12 h after withdrowal. LF-LO: bipolar record-
          ing from the left frontal electrode (LF) to the left
          occipital electrode (LO), RF-RO: bipolar recording from
          the right frontal electrode (RF) to the right occipital
          electrode (RO), LF-RF: bipolar recording from LF to RF,
          LO-RO: bipolar recording from LO to RO, LF-E: unipolar
          recording from LF, LO-E: unipolar recording from LO,
          RF-E: unipolar recording from RF, RO-E: unipolar
          recording from RO.

EEGs of control-rats showed no Sp-D during the recording period.

As only 3 of 10 rats which were not given 2.5 g/kg of EtOH as
a priming dose before inhalation of EtOH vapor showed Sp-D, about
9 to 11 h after withdrowal, rats were given a priming dose in this
series of experiments.

Alcohol, GOT and GPT concentrations in blood

Rats inhaling EtOH for 5 days gained no weight (101±1 %,
mean±S.D., N=13), whereas normal rats gained about 15 % of their
initial body weight. The EtOH concentration in blood was 0.448

Table 1. The effect of chronic alcohol administration on the
guanidino compounds in the brain of rats

	Control	Alcohol	
GAA	7.26± 0.87	4.02± 0.42	#
NAA	3.41± 0.51	2.79± 0.27	
CRN	686 ±151	307 ±80	
Arg	152 ± 19	109 ± 9	
MG	0.49± 0.09	0.23± 0.03	#

(mean±S.E.M. nmol/g tissue, N=6)

#, P<0.05;
GAA, guanidinoacetic acid; NAA, N-acetylarginine;
CRN, creatinine; Arg, arginine; MG, methylguanidine.

0.025 mg/ml (mean±S.D., N=6) at the withdrawal from EtOH. The
serum GOT concentration of EtOH-rats was 144±51 Karmen unit (KU),
whereas that of normal rats was 100±9 KU, and the serum GPT concentration of EtOH-rats was 37±16 KU, whereas that of normal rats
was 25±3 KU.

Concentration of guanidino compounds in various organs

Taurocyamine, guanidinoacetic acid (GAA), guanidinoglutaric
acid, guanidinosuccinic acid (GSA), N-acetylarginine (NAA),
β-guanidinopropionic acid (GPA), γ-guanidinobutyric acid (GBA),
creatinine (CRN), arginine (Arg), homoarginine (HArg), guanidine
and methylguanidine (MG) were identified and measured using the
guanidino compound analyzer.

In the brain, GAA, NAA, CRN, Arg and MG were detectably
present prior to EtOH administration. After EtOH administration,
GAA and MG decreased in concentration, and CRN tended to be lower
than in control-rats, but NAA and Arg did not change (Table 1).

In the liver, GSA, GAA, GPA, CRN, GBA, Arg, HArg and MG were
also measured prior to EtOH. GAA and Arg increased and GSA decreased after EtOH administration. However, EtOH administration
caused no change in the contents of GPA, CRN, GBA, HArg and MG
(Table 2).

In the kidney, GAA, GPA, CRN, Arg and MG were present before
EtOH. In the control rats, only a trace of HArg was detected, but
after EtOH the renal content of HArg markedly increased to 6.2
nmol/g tissue. However, the other 5 guanidino compounds showed
no change in concentration (Table 3).

Table 2.   The effect of lysine on the guanidino compounds in
           the liver of rats administered alcohol chronically

	Control	Alcohol	Alcohol + Lysine
GSA	7.92± 0.97	2.37± 0.48 #	1.60± 0.49 *
GAA	3.65± 0.44	5.58± 0.59 #	6.04± 0.86 *
GPA	30.2 ± 2.0	28.4 ± 2.4	20.8 ± 1.5 *+
CRN	47.8 ± 6.0	32.4 ± 4.3	101 ±24
GBA	144 ±35	104 ±10	115 ± 7
Arg	17.5 ± 1.2	22.5 ± 1.1 #	15.4 ± 1.6 +
HArg	2.14± 0.56	3.18± 0.23	3.41± 0.47
MG	0.35± 0.02	0.47± 0.10	0.26± 0.02 *
	(N=6)	(N=6)	(N=5)

(mean±S.E.M. nmol/g tissue)

#: $P<0.05$ (Control vs Alcohol)
*: $P<0.05$ (Control vs Alcohol + Lysine)
+: $P<0.05$ (Alcohol vs Alcohol + Lysine)
GSA, guanidinosuccinic acid;   GAA, guanidinoacetic acid;
GPA, β-guanidinopropionic acid;   CRN, creatinine;
GBA, γ-guanidinobutyric acid;   Arg, arginine;
HArg, homoarginine;   MG, methylguanidine.

     GAA, CRN, Arg, HArg and MG were present in the blood. GAA,
Arg and HArg increased whereas CRN decreased, and there was no
change in MG (Table 4).

Identification of HArg in the kidney of EtOH-rats

     A substance having the same retention time as HArg, was iden-
tified by the GC/MS technique in the kidney of EtOH-rats. Fig. 3
shows an example of a chromatogram of guanidino compounds from the
kidney of EtOH-rats. The HArg peak is observed between the Arg
peak and the buffer change peak.

     In another experiment, the eluate from a Dowex 1x8 column was
analysed by paper chromatography, using Toyo-Roshi No.526 filter
paper (40 x 40 cm) and n-butanol / acetic acid / water (12:3:5 by
vol.) as the solvent. The area having the same Rf value as standard HArg was cut from the filter paper. The substance was
extracted in water, dried in vacuo, converted into a dimethylpyrimidyl butyl ester derivative by reaction with acetylacetone
and butyl alcohol, and analysed by the GC/MS technique.

Table 3. The effect of lysine on the guanidino compounds in the kidney of rats administered alcohol chronically

	Control	Alcohol	Alcohol + Lysine
GAA	217 ±23	200 ±19	209 ±15
GPA	4.41± 0.67	3.94± 0.34	4.40± 0.62
CRN	122 ±27	104 ±16	90 ±13
Arg	258 ±16	253 ±26	386 ±47
HArg	Trace	6.01± 0.87 #	6.31± 1.01 *
MG	0.95± 0.06	1.46± 0.37	1.33± 0.09 *
	(N=6)	(N=6)	(N=5)

(mean±S.E.M. nmol/g tissue)

#: P<0.05 (Control vs Alcohol)
*: P<0.05 (Control vs Alcohol + Lysine)
GAA, guanidinoacetic acid;   GPA, β-guanidinopropionic acid;
CRN, creatinine;   Arg, arginine;
HArg, homoarginine;   MG, methylguanidine.

Fig. 4 shows the mass spectra of the substance from kidney and standard HArg, the former being identical to the dimethyl derivative of the HArg butylester (M^+ = 500).

Effect of lysine on the concentration of guanidino compounds

Intraperitoneal injections of Lys every morning during EtOH administration, normalized the Arg concentration in the liver which was increased by EtOH administration, but had no effect on either the GSA level, which remained low, or the GAA level, which remained high. GPA and MG concentrations, which did not change during EtOH administration, decreased with Lys loading. CRN, GBA and Arg contents, which showed no change during EtOH administration, were not influenced by Lys administration (Table 2).

In the kidney, HArg and MG concentration remained high after the addition of lysine and there was no influence on GAA, GPA, CRN and Arg levels (Table 3).

DISCUSSION

Whereas there are many reports about the effect of EtOH on neurotransmitter metabolism[7,8,9], there are none on the effect of EtOH on those guanidino compounds that profoundly effect brain, liver and kidney function physiologically and pharmacologically.

Table 4. The effect of chronic alcohol administration on
          the guanidino compounds in the serum of rats

	Control	Alcohol
GAA	4.36±0.20	6.58± 0.49 #
CRN	29.4 ±0.9	21.0 ± 1.0 #
Arg	233 ±9	317 ±16 #
HArg	0.73±0.04	1.67± 0.14 #
MG	0.07±0.01	0.04± 0.01

(mean±S.E.M. nmol/ml, N=6)

#: P<0.05
GAA, guanidinoacetic acid;   CRN, creatinine;
Arg, arginine;   HArg, homoarginine;
MG, methylguanidine.

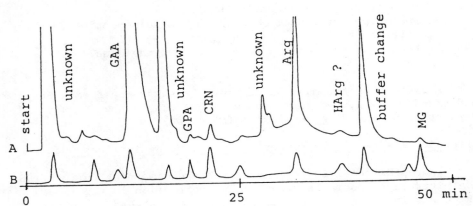

Fig. 3. Chromatogram of guanidino compounds in the kidney of rats
         administrered alcohol for 5 days (A), and of the standard
         sample (B). GAA, guanidinoacetic acid; CRN, creatinine;
         GPA, β-guanidinopropionic acid; Arg, arginine; HArg, homo-
         arginine; and MG, methylguanidine.

    A peak appeared in the chromatogram of the guanidino compound
analysis of the kidney of rats administered EtOH identical to that
of HArg, which induces spike discharges when applied topically to
the sensorimotor cortex[10]. Although a product identified as HArg
by chromatography and nuclear magnetic resonance can be synthesized
in rat kidney homogenates from Lys[11], HArg has not been identified
before in kidney. Therefore we applied the GC/MS technique to sub-
stantiate its presence (M^+ = 500). In the mass spectra, fragments
m/e 123 and 124 could represent the 2'-amino-4', 6'-dimethyl-
pyrimidine residue with a difference of one or two hydrogen atoms.

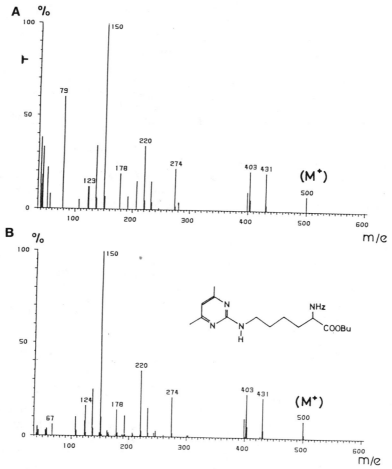

Fig. 4. Mass spectra of homoarginine in the trifluoroacetylated
dimethylpyrimidyl derivative form from the kidney of rats
administered alcohol for 5 days (A), and of the standard
sample (B).

Fragment m/e 150 could represent the 2'-aminoethyl-4', 6'-dimethyl-
pyrimidine residue. Therefore, the substance isolated from kidney
by chromatography could be HArg.

In uremic patients, the serum MG level is elevated[12] and GPA,
possibly related to hemolysis, appears in the urine[13]. EtOH did
not cause a change in the serum MG content, so the increase in
HArg in the kidney and serum is probably not related to a renal

disorder such as uremia. HArg has been identified in the brain, liver and serum[6], and Ryan and Wells reported that an injection of 2.2 g/kg of Lys caused 300 nmol/g tissue of HArg in rat kidney and 90 nmol/g tissue in liver[14]. We loaded 400 mg/kg of Lys every morning for 6 days, but the HArg content in kidney and liver of EtOH administered rats did not increase. The pig kidney argininosuccinic acid cleavage enzyme can synthesize homoargininosuccinic acid from HArg and fumaric acid. The reverse reaction, which may be possible in mammals[15], is another proposed metabolic pathway for the origin of HArg, but the specific activity of this enzyme for HArg is 1/100 that for Arg. Our findings suggest that there is another metabolic pathway for HArg that is activated or enhanced by EtOH administration, and that 400 mg/kg of Lys might be used completely in the so called "lysine-urea" cycle[16], since in liver, Lys normalized the Arg content depressed by EtOH administration.

Since GAA showed an inhibitory effect on the Tonic Autoactive Neuron identified in the subesophageal ganglia of the African giant snail at high concentration[17], the crayfish stretch receptor[18] and the mammalian cortical neuron[19,20], GAA may act as an agonist of GABA. Although the GAA content in brain was decreased by EtOH administration, the relation between the GAA content and convulsions during withdrawal from EtOH is still obscure.

Recently, when we made a preliminary investigation of the effects of EtOH on the guanidino compounds in brain and serum[21], the Arg content in brain was found to decrease significantly after EtOH administration from 161 ± 45 to 104 ± 20 nmol/g tissue (mean$\pm$S.D. of 5 rats, $p < .05$). In this investigation, however, the change in the Arg content in the brains of 6 rats was not significant (from 152 ± 47 to 109 ± 22) because of a high standard deviation, so that a change occuring in the Arg content after EtOH administration seems rather doubtful.

GAA, CRN, Arg and MG were all observed in the serum, brain, liver and kidney. GAA decreased in the brain but increased in the serum and liver and did not change in the kidney. CRN increased in the serum, Arg increased in the serum and liver, and MG decreased in the brain. These differences in the changes in concentrations of guanidino compounds between organs indicates that the metabolism of these compounds in the respective organs changed after EtOH administration and that some of these compounds might be transported selectively across the blood-brain barrier.

CONCLUSION

Guanidino compounds in the brain, liver, kidney and blood of rats, housed for 5 days in chambers supplied with 20 mg/1 of ethanol (EtOH) vapor, were fluorometrically analysed using a JASCO G-520 guanidino compound analyzer.

In the brain, guanidinoacetic acid (GAA), N-acetylarginine (NAA), creatinine (CRN), arginine (Arg) and methylguanidine (MG) were detected. After withdrawal for 12 h from EtOH, GAA and MG levels were found to be significantly decreased and CRN tended to decline.

In the liver, guanidinosuccinic acid (GSA), GAA, β-guanidino-propionic acid (GPA), CRN, γ-guanidinobutyric acid (GBA), Arg, homoarginine (HArg) and MG were detected. GAA and Arg increased and GSA decreased after the EtOH administration.

In the kidney, GAA, GPA, CRN, Arg and MG were detected. HArg identified by the GC/MS technique, greatly increased (from a trace to 6.2 nmol/g tissue) when EtOH was administered.

GAA, CRN, Arg, HArg and MG were present in the blood. GAA, Arg and HArg increased whereas CRN decreased after the EtOH administration.

When 400 mg/kg of lysine (Lys) was administered every morning during EtOH administration, the Arg level was normalized, but GSA remained low and GAA high in the liver. In the kidney, HArg and MG contents remained high. Our findings suggest that there is a metabolic pathway for HArg production that is activated or enhanced by EtOH administration.

ACKOWLEDGEMENTS

We thank Mr. Hisao Matsumoto, Institute of Research and Development, Yamanouchi Pharmaceutical Company for his invaluable help with the GC/MS analysis.

This investigation was supported by a Grant-in-Aid from the Ministry of Health and Welfare of Japan.

REFERENCES

1. M. Matsumoto, K. Kobayashi, H. Kishikawa and A. Mori, Con-
    vulsive activity of methylguanidine in cats and rabbits,
    IRCS Med. Sci., 4:65 (1976).
2. H. Shiraga and A. Mori, Convulsive activity of α-guanidino-
    glutaric acid in rats, IRCS Med. Sci., 10:855 (1982).
3. A. Mizuno, J. Mukawa, K. Kobayashi and A. Mori, Convulsive
    activity of taurocyamine in cats and rabbits,
    IRCS Med. Sci., 3:385 (1975).

4. A. Mori, M. Akagi, Y. Katayama and Y. Watanabe, α-Guanidino-glutaric acid in cobalt-induced epileptogenic cerebral cortex of cats, J. Neurochem., 35:603 (1980).

5. A. Mori, M. Hosotani and L. C. Tye, Studies on brain guanidi-no compounds by automatic liquid chromatography, Biochem. Med., 10:8 (1974).

6. A. Mori, I. Ichimura and H. Matsumoto, Gas chromatography-mass spectrometry of guanidino compounds in brain, Anal. Biochem., 89:393 (1978).

7. W. A. Hunt and E. Majchrowicz, Turnover rates and steady state levels of serotonin in alcohol-dependent rats, Brain Res., 72:181 (1974).

8. S. W. French, D. S. Palman, M. E. Narod, P. Z. Reid and W. Ramey, Noradrenergic sensitivity of the cerebral cortex after chronic ethanol injection and withdrawal, J. Pharmacol. Exp. Ther., 194:319 (1975).

9. P. J. Griffith, J. M. Littleton and A. Dritz, Changes in monoamine concentrations in mouse brain associated with ethanol dependency, Br. J. Pharmacol., 50:489 (1974).

10. I. Yokoi, J. Toma and A. Mori, The effect of homoarginine on the EEG activity of rats, Neurochemical Pathology (in press)

11. W. L. Ryan, R. J. Johnson and S. Dimari, Homoarginine synthesis by rat kidney, Arch. Biochein. Biophys., 131:521 (1969).

12. M. Matsumoto, H. Kishikawa and A. Mori, Guanidino compounds in the sera of uremic patients and in the sera and brain of experimental uremic rabbits, Biochem. Med., 16:1 (1976).

13. R. Shainkin, Y. Giatt and G. M. Berlyne, The presence and toxicity of guanidinopropionic acid in uremia, Kidney International, 7:302 (1975).

14. W. L. Ryan and I. C. Wells, Homocitrulline and homoarginine synthesis from lysine, Science., 144:1122 (1964).

15. J. J. Strandholm, N. R. M. Buist and N. G. Kennaway, Homo-argininosuccinic acid synthesis by an enzymes from pig kidney, Biochim. Biophys. Acta., 237:293 (1971).

16. A. Scott-Emuakpor, J. V. Higgins and A. F. Kohrman, Citrullinemia: A new case, with implications concerning adaptation to defective urea synthesis, Pediat. Res., 6:626 (1972).

17. M. Matsumoto, I. Yokoi, H. Takeuchi and A. Mori, Effects of guanidino compounds on the electrical activity of giant neurons identified in subesophageal ganglia of the African giant snail, Achatina fulica Ferussac, Comp. Biochm. Physiol., 54C:123 (1976).

18. C. Edwards and S. W. Kuffler, The blocking effect of γ-amino butyric acid (GABA) and the action of related compounds on single nerve cells, J. Neurochem., 4:19 (1959).

19. K. Krnjevic and J. W. Phillis, Iontophoretic studies of neurones in the mammalian cerebral cortex, J. Physiol., Lond., 165:274 (1963).

20. J. M. Crawford and D. R. Curtis, The excitation and depres-
     sion of mammalian cortex neurones by amino acids, <u>Br. J.
     Pharmac</u>., 23:313 (1964).
21. I. Yokoi, J. Toma and A. Mori, Effects of taurine on
     guanidino compounds in the brain and serum of rats
     administered alcohol chronically, <u>Neurosciences</u>,
     9:177 (1983).

IV: INVOLVEMENT OF GUANIDINO COMPOUND

IN ACUTE AND CHRONIC RENAL FAILURE

GUANDINO COMPOUNDS: IMPLICATIONS IN UREMIA

Burton D. Cohen

Bronx-Lebanon Hospital
Bronx, N.Y., U.S.A.

Guanidine, which may be described as an "aminated" urea (Fig.1), was first isolated in uremia as far back as 1927[1]. It is an extremely toxic material, producing effects when injected into dogs that mimic many of the clinical symptoms of uremia, such as encephalopathy and gastroenteropathy. Its structural similarity to urea makes it an attractive candidate for the role of the enigmatic "uremic toxin". It makes economic sense, in states of nitrogen excess, to assemble a molecule which binds three, in place of two, amino groups to carbon.

While the original observation over a half century ago relating uremia and hyperguanidinemia was never confirmed, numerous analogues and derivatives have been isolated and implicated in uremia[2,3,4,5] including: guanidinoacetic acid (GAA), guanidinosuccinic acid (GSA), guanidinobutyric acid (GBA), guanidinopropionic acid (GPA), methylguanidine (MG) and dimethylguanidine (DMG). Moreover, a variety of toxic manifestations of uremia have been attributed to guanidines although Koch's postulates relating agents to pathophysiology have never been conclusively satisfied. These include thrombaesthenia, erythrohemolysis, encephalopathy, neuropathy, pericarditis, ascites, hyperlipidemia, pseudodiabetes and atherosclerosis.

This concept, that a substance, normally excreted by the kidney, is retained in renal insufficiency, accumulates and interferes with cellular function, is the classic or traditional formulation of the problem of toxicity in uremia. There is an opposing and somewhat more sophisticated view which, while different, is not necessarily in conflict with the traditional hypothesis. This new theory proposes that toxicity in uremia is an effect of over-

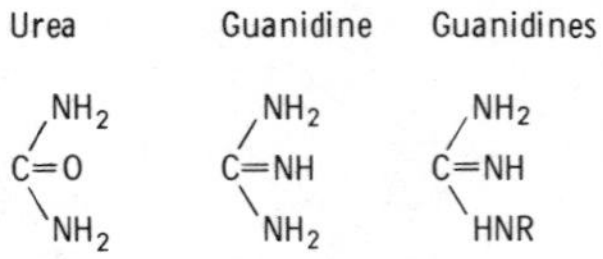

Urea   Guanidine   Guanidines

$$
\begin{array}{ccc}
\underset{\displaystyle NH_2}{\overset{\displaystyle NH_2}{C=O}} & \underset{\displaystyle NH_2}{\overset{\displaystyle NH_2}{C=NH}} & \underset{\displaystyle HNR}{\overset{\displaystyle NH_2}{C=NH}}
\end{array}
$$

Fig. 1.   Structural formulae showing the relationship of
guanidines to urea.

zealous compensating events and misguided efforts to restore extracellular homeostasis in the face of a diminishing nephron population.   When toxicity,however, becomes a compensation, there remains
an uncertainty whether the effect should be combatted or encouraged.

Let us begin with the traditional hypothesis and let me in-
troduce the subject by quoting from Richard Bright writing in
1833[6] : "What seems to account for the general derangement and
suffering (in uremia) is the fact that urea.... or the elements of
which it is formed are abundant in the blood."

It is clever to quote antiquity.   It immediately allies the
audience and the speaker.   We both want Richard Bright to be
correct in his primitive assessment 150 years ago.   Moreover, it
establishes a connection with men of great renown.   It links the
present effort to the past in a sort of "passing the torch".

Unhappily, the torch enlightening uremia and its causes has
been passed for well over a century without much increase in illu-
mination.   Uremia and the uremic toxin remain as elusive today as
in the past despite the miracle of dialysis and other marvelous
mechanisms for managing endstage renal disease.   We have, in fact,
stumbled on the cure before discoveing the cause and this is
decidedly in the wrong order of things.

Recently, a new and intersting development has enlivened the
"toxin game".   It has become fashionable to quote the afore-
mentioned Koch's postulates with the demand that a potential candidate meet these criteria before being certified as legitimate.
The substance must be isolated from the blood or tissue of uremic
persons and, injected into animals, must induce the signs, symptoms or chemical aberrations of uremia.

This, again, establishes an immediate connection with scien-
tific greatness and past success.   There is an almost magic
feeling that, what Koch's postulates did for plague and pesti-
lence, they will again accomplish for uremia.   There are, however,
fundamental differences between infections and uremia.   If there
is anything axiomatic in uremia, it is the very antithesis of
Koch's postulstes:

1. Nothing, which can be measured, is normal in uremia.

2. There is nothing which can be injected into animals that will not produce the signs, symptoms or chemical aberrations of uremia.

With these cautions in mind let us look at the various claims advanced for guanidines in uremia. Two observations, thrombaesthenia and atherogenesis, appear to present a paradox. Untreated, roughtly 10 % of uremics die from hemorrhagic catastrophes while some 40 %, even with dialysis therapy, die as a result of cerebral or coronary vascular thromboses.

Fifteen years ago, Stewart and Castaldi[7] summarized the bleeding diathesis in uremia as an acquired thrombaesthenia reversible by dialysis which suggests, therefore, a small molecular weight, circulating platelet toxin. In the same year[8] we reported that GSA mimicked the effect of the nucleotide, adenosine, in inhibiting the action of adenosine diphosphate on platelet factor 3 activation.

In 1972[9], we reported the startling finding that methionine given to both rats and humans lowered serum and urinary levels of GSA. The mechanism of this remains obscure but one possibility is that both guanidinosuccinic acid, and guanidine, produced in the liver in response to increased serum urea, act as methyl acceptors giving methyl guanidinosuccinic acid (MGSA) and methylguanidine (MG). The reaction proposed would be analogous to that occurring in the liver between S-adenosyl methionine (SAM) and guanidino-acetic acid to produce creatine:

GSA + SAM $\longrightarrow$ MGSA + homocysteine + adenosine

Using a technique based upon that of Bessman, et al[10], we measured homocysteine in the plasma of uremic subjects[11]. Results are shown in Table 1.

To date a total of 16 studies report values of cysteine in the plasma of uremic subjects in various stages of treatment[12-16]. In only 3 were normal values found, and in all 13 remaining the increases reported were comparable to those in Table 1. Hypercysteinemia is, in fact, the most consistent amino acidemia reported in uremic subjects.

Homocystinuria is a disease characterized by circulating homocysteine, an amino acid not present in normal plasma, and clinically by mental deficiency, skeletal abnormalities and thromboembolism. The defect is thought to reside in the enzyme cystathionine synthetase which normally converts homocysteine to cystathionine. In one case[17], a defect in tetrahydrofolate methyl transferase, which converts homocysteine back to methionine, was observed, with similar embolic complications, which suggests that homocysteinemia

Table 1.   Plasma concentration of sulphur-
           containing amino acids in uremia

	Homocysteine mg/dl	Cysteine mg/dl
Controls		
BC	0.11	1.0
HP	0.00	1.2
RR	0.00	1.4
VA	0.06	1.4
SM	0.06	2.0
RS	0.00	1.4
EJ	0.10	1.5
Mean	0.05	1.4
SD	0.05	0.3
Uremics		
JS	0.24	3.9
JH	0.10	4.0
AH	0.23	3.1
JN	0.31	3.8
WP	0.19	3.5
SP	0.57	4.1
ED	0.19	4.1
BS	0.19	2.1
DD	0.30	2.3
TM	0.19	3.5
Mean	0.25	3.4
SD	0.13	0.7
P	0.0025	0.0005

is the common element in the vascular damage.   Several subsequent
studies in animals have confirmed the potent endothelial necrot-
izing effect of injected homocysteine[18,19].   Earlier case repotes
depended largely upon the demonstration of homocystine in urine.
Later studies include reliable plasma levels, the most recent of
which[20] notes a range in 6 cases from 0.27 to 2.39 mg/dl.   The
uremic patients reported here fall into the lower limits of that
range.

In 1964, Welt, et al[21] reported an ion transport defect in red cells from uremic patients which led to an elevated sodium content and an increased intracellular ATP. Incubation of normal cells in plasma from uremic patients induced a defect in the glycoside inhibitable ATPase of the cell membrane. Van den Noort, et al[22] described similar increases in intracellular ATP in brain tissue of uremic rats. Minkoff and coworkers[23] showed that comparable alteration in brain cell ATPase may be induced by injection of methyl guanidine.

Together with our finding that GSA inhibits the ADP-induced platelet release reaction in a manner identical to uremic serum, these data suggest that inhibition of ATPase and ADPase is a generalized phenomenon in uremia and that adenosine, one of the byproducts of the demethylation of methionine and a potent inhibitor of both enzymes, may be implicated in this process leading to diffuse toxicity.

As for homocysteine, the other product of the interaction of guanidino acids and methionine, its tendency to induce vasculitis has already been discussed. In 1974, Warnock, et al[24] noted a decreased SGOT activity in the serum of uremics and suggested that this might reflect a deficiency in vitamin B6. In the same year Dobbelstein, et al[25] followed by several investigators, using different enzyme substrates, demonstrated inhibition of a number of pyridoxal-dependent reactions by uremic serum or guanidines[26]. Since homocysteine is among the known inhibitors of pyridoxine[27,28] the apparent deficiency of the vitamin noted in uremia may be another manifestation of the guanidine-induced demethylation of methionine.

So much for the traditional hypothesis and its relationship to guanidines. Let us turn to the more complex hormonal hypothesis first advanced by Bricker[29] although Charcot[30] writing a century ago expressed a vague insight: "If the secretion of water and the excretion of urea still go on in the kidney, even when it seems to be much altered, it is because some of the glandular portions have preserved their physiological structure." Phosphate is the prototype (Fig.2), the retention of which leads to increased circulating parathormone which results, then, in reduced tubular reabsorption and restoration of phosphate balance. The hormone compensates for reduced filtration by encouraging tubular rejection, a salutory effect restoring homeostasis, the price for which, or the "trade-off" is uremic osteodystrophy and metastatic calcification.

There is no tubular reabsorption of nitrogen which is excreted wholly via filtration, the human organism having lost the capacity to reutilize nitrogen in the form of ammonia or urea. But has he? Let us take another look at guanidines and where they originate. Are they, perhaps, a dormant mechanism for urea fixation?

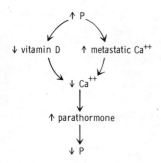

Fig. 2.   In renal insufficiency parathormone serves to maintain
           phosphate balance.

First, let us reexamine the problem of nitrogen in uremia.
It represents, along with sodium and phosphate, the dominant
urinary solute and its retention in renal insufficiency is the
hallmark of the disease. What can we do with nitrogen?

1. REDUCE INTAKE

Twenty years ago Giordano[31] and Giovannetti[32] reported
clinical improvement and a marked reduction in symptomatology in
uremics on a low protein diet provided the intake of essential
amino acids was maintained and the diet was isocaloric. More
recently, Brenner[33] proposed that protein restriction will prolong
the life of even the normal kidney and halt progression of renal
failure in individuals with renal damage and subclinical re-
ductions in nephron population.

2. INCREASE PROTEOGENESIS

There are two mechnisms which control the relative rate of
proteolysis to proteogeneses in muscle. The first is short term
and hormonal and the hormone involved is insulin which stimulates
amino acid uptake and protein synthesis. Thus, insulin is a
nitrogen-sparing hormone which would serve in uremia the same role
as parathormone does in relation to phosphate. Numerous studies
report hyperinsulinism in uremia beginning with the primitive
observations of insulinlike activity[34] and progressing to immuno-
reactive insulinemia proportionate to the level of blood urea[35].

During proteolysis, however, branched-chain or ketogenic
amino acids such as valine, leucine and isoleucine are locally
catabolized so that they are no longer available for proteogenesis
irrespective of the level of plasma insulin. They become, then,

the long term limiting factor in proteogenesis and must be re-
stored exogenously as was empirically discovered by both Giordano
and Giovannetti in their earlier studies in uremia. Concurrent
with this, several investigators[36,37] discovered that the keto-
analogues of valine, leucine and isoleucine (along with that of
methionine) could be substituted for the amino acids in the diet
with a clinically salutory effect though there is no evidence that
these keto acids can be significantly transaminated[38,39]. It is
likely that the keto acids (KA) are nitrogen-sparing by retro-
inhibition of the ammoniolytic breakdown of the branched chain
amino acids (BCAA):

$$BCAA \rightarrow KA + NH_3 \rightarrow CO_2 + NH_3 \rightarrow urea$$

In summary, the combination of protein restriction with adequate maintenance of branched-chain amino acids (or ketoanalogues)
is one mechanism for increasing storage of nitrogen and compensating for reduced filtration. The organism cooperates physiologically in this process by increasing production of insulin, the
price of which is an increased sensitivity to hypoglycemic agents
which is most apparent in diabetics with failing kidneys[40].

3. INCREASE NON-PROTEIN NITROGEN STORES

Most forms assumed by non-potein nitrogen are poorly stored.
Urea diffuses freely through all the body pools where it can potentially disrupt osmotic relationships. Uric acid is poorly soluble.
Free amino acids stimulate gluconeogenesis resulting in an increased urea and glucose production such as is seen in uremia[41,42],
leading to the condition known as azotemic pseudodiabetes. Of all
the alternatives to protein only one seems to offer a possibility
of storage without toxicity and that is creatine. Elsewhere in
this text we offer evidence in support of Natelson's hypothesis
that guanidine (G) and guanidinosuccinic acid (GSA) are byproducts
of a pathway for the fixation of urea in the form of creatine, the
principle endocrine stimulus for which is the hormone, glucagon[43].
Whether this is quantitatively significant is unknown but G and
GSA would appear to be trade-offs for this compensatory pathway.

4. INCREASE FILTRATION

If phosphate is excreted largely via tubular rejection then a
loss in the total number of nephrons or conduits for phosphate excretion is logically compensated for by increased tubular rejection
mediated by parathyroid hormone. If nitrogen is excreted wholly
via glomerular filtration then loss of nephrons is reasonably
compensated by increased filtration. Indeed, Brenner[33] points
out that hyperperfusion and ultrafiltration are a characteristic
finding following episodes of renal damage and ultimately the

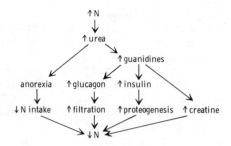

Fig. 3.   Maintenance of nitrogen balance in uremia is more
           complex. The production of guanidines from urea appears
           to be an important step leading to increased filtration,
           proteogenesis and the generation of non-protein nitrogen
           stores.

increased stress placed upon the filtering membrane leads to
progressive insufficiency. The mechanism for this is unknown but
for years at least one hormone, glucagon, has been recognized as
natriuretic as a consequence of glomerular hyperperfusion[44].
Glucagon is known to be measurably elevated in uremia[45] and,
indeed, contributes to the gluconeogenesis and glycogenolysis that
characterizes azotemic pseudodiabetes.

    Of all the substances which stimulate the islet cells of the
pancreas by far the most potent are the guanidines. Arginine and
GAA are known activators[46] and we reported similar sensitivity to
GSA and MG[47].

    Figure 3 is an effort to integrate all of these proposals.
The cost of all this is gastroenteritis, pseudodiabetes, thrombaesthenia, atherogenesis, encephalopathy, erythrohemolysis, neuropathy, hypoglycemia, and progressive renal insufficiency. It
would seem that the trade-off in this instance is both destructive
and costly. Clinical studies are currently under way in our
laboratory using phenobarbitol and methionine to inhibit microsomal oxidation and depress activity of the guanidine cycle.

REFERENCES

1.   R. H. Major and C. J. Weber, The probable presence of
      increased amounts of guanidine in the blood of patients
      with arterial hypertension, Bull. J. Hopkins Hospital,
      40:85 (1927).
2.   B. D. Cohen, I. M. Stein and J. E. Bonas, Guanidinosuccuinic
      acidemia in uremia: A possible alternate pathway for urea
      synthesis, Am. J. Med., 45:63 (1968).

3. G. Perez, A. Rey, M. Micklus and I. Stein, Cation-exchange chromatography of guanidine derivatives in plasma of patients with chronic renal failure, Clin. Chem., 22:240 (1976).

4. R. Shinkin, Y. Giatt and G. M. Berlyne, The presence and toxicity of guanidinopropionic acid in uremia. Kidney Int., 7:S301 (1975).

5. S. Giovannetti, M. Biagine and L. Cioni, Evidence that methylguanidine is retained in chronic renal failure, Experientia, 24:341 (1968).

6. R. Bright, On the functions of the abdomen and some of the diagnostic marks of its disease, London Medical Gazette, 12:150 (1833).

7. J. H. Stewart and P. A. Castaldi, Uremic bleeding: a reversible platelet defect corrected by dialysis, Quart. J. Med., 36:409 (1967).

8. H. I. Horowitz, B. D. Cohen, P. Martinez and M. F. Papayoanou: Defective ADP-induced platelet factor 3 activation in uremia. Blood, 30:331 (1967).

9. B. D. Cohen, Aberrations of the urea cycle in uremia, in: "Uremia: An International Conference on Pathogenesis, Diagnosis and Therapy, " R. Kluthe, G. Berlyne and B. Burton, eds., Georg Thieme Verlag, Stuttgart (1972).

10. S. P. Bessman, Z. H. Koppanyi and R. A. Wapnir, A rapid method for homocysteine assay in physiological fluids, Anal. Biochem., 18:213 (1967).

11. B. D. Cohen, H. Patel and R. S. Kornhauser, Alternate reasons for atherogenesis in uremia, Proc. Dialysis Transplant Forum, 7:178 (1977).

12. J. D. Kopple, Metabolic and endocrine abnormalities, in: "Clinical Aspects of Uremia and Dialysis," A. Sellers and S. Massry, eds., Charles C. Thomas, Springfield (1976).

13. R. P. Betts and A. Green, Plasma and urine amino acid concentrations in children with chronic renal insufficiency, Nephron, 18:132 (1977).

14. J. R. Condon and A. M. Asatoor, Amino acid metabolism in uremic patients, Clin. Chem. Acta, 32:333 (1971).

15. E. Held, W. Winkelmann, K. Finke, H. Dehn, G. Seyffart and H. J. Gurland, Plasma aminosauren bei chronischer miereninsuffizienz, Klin. Wschr., 52:948 (1974).

16. D. Muting and B. D. Dishuk, Free amino acids in serum, cerebrospinal fluid, and urine in renal disease with and without uremia, Proc. Soc. Exp. Biol. Med., 126:754 (1967).

17. K. S. McCully, Vascular pathology of homocysteinemia: implication for the pathogenesis of arteriosclerosis, Am. J. Path., 56:111 (1969).

18. K. S. McCully and B. D. Ragsdale, Production of arterio-sclerosis by homocysteinemia. Am. J. Path., 61:1 (1970).

19. L. A. Harker, S. J. Slichter, C. R. Scott and R. Ross, Homocysteinemia: vascular injury and arterial thrombosis.

N. Engl. J. Med., 291:537 (1974).

20. E. R. Uhlemann, J. H. Ten Pas, A. W. Lucky, J. D. Schulman, S. H. Mudd and N. R. Shulman, Platelet survival and morphology in homocystinuria due to cystathionine synthase deficiency, N. Engl. J. Med., 295:1283 (1976).

21. L. G. Welt, J. R. Sachs and T. J. McManus, An ion transport defect in erythrocytes from uremic subjects. J. Assoc. Am. Phys., 77:169 (1964).

22. S. Van den Noort, R. E. Eckel, K. L. Brine and J. Hrdlicka, Brain metabolism in experimental uremia, Arch. Intern. Med., 126:831 (1970).

23. F. Monkoff, G. Gaertner, M. Darab, C. Mercier and M. L. Levin, Inhibition of brain sodium potassium ATP-ase in uremic rats, J. Lab. Clin. Med., 80:71 (1972)

24. L. G. Warnock, W. J. Stone and C. Wagner, Decreased aspartate aminotransferase (SGOT) activity in serum of uremic patients. Clin. Chem., 20:1213 (1974).

25. H. Dobbelstein, W. F. Korner, W. Mempel, H. Grosse-Wilde and H. H. Edel, Vitamin B6 deficiency in uremia and its implications for the depression of immune responses, Kidney Int., 5:233 (1974).

26. V. Gang, H. Berneburg, H. Hennemann and G. Hevendahl, Diamine oxidase (histaminase) in chronic renal disease and its inhibition in vitro by methylguanidine, Clin. Nephrol., 5:171 (1976).

27. A. S. M. Selim and D. M. Greenberg, An enzyme that synthesizes cystathionine and deaminates L-serine, J. Biol. Chem., 234:1474 (1959).

28. A. Pestana, I. V. Sandoval and A. Sols, Inhibition by homocysteine of serine dehydratase and other pyridoxal S-phosphate enzymes of the rat through cofactor blockage, Arch. Biochem. Biophys., 146:373 (1971).

29. N. S. Bricker and L. G. Fine, The Trade-off hypothesis: current status, Kidney Int., 13:55 (1978).

30. J. M. Charcot, "Lectures on Bright's Disease of the Kidneys," Wm. Wood & Co., New York (1878).

31. C. Giordano, Use of exogenous and endogenous urea for protein synthesis in normal and uremic subjects, J. Lab. Clin. Med., 62:231 (1963).

32. S. Giovannetti and Q. Maggiore, A low-nitrogen diet with proteins of high biological value for severe chronic uremia, Lancet, 1:1000 (1964).

33. B. M. Brenner, T. W. Meyer and T. H. Hostetter, Dietary protein intake and the progressive nature of kidney disease, N. Engl. J. Med., 307:652 (1982).

34. J. M. Cerletty and H. H. Engbring, Azotemia and glucose intolerance, Ann. Intern. Med., 66:1097 (1967).

35. J. D. Bagdade, Uremic lipemia, an unrecognized abnormality in triglyceride production and removal, Arch. Intern. Med., 126:875 (1970).

36. P. Richards, A. Metcalfe-Gibson, F. E. Ward, O. Wrong and B. J. Houghton, Utilization of ammonia nitrogen for protein synthesis in man and the effect of protein restriction in uremia, Lancet, 2:845 (1967).

37. M. Walser, A. W. Coulter, S. Dighe and F. R. Crantz, The effect of keto-analogues of essential amino acids in severe chronic uremia, J. Clin. Invest., 52:678 (1973)

38. S. Ell, M. Fynn, P. Richards and D. Halliday, Metabolic studies with keto acid diets, Am. J. Clin. Nutr., 31:1776 (1978).

39. M. Walser, Nutritional management of chronic renal failure, Am. J. Kid. Dis., 1:261 (1982).

40. C. G. Zubrod, S. L. Eversole and G. W. Dana, Amelioration of diabetes and striking rarity of acidosis in patients with Kimmelstiel-Wilson lesions, N. Engl. J. Med., 245:518 (1951).

41. A. L. Sellers, J. Katz and J. Marmorston, Effect of bilateral nephrectomy on urea formation in rat liver slices, Am. J. Physiol., 191:345 (1957).

42. R. Dzurik, T. R. Niederland and P. Cernacek, Carbohydrate metabolism by rat liver slices incubated in serum obtained from uremic patients, Clin. Sci., 37:409 (1969).

43. B. D. Cohen and H. Patel, Guanidinosuccinic acid and the alternate urea cycle in: "Urea Cycle Diseases", A. Lowenthal, A. Mori, B. Marescau, eds., Plenum Publ. Co., New York (1983).

44. M. Levy and N. L. Starr, The mechnism of glucagon-induced natriuresis in dogs, Kidney Int., 2:76 (1972).

45. G. L. Bilbrey, G. R. Faloona, M. G. White and J. P. Knochel, Hyperglucagonemia of renal failure, J. Clin. Invest., 53:841 (1947).

46. R. N. Alsevor, R. H. Georg and K. E. Sussman, Stimulation of insulin secretion by guanidinoacetic acid and other gua- nidine derivatives, Endocrinology, 86:332 (1970).

47. B. D. Cohen, D. G. Handelsman and B. N. Pai, Toxicity arising from the urea cycle, Kidney Int., 7:S285 (1975).

A STUDY OF CREATINE METABOLISM IN CHRONIC RENAL FAILURE RATS

S. Owada, S. Ozawa, M. Inouchi, Y. Kimura
and M. Ishida

Ist Department of Internal Medicine
St. Marianna University School of Medicine
Sugao, Miyamae-ku, Kawasaki, Japan

INTRODUCTION

Creatine is formed by the methylation of guanidinoacetic acid (GAA). GAA is formed from arginine and glycine in a reaction catalyzed by glycine amidinotransferase (GAT). In chronic uremic subjects, the creatine level is increased in the plasma and normal in the urine[1,2]. The cause of the elevated creatine is not entirely clear. Serum GAA levels are elevated and GAA in the urine is decreased in renal failure[1,2]. The mechanism of these findings is also unknown. The present report describes a study of creatine metabolism in chronic uremic rats. Creatine and GAA levels in plasma, liver and kidney were measured. GAT activity in the kidney was determined. Conversion of L-[guanidino-^{14}C] arginine to creatine, GAA and urea was also measured in the plasma, kidney, liver and muscle.

MATERIALS AND METHODS

Animals

Male Wister rats, about 100 g body weight, were used for a model of chronic renal failure (CRF). One week after resection of 2/3 to 3/4 of the right kidney, the contralateral kidney was removed. Three to six months later, the animals were used for the experiment.

277

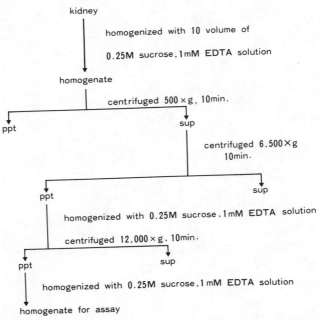

Fig. 1. Preparation of the mitochondrial fraction for kidney
glycine amidinotransferase assay.

Chemical analysis

Plasma creatinine and urea nitrogen (BUN) were measured by
the Jaffe reaction and urease-indophenol method, respectively.
Creatine, GAA and arginine levels in the plasma and tissues were
measured by HPLC (ninhydrin method) employing Shimadzu LC-3A.
Tissue homogenates and plasma were deproteinized by ultrafiltration using YMT membrane (Micropartition System-1, Amicon). Eluting buffers were as follows: 1st buffer; 0.15 N Na^+ (citrate), pH
3.3, 2nd buffer; 0.35 N Na^+ (citrate), pH 4.8, 3rd buffer; 0.35 N
Na^+ (citrate), pH 6.0, 4th buffer; 0.35 N (Na^+) citrate, 0.30 N
(Na^+) chloride, 0.20 N (Na^+) borate, pH 11.4, 5th; 0.20 N NaOH and
6th water. The gradient time program was as follows:

$$\begin{array}{cccccc} & 8 \text{ min} & 8 \text{ min} & 2 \text{ min} & 22 \text{ min} & 2 \text{ min} & 3 \text{ min} \\ \text{1st} & \rightarrow \text{2nd} & \rightarrow \text{3rd} & \rightarrow \text{4th} & \rightarrow \text{5th} & \rightarrow \text{6th} & \rightarrow \text{1st} \end{array}$$

Determination of enzyme activity

Glycine amidinotransferase (GAT) activity in kidney was measured according to the method of Van Pilsum[3] . Fig. 1 shows the
method for the preparation of the mitochondrial fraction of the

Table 1. Reagents for GAT activity assay

A. 0.1 M Phosphate buffer, pH 7.4

B. Arginine-glycine substrate
    40 mM L-arginine-HCl ⎫
    40 mM glycine   ⎬ in 0.1 M phosphate
    30 mM sodium fluoride ⎭ buffer, pH 7.4

C. Arginine substrate
    40 mM L-arginine-HCl ⎫ in 0.1 M phosphate
    30 mM sodium fluoride ⎭ buffer, pH 7.4

D. Ornithine standard solutions
    0.2, 0.4, 0.6, 0.8, 1.0 mM ornithine

E. Protein precipitant
    0.6 N perchloric acid

F. Ninhydrin color reagent
    6 % ninhydrin in methyl cellosolve

---

kidney. The assay procedure for renal GAT activity is shown in Table 1. and Table 2. Enzyme activity is expressed as the micromoles of product per gram wet weight and milligram protein per hour.

Radioactive assay

Two control rats and two CRF rats were used for the experiment. After 15 hours fasting, 2.5 µCi of L-[guanidino-^{14}C] arginine monohydrochloride (specific activity; 56.0 mCi/mmol, Amersham) per 100 g body weight was injected into the peritoneal cavity. The animals were sacrificed 60 and 120 minutes after injection and blood were taken with heparin and then the kidney, liver and psoas muscle were rapidly removed to determine the radioactivities of each.

Analysis of the metabolites of ^{14}C-arginine was performed by the method of Funahashi et al[4]. Tissues were homogenized with 5 volume of cold water, centrifuged for 30 min. at 25,000 rpm and then radioactivity of the supernatants was counted. Supernatants of tissue homogenates and plasma were quickly acidified by the addition of cold 10 % perchloric acid (PCA) to a final concentration of 5 % PCA. These samples were centrifuged and the precipitates were washed twice with cold 5 % PCA. The supernatant and wash fluid were combined and neutralized with 10 N KOH. The precipitate was centrifuged down and washed once with cold water.

Table 2. Assay procedure of kidney GAT activity

	Sample	Blank	Standard
Enzyme solution	0.5 ml	0.5 ml	–
Arg-gly subsrtate	0.5 ml	–	0.5 ml
Arg substrate	–	0.5 ml	–
Ornithine standard	–	–	0.5 ml

Incubation, 37 °C, 60 min, with shaking

0.6N Perchloric acid	1.5 ml	1.5 ml	1.5 ml

Filtration (Whatman No. 1 filter paper)

Protein free filtrate	1.0 ml	1.0 ml	1.0 ml
Ninhydrin color reagent	3.0 ml	3.0 ml	3.0 ml

Boiling in water bath, 100 °C, 25 min.

Cooling in cold water
Colorimetry at 505 nm, within 30 min.

The supernatant and wash fluid were combined, adjusted to pH 2 by adding one tenth volume of 0.1 N HCl and applied to a column (0.7 x 4 cm) of AG 50W-X8 (pyridine form, 200-400 mesh). The column was eluted successively with 10 ml of water, 7 ml of 0.08 N pyridine acetate, pH 4.19, 5 ml of 0.3 N pyridine acetate, pH 4.65, 2 ml of 1.0 N pyridine acetate, pH 4.96 and 8 ml of 2.5 N pyridine acetate, pH 5.18. Urea was found in the void volume. Creatine, GAA and arginine were eluted with 0.08 N, 0.3 N and 2.5 N pyridine acetate buffer respectively. Samples of 0.5 ml of each fraction were transfered to scintillation vials and mixed with 7 ml of scintillator (Biofluor, New England Nuclear). Radioactivities were counted by Aloka-LSC-753 liquid scintillation counter. By this method, creatine and GAA fractions were not completely separated. Significant amounts of creatine were found in 0.3 N pyridine acetate buffer (GAA fraction). To calculate the creatine concentration exactly, 0.08 N and 0.3 N pyridine acetate fractions were assayed by HPLC and then the creatine and GAA counts were corrected.

Table 3. Plasma creatinine and BUN in control and CRF rats

	Control rats	CRF rats
Creatinine (mg/dl)	0.6 ± 0.1	2.7 ± 0.7
BUN (mg/dl)	15.5 ± 2.5	117.2 ± 33.3

(n=3, mean ± SD)

Table 4. Creatine, GAA and arginine levels in plasma and tissues in control and CRF rats (n=3, mean)

	Creatine		GAA		Arginine	
	Cont	CRF	Cont	CRF	Cont	CRF
Plasma (μmol/l)	175.0	34.8	7.6	4.6	156.4	124.6
Liver (μmol/100g-tissue)	49.0	11.0	3.6	0.9	17.8	14.4
Kidney (μmol/100g-tissue)	234.0	42.6	46.8	43.7	231.8	203.9

Table 5. Specific radioactivities of urea, creatine, GAA and arginine in kidney after the injection of labelled arginine into the peritoneal cavity. Plasma creatinine levels in control and CRF-rats.

	Urea	Creatine	GAA	Arginine	Creatinine*
Cont-1	18,316	3,779	5,083	2,433	0.3
Cont-2	8,839	593	717	2,010	0.4
CRF-1	3,583	877	331	1,982	7.1
CRF-2	7,807	637	407	1,110	5.2

Values are given as dpm/g-tissue. Values of Cont-1 and CRF-1 are measured 60 min. after injection, Cont-2 and CRF-2 are measured 120 min. after injection. *Values are expressed as mg/dl.

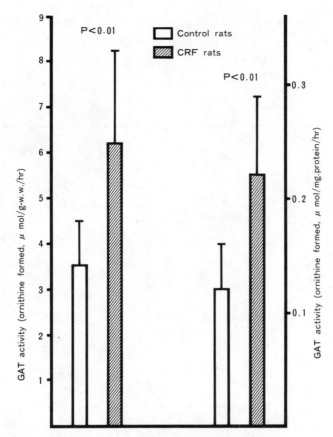

Fig. 2. Kidney glycine amidinotransferase activities in control
         (n=10) and CRF (n=20) rats.

RESULTS

1) Table 3 shows plasma creatinine and BUN levels in control
and CRF rats. Both creatinine and BUN were higher in CRF rats.

2) Table 4 shows the levels of creatine, GAA and arginine in
the plasma and tissues in control and CRF rats. Concentrations of
plasma, liver and kidney creatine were lower in the CRF rats.
Though GAA was lower in the plasma and liver of CRF rats, there
was no significant difference in the kidney.

3) Fig. 2 shows the specific activities of kidney GAT in the
control and CRF rats. GAT activity in the kidney of CRF rats was
significantly higher.

Table 6. Specific radioactivities of urea, creatine,
          GAA and arginine in liver, muscle and plasma
          after the injection of labelled arginine
          into the peritoneal cavity.

		Urea	Creatine	GAA	Arginine
Cont-1	Liver	4,513	1,401	402	2,284
	Muscle	3,903	222	357	2,757
	Plasma	11,347	678	369	3,323
Cont-2	Liver	5,577	854	237	2,699
	Muscle	6,789	557	198	1,299
	Plasma	15,618	775	243	1,083
CRF-1	Liver	12,351	1,827	1,345	1,631
	Muscle	4,806	463	170	1,525
	Plasma	3,313	179	302	2,224
CRF-2	Liver	6,903	274	327	2,538
	Muscle	8,209	484	174	1,447
	Plasma	20,073	150	268	1,775

Values are given as dpm/g-tissue and dpm/ml-plasma.
Values of Cont-1 and CRF-1 are measured 60 min.
after injection, Cont-2 and CRF-2 are measured 120
min. after injection.

4) Table 5 and Table 6 show the radioactivities of urea,
creatine, GAA and arginine. Specific radioactivity of GAA in the
kidney was lower in CRF rats. Plasma creatine activity was lower
in the CRF rats.

DISCUSSION

Creatine is formed by methylation of GAA. GAA is formed from
arginine and glycine in a reaction catalyzed by GAT. In chronic
uremic subjects, the creatine level is increased in the plasma
and normal in the urine. The serum GAA is elevated and urine GAA
decreased in renal failure. In contrast to the reports of elevated
GAA levels in uremic subjects, Tsubakihara[5] found a tendency for
decreased GAA levels. He studied the levels of GAA and creatine

in serum, liver and muscle and also the effects of GAA on creatine
metabolism in normal and CRF rats. In CRF rats, levels of GAA in
serum and liver were significantly lower than that of normal rats.
The GAA and creatine content was significantly elevated when CRF
rats were fed a diet containing GAA. From the above findings he
concluded that GAA production is decreased in the kidney. Therefore, creatine synthesis is decreased in the liver and creatine
concentration in muscle is lowered in CRF rats. From our study,
creatine and GAA concentrations were found to be low in plasma and
liver. In the kidney creatine was low and there was no significant
change in the GAA concentration in the kidney tissues. However,
in relation to the total renal mass, GAA was decreased in the CRF
rats. These findings suggest that creatine production is decreased
in CRF rats.

Activity of GAT was elevated in the uremic rats in comparison
with the control rats. However GAT activity in the total kidney
mass was low. These findings are similar to the data described by
Wang et al.[6]. They found that GAT activity per gram of kidney
tissue was always elevated in uremic rats as compared to control
rats except when arginine-free diets were fed. However, in the
total renal mass, GAT activity was lower in uremic rats suggesting
the possibility that there is a decreased capacity to form creatine in uremia.

Conversion of ^{14}C-labelled arginine to creatine in the plasma
and GAA in the kidney was lower in CRF rats. This findings also
supports the possibility of a decreased creatine production. In
conclusion, creatine production may be decreased in chronic renal
failure in rats.

REFERENCES

1. B. D. Cohen, Guanidinosuccinic acid in uremia, Arch. Intern.
    Med., 126:846 (1970).
2. M. Sasaki, K. Takahara and S. Natelson, Urinary guanidino-
    acetate/guanidinosuccinate ratio, an indicator of kidney
    dysfunction, Clin. Chem., 19:315 (1973).
3. J. F. Van Pilsum, D. Taylor, B. Zakis and P. McCormic,
    Simplified assay for transamidinase activities of rat
    kidney homogenates, Anal. Biochem., 35:277 (1970).
4. M. Funahashi, H. Kato, S. Shiosaka and H. Nakagawa, Formation
    of arginine and guanidinoacetic acid in the kidney in vivo.
    Their relations with the liver and their regulation,
    J. Biochem., 89:1347 (1981).
5. M. Tsubakihara, Guanidinoacetic acid (GAA)-creatine metabo-
    lism in chronic renal failure rats, Proceedings of the 1st
    Meeting of Research Society for Pathogenesis of Uremia,
    Japan, 54 (1982).

6. M. Wang, W. Chan, I. Schutz, J. D. Kopple and M. E. Swendseid, Plasma arginine and ornithine levels and selected enzyme activities in uremic rats fed various amounts of arginine, Life Sciences, 22:2129 (1978).

GUANIDINO COMPOUNDS IN PATIENTS WITH ACUTE RENAL FAILURE

Yoshinori Kosogabe, Yoji Ochiai, Rikiya Matsuda,
Kyoko Nishitani*, Shinya Abe*, Yoshitaro Itano*,
Teruo Yamada* and Futami Kosaka*

Division of Intensive Care Unit, Okayama University
Hospital, and *Department of Anesthesiology, Okayama
University Medical School, Okayama 700, Japan

INTRODUCTION

Alterations in urea cycle activity in acute renal failure
(ARF) may have important metabolic consequences. There are many
reports which study the concentration of guanidino compounds (GC)
in the sera of patients with chronic renal failure (CRF) and the
relationship of these substances to the degree of renal failure[1].
In a recent publication[2], we pointed out that there were very few
reports on GC in ARF and that the relative importance of the
plasma level of GC in the pathogenesis of ARF was uncertain. To
predict the clinical course of ARF, it might be useful clinically
to know of changes in GC, many of which are potential uremic
toxins. As reported at the fifth Symposium on the Analysis of GC
in Japan (Osaka, 1982), it is of interest to trace variations in
GC for determining the course and prognosis of ARF. The purpose
of this study is to trace the clinical meaning of GC in ARF, and
to observe changes in GC by making circle diagrams called guanidinograms. We describe our cumulative studies on GC in 50 patients with ARF.

MATERIALS AND METHODS

Patients

Fifty oliguric and anuric patients with ARF were studied from
September 1981 to April 1983; 30 were males and 20 were females
(average age, 51.2). They received hemodialysis and hemoperfusion

at the Intensive Care Unit of Okayama University Hospital. As
etiologies of ARF there were 11 cases of postoperative hepatorenal
syndrome after gastrointestinal surgery, 7 cases of postoperative
low output syndrome after cardiac surgery, 4 cases of obstetrical
shock, 2 cases of endotoxin shock, 4 cases involving nephrotoxic
substances such as antibiotics and contrast media, 1 case of
gramoxon poisoning, 4 cases of obstruction of the urinary tract
and 17 cases of unknown cause. Patients on repeat dialysis treatment were on unrestricted protein intake.

Analysis of guanidino compounds

Blood samples were obtained from each patient several times
during the initial and oliguric stages of ARF. Immediately following collection, plasma was deproteinized by adjusting the final
concentration to approximately 10 % with 50 % trichloroacetic acid.
Plasma was centrifuged (800 x g, 20 min., 0 °C) and the supernatant
was filtered through a microfilter. The filtrate obtained was
used as the specimen and was analysed by HPLC LC-3A (Shimadzu,
Japan)[2]. Analyses were made by the step gradient system using a
strong acidic cation exchanger ISC-05/SO504, at a reaction temperature of 55 °C and a flow rate of 0.8 ml/min. For reagents,
sodium citrate, boric acid and NaCl were used as eluants and 0.6 %
ninhydrin in 0.5N NaOH for detection.

RESULTS

1) At the initial stage of ARF, that is, at the time of inducing hemodialysis, the concentrations of each GC in plasma are
shown in Table 1. These were classified on the basis of the
underlying diseases into 5 types: the parenchymal ARF group, the
rapidly deteriorated group, pre- and post-renal groups and the
endotoxin group. Nine types of GC were studied: guanidinosuccinic
acid (GSA), creatine (CR), guanidinoacetic acid (GAA), N-acetylarginine, guanidinopropionic acid (GPA), guanidinobutyric acid
(GBA), arginine (ARG), guanidine (G) and methylguanidine (MG).
The plasma concentrations of GSA and MG were higher than those
observed in normal subjects. The CR concentration was also elevated except for the post-renal group. The concentrations of GC
in normal subjects were as follows (in μM/1): CR 131.0±58.7, GAA
2.0±0.3, NAA 0.7±0.1, ARG 84.5±0.8. GSA, GPA, GBA, G and MG were
not present in normal subjects in detectable quantities.

2) At the oliguric stage of ARF, the plasma concentrations
of GC were obtained (Table 2). GSA, GBA, ARG, G and MG increased
compared with that at the initial stage except for the rapidly
deteriorated and post-renal groups. In these two groups, GSA and
GBA were reduced by hemodialysis.

Table 1. Guanidino compounds in initial stage of acute renal failure.

	parenchymal ARF group (n=15)	rapidly deteriorated group (n=4)	post-renal group (n=3)	endotoxin group (n=8)	pre-renal group (n=10)
GSA	10.0±2.2	28.1±6.4	27.3±16.6	22.5±2.2	5.6±1.1
CR	669.1±178.8	1070.4±71.8	82.6±45.1	825.4±261.3	290.4±106.0
GAA	1.4±0.4	1.3±1.0	0.7±0.3	1.2±0.2	1.3±0.7
NAA	7.9±2.3	4.9±4.1	2.1±1.0	4.2±1.1	8.7±3.5
GPA	2.5±0.7	0.1±0.1	0.6±0.6	1.2±0.7	0.9±0.5
GBA	4.3±2.4	2.7±2.7	1.7±0.9	2.8±0.6	0
ARG	68.4±9.2	76.1±17.9	36.6±8.7	109.8±25.5	57.2±10.7
G	1.2±0.2	0	0.9±0.9	0.6±0.2	0
MG	0.7±0.2	1.9±1.5	3.5±2.5	0.5±0.2	0.2±0.2

(μmol/l, Mean±S.E.)

3) A guanidinogram is a circle diagram of eight components of GC. This was devised to display graphically the changes in GC. During the oliguric stage, compared with the initial stage, marked increases in GSA, GBA, ARG, G and MG were noted (Fig. 1a). In pre-renal renal failure, at the oliguric stage, increases in GSA, ARG, G and MG are noted compared with those seen in the initial stage (Fig. 1b). In the cases which ultimately developed CRF, marked increases of MG and G were noted. In the group for which dialysis was required because of rapid deterioration in renal function, the guanidinogram at the initial stage of ARF was noted to respond to hemodialysis (Fig. 1c) with a reduction GC. At the same time in cases of post-renal renal failure, GSA and MG were reduced and the shaded area on the guanidinogram was also reduced by hemodialysis (Fig. 1d).

4) In ARF with endotoxemia, an unusual pattern on the guanidinogram was noted. GSA, GBA, G and MG increased markedly during the oliguric stage compared with values at the initial stage, suggesting an influence of endotoxin on metabolism (Fig. 2a). When a comparison is made between those who survived and those who died on the guanidinogram, a marked increase of GSA, CR and G are noted in the group with fatal outcome (Fig. 2b).

5) At the initial stage of ARF, good correlation between GSA and creatinine (CRN), GSA and blood urea nitrogen (BUN), and MG and CRN. On the other hand, at the oliguric stage, no correlations are noted (Table 3).

Table 2.   Guanidino compounds in oliguric stage of acute renal
           failure.

	parenchymal ARF group (n=15)	rapidly deteriorated group (n=4)	post-renal group	endotoxin group (n=5)	pre-renal group (n=10)
GSA	24.7±4.8	12.3±0.9	8	36.6±8.4	11.7±2.0
CR	609.2±145.6	578.9±505.4	234	764.1±356.3	378.5±94.0
GAA	1.9±0.6	0.07±0.07	0.7	3.3±2.1	1.3±0.7
NAA	8.5±1.3	3.0±2.7	6.7	7.7±3.3	10.6±2.0
GPA	2.1±0.2	0.6±0.3	0	0	1.3±0.4
GBA	6.4±4.3	0.9±0.9	0.2	8.9±7.7	1.1±0.8
ARG	136.9±31.5	94.0±51.0	46.7	163.2±49.2	40.2±57.1
G	2.5±0.6	0	0	2.5±0.9	0.9±0.6
MG	3.4±0.9	0.8±0.5	4.1	2.6±1.3	2.9±1.4

Post-renal group is only one case.   (μmol/1, Mean±S.E.)

6)   However, the correlation (Fig. 3) between GAA and ARG was
not as significant in the initial stage (closed circles) as at the
oliguric stage (open circles). The ARG/GAA ratio and CR/GAA ratio
(Fig. 4) were highly correlated at the initial stage as shown by
the closed circles. Correlation at the oliguric stage between CR
and GAA (data not shown) was noted to be poor (r=0.40).

DISCUSSION

     Guanidino compounds (GC) are normal metabolic products of
protein catabolism[3]. Plasma concentrations of GC, especially GSA
and MG are increased in the uremic state. For example, guanidine
was found to increase up to ten times normal in the sera of dogs
on the fourth day after bilateral nephrectomy and GAA, as well as
GSA, are known to accumulate in renal failure[4]. However, the
etiologic role of GC in the symptoms of acute renal failure is
uncertain. The combined effects of GC now known to be present in
uremic body fluids, rather than the toxicity of any one of them
singly, must be considered, therefore, as a very likely cause of
uremic symptoms and complications[1].

     In this study we used high-pressure liquid chromatography
(HPLC) to estimate levels of GC. Cohen first described a method

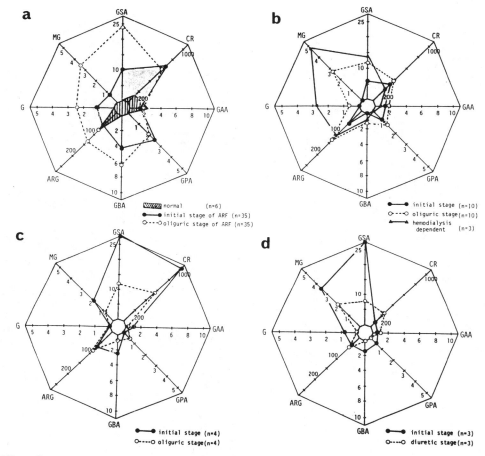

Fig. 1.   Guanidinograms: Circle diagrams of guanidino compounds in
          acute renal failure.
          1-a; at the initial and oliguric stage,
          1-b; pre-renal acute renal failure,
          1-c; rapidly deteriorated renal diseases,
          1-d; post-renal acute renal failure.

for GC measurement using ion exchange chromatography[5]. We used
fluorimetry, a method of detection which is simpler and more sensitive than the Sakaguchi reaction to assay for GC. This technique for the determination of GC has been recently described
using HPLC and fluometric detection with 0.5N NaOH and 0.6% ninhydrin as detection reagents[2]. A rapid, more sensitive determination of GC with a small amount of sample became possible by this
method. The analysis of GC in biological fluids provides us with
a great deal of useful information on the pathophysiology and

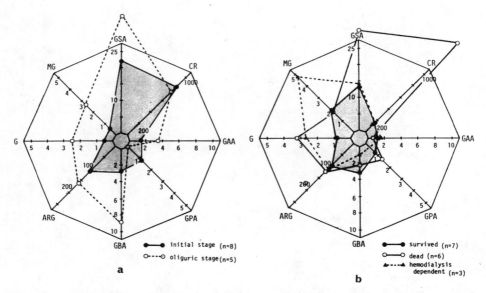

Fig. 2.   Guanidinograms: Circle diagrams of guanidino compounds in
          acute renal failure (ARF).
          2-a; ARF with endotoxemia,
          2-b; groups of survived, dead and hemodialysis dependent.

Table 3.   Correlation coefficient at two stages
           in acute renal failure.

		BUN	Creatinine
initial stage of ARF	GSA	r=0.691 (n=13)	r=0.805 (n=10)
	MG	r=0.110 (n=7)	r=0.886 (n=10)
oliguric stage of ARF	GSA	r=0.084 (n=20)	r=0.110 (n=20)
	MG	r=0.173 (n=20)	r=-0.055 (n=13)

GSA: guanidinosuccinic acid
MG: methylguanidine

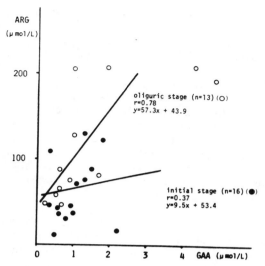

Fig. 3.   Relationship between GAA and ARG in acute renal failure.

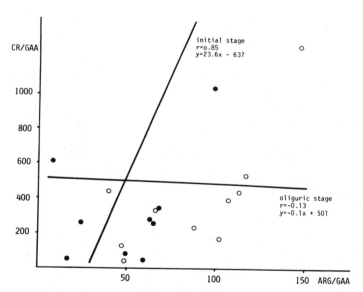

Fig. 4.   Relationship between ARG/GAA and CR/GAA in acute renal
          failure (n=12).

treatment of patients with ARF.   In this report ARF was investigated by analyzing GC^2.   The purpose of the study is to find a
clinical meaning for GC in ARF.

Plasma GSA concentrations in uremia have been shown to interfere with the activation of ADP-induced platelet factor 3 in vitro[6]. G and MG are known to inhibit oxygen and phosphate uptake by mitochondria from rabbit kidney cortex[1] and to induce similar acute toxic effects. GAA and GSA are known to exert enzymatic inhibitions in vitro[7]. A recent study demonstrated the presense of GPA in the plasma of uremic patients and showed that in physiologic concentrations it depressed glucose-6-phosphate dehydrogenase activity in vitro. GBA has been previously demonstrated in plasma of normal and uremic subjects[8,9], but its toxicity is uncertain. We propose that the effect of GC must be considered from the total quantity of GC, especially in ARF where the causative factors are different so there are differences in GC concentrations as reflected in the guanidinogram.

In conclusion, it is suggested that variations of GC are easily recognized by means of guanidinograms and that these are useful indices to evaluate the clinical course and prognoses in ARF.

REFERENCES

1. S. Giovannetti, P. L. Balestri and G. Barcotti, Methylguanidine in uremia, Arch. Intern. Med., 131:709 (1973).
2. Y. Ochiai, S. Abe, T. Yamada, K. Tada and F. Kosaka, Guanidino compounds and hemodialysis, in:"Urea Cycle diseases," A. Lowenthal, A. Mori and B. Marescau, eds., Plenum press, New York and London (1982).
3. J. Sawynok and J. K. Dawborn, Plasma concentration and urinary excretion of guanidine derivates in normal subjects and patients with renal failure, Clin. Exp. Pharmacol. Physiol., 2:1 (1975).
4. M. H. Carr and P. R. Schloerb, Analysis for guanidine and methylguanidine in uremic plasma, Anal. Biochem., 1:221 (1960).
5. B. D. Cohen, Guanidinosuccinic acid in uremia, Arch. Intern. Med., 126:846 (1970).
6. H. I. Horowitz, I. M. Stein, B. D. Cohen and J. G. White, Further studies on the platelet-inhibitory effect of GSA and its role in uremic bleeding, Am. J. Med., 49:336 (1970).
7. E. T. Lonergan, M. Semar and R. Lange, A dialysable toxic factor in uremia, Trans. Am. Soc. Artif. Intern. Organs, 16: 269 (1970).
8. G. Perez, A. Rey, M. J. Micklus and I. Stein, Cation-exchange chromatography of guanidine derivatives in plasma of patients with chronic renal failure, Clin. Chem., 22:240 (1976).
9. M. Sasaki, K. Takahara and S. Natelson, Urinary guanidinoacetate/ guanidinosuccinate ratio: An indicator of Kidney dysfunction, Clin. Chem., 19:315 (1973)

POSSIBILITY OF A COMMON METABOLIC PATHWAY FOR THE PRODUCTION OF METHYLGUANIDINE AND DIMETHYLAMINE IN CHRONIC RENAL FAILURE

Fumitake Gejyo, Sadao Baba*, Yoko Watanabe*
Bellamkonda K. Kishore, Yasushi Suzuki
and Masaaki Arakawa

Department of Medicine (II), Niigata University
School of Medicine, and Department of Medical
Technology*, College of Biomedical Technology
Niigata University, Niigata 951, Japan

INTRODUCTION

It is well known that various nitrogenous metabolites accumulate in the blood of uremic patients[1,2]. These substances, which occur in uremic blood in elevated concentrations, are considered to be toxic and to give rise to many uremic symptoms.

Of these substances, guanidino compounds and aliphatic monoamines are found to be particularly toxic. However, the metabolic pathways of the guanidino compounds and aliphatic monoamines remain to be clarified.

Creatinine is known to be a precursor of methylguanidine, and the increased methylguanidine in chronic renal failure may be due to the elevated creatinine concentration[3]. The conversion of creatinine to methylguanidine in vitro was first postulated by Pfiffner and Myers[4]. Gonella et al.[5] reported that methylguanidine arises from the degradation of creatinine by gastrointestinal bacteria. Jones and Burnett[6] have recently shown that creatinase activity is enhanced in the gut flora of rats fed with creatinine, suggesting that the high concentration of creatinine in body fluids is responsible for such "induction". This finding explains why gut flora from uremics is more efficient in metabolizing creatinine than that from normal subjects.

The source of aliphatic monoamines such as methylamine and dimethylamine is both exogenous and endogenous[7]. A significant

295

generation of these amines is known to occur in the bowel by bacterial metabolism: trimethylamine from lecithin and choline, and
methylamine from sarcosine and creatinine. Both trimethylamine and
methylamine have been suggested as precursors of dimethylamine.
Simenhoff et al.[8] have carried out biochemical studies on aliphatic
amines in uremia in relation to their site of generation by
bacterial transformation, and their clinical significance and
toxicity. They have demonstrated that creatinine is an endogenous
precursor of dimethylamine in chronic renal failure, and that
bacterial overgrowth and increased production of duodenal dimethylamine in the small intestine become apparent at a serum creatinine
level of 8 mg/dl or more.

    These reports suggest the possibility of a common metabolic
pathway from creatinine to both methylguanidine and dimethylamine
in chronic renal failure. Or it may at least point to the involvement of the same type of gut flora in the production of methylguanidine and dimethylamine from creatinine. However, no studies
are available correlating the values of these two compounds, viz.,
methylguanidine and dimethylamine in patients with chronic renal
failure.

    The present study was undertaken to investigate the degree of
elevation of guanidino compounds and aliphatic monoamines in serum,
and the correlations among the serum levels of creatinine, guanidino
compounds and aliphatic monoamines in chronic renal failure. It
was also attempted to examine the possibility of a common metabolic
pathway from creatinine to both methylguanidine and dimethylamine
in chronic renal failure.

MATERIALS AND METHODS

Patients

    Thirty patients with chronic renal failure (17 males and 13
females) selected randomly from the Nephrology Uint of Niigata
University Hospital constituted the study material. The glomerular filtration rate of each patient was under 30 ml/min. Twenty-
four patients had mainly chronic glomerulonephritis. One patient
had renal amyloidosis, and the remainder had polycystic kidney,
polyarteritis nodosa, diabetic nephropathy, Alport's syndrome and
interstitial nephritis respectively. Prior to study the patients
had been treated with conventional conservative methods excluding
hemodialysis, and had not been given any antibitics. Protein
intake was restricted to 0.8 g/kg body weight/day.

    Blood specimens were obtained before the first meal of the
day. The blood was centrifuged and the separated sera were kept
at -20 °C until use.

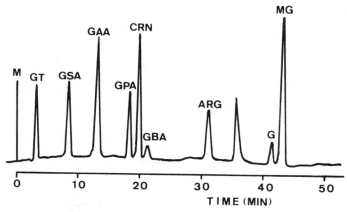

Fig. 1. A chromatogram of a standard mixture of guanidino compounds: taurocyamine (GT) 1 µmole/dl, guanidinosuccinic
acid (GSA) 1 µmole/dl, guanidinoacetic acid (GAA) 0.5
µmole/dl, guanidinopropionic acid (GPA) 0.5 µmole/dl,
creatinine (CRN) 5 µmole/dl, guanidinobutyric acid (GBA)
1 µmole/dl, arginine (ARG) 2 µmole/dl, guanidine (G) 2
µmole/dl and methylguanidine (MG) 0.5 µmole/dl.
M shows an injection mark.

Quantitative determination of guanidino compounds in serum

        The concentration of guanidino compounds in serum was deter-
mined according to the slightly modified method of Yamamoto et al.[9]
and Ando et al.[10]. The principle steps were as fllows: 0.5-1.0 ml
of serum was deproteinized with its half volume of 30 % trichloroacetic acid. Two hundreds-µl aliquots of the supernatant were
subjected to chromatography. High-performance liquid chromatography was performed using a Guanidino Analyzer G-520 (Japan Spectroscopic Co., Hachioji, Tokyo, Japan). The 3.0 x 150 mm chromatographic column was packed with CM-10S (a styrene-divinylbenzene
co-plymer), cross-linked at 10 %, with a particle diameter of
11.5 µm. The column eluent was mixed with streams of 2 N NaOH and
9,10-phenanthrenequinone reagent, and the guanidino compounds in
the sample were converted to the fluorescent product at 60 °C.
Detection was carried out with the excitation maximum at 365 nm
and the emission maximum at 475 nm using a Jasco Model FP-110C
fluorescence photometer. The time for one analysis of individual
guanidino compounds was within 50 min (Fig. 1).

Quantitative determination of aliphatic monoamines in serum

        Aliphatic monoamines in serum were estimated by a high-
performance liquid chromatographic method. Details of the method
have been described previously[11]. The outline of the experiment is

as follows: 0.5-2.0 ml of serum was diluted with the 4-fold volume
of water and then deproteinized by adding 0.25-1.0 ml of 20 % zinc
sulfate and 0.25-1.0 ml of N NaOH. To the deproteinized sample in
a glass-stoppered test tube, a small quantity (100 mg or more) of
crystalline sodium borate and 0.5 ml of 0.5 M 2,4-dinitrobenzene-
sulfonate were added. The mixture was then heated in a boiling
water bath for 40 min, and after cooling 8.0 ml of benzene were
added. The mixture was vigorously shaken for 2 min. The benzene
layer was washed with 2.0 ml of saturated sodium chrolide and dried
over anhydrous sodium sulfate. The benzene extract (6.0 ml) obtained was concentrated to dryness under nitrogen, and the residue
was dissolved in 0.2 ml of tetrahydrofuran. Ten-μl aliquots of the
solution were injected to the apparatus. The apparatus used was a
Waters ALC-201 equipped with a Model 6000A solvent delivery system,
a Model U6K injector, and a μ Bondapak C_{18} reversed-phase column
(3.9 mm i.d. x 30 cm), (Waters Associates, Inc., Milford, Mass.,
USA). The detector was a Jasco Model UVIDEC-100-111 variable wavelength ultraviolet spectrophotometer. The mobile phase was ethanol-
water (1:1 v/v) and its flow rate was 1.0 ml/min. The effluent was
monitored at 357 nm. The 2,4-dinitrophenyl derivatives of ethanolamine, methylamine and dimethylamine were separated satisfactorily
within 6 min (Fig. 2).

Serum levels of urea nitrogen and creatinine were measured by
methods based on the urease-glutamate dehydrogenase-coupled reaction and the Jaffe reaction, respectively, on an Autoanalyzer
(JEOL, JCA-MS24).

RESULTS

The analytical methods

The chromatographic system used for guanidino compounds was
designed to estimate the sensitive 9,10-phenanthrenequinone derivatives. This method is more sensitive than the conventional
method utilizing Sakaguchi's colorimetric procedure[12]. Fig.1
shows a chromatogram of a standard mixture of guanidino compounds
obtained from Wako Pure Chemical Ind. LTD, Osaka, Japan. Typical
chromatograms of serum guanidino compounds of a normal subject and
a uremic patient are shown in Figs.3 and 4, respectively.

A high-performance liquid chromatographic method for the
determination of aliphatic monoamines (ethanolamine, methylamine
and dimethylamine) in serum was developed. The accuracy of the
method was assessed by measuring four different levels of three
amines spiked to serum in replicates. The representative chromatograms obtained from a normal subject and a uremic patient are
illustrated in Figs. 5 and 6, respectively.

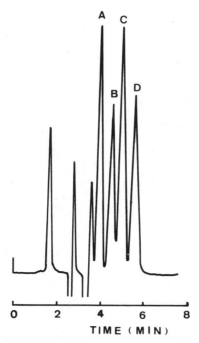

Fig. 2. A chromatogram of a standard mixture of 2,4-dinitrophenyl derivatives of ethanolamine (A), ammonia (B), methylamine (C) and dimethylamine (D).

Serum levels of guanidino compounds and aliphatic monoamines in 30 patients with chronic renal failure

Table 1 shows the concentrations of guanidino compounds, aliphatic monoamines, urea nitrogen and creatinine in the serum of 30 patients with chromic renal failure. These parameters except guanidinoacetic acid and ethanolamine showed significant elevations as compared to the normal values. Table 2 gives the zero order correlation coefficients between the pairs of parameters studied. A highly significant correlation was found between creatinine and dimethylamine (Fig. 7) or methylguanidine (Fig. 8). A significant correlation also existed between methylguanidine and dimethylamine. Of the parameters studied, guanidinosuccinic acid had an excellent correlation with urea only. There was no correlation between ethanolamine and the other parameters.

DISCUSSION

We have developed a relatively rapid and sensitive method for the quantitative determination of serum aliphatic monoamines using

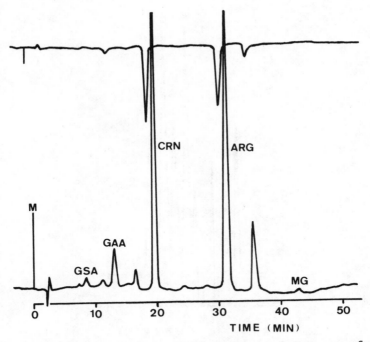

Fig. 3. A chromatogram of guanidino compound in serum from a normal control.

high-performance liquid chromatographic apparatus. Dunn et al.[13] have described a method for measurement of dimethylamine and trimethylamine in ultrafiltered serum by gas liquid chromatography. However, their method was not sufficiently sensitive to confirm the existence of methylamine in serum. They have reported that serum concentration of dimethylamine in uremic patients ranged from 100 to 600 µg/dl, which was about three times higher than our data.

Simenhoff et al.[14] demonstrated that there was a significant correlation between serum dimethylamine level and choice reaction time, which is a marker of neurophysiological function in uremic patients on hemodialysis. Serum trimethylamine levels were also related to choice reaction time and electroencephalographic abnormality. Clinical toxicity of aliphatic monoamines in uremia seems to be emphasized by the correlation of abnormal neurobehavioral parameters with these levels of serum and to be improved by administration of nonabsorbable broad spectrum antibiotics.

Since methylguanidine is considered to be a very important uremic toxin among guanidino compounds, the clarification of metabolic pathway(s) of methylguanidine as well as aliphatic monoamines is important especially in the context of the treatment of patients with chronic renal failure.

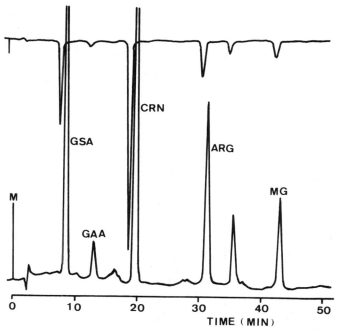

Fig. 4. A chromatogram of guanidino compound in serum from
a uremic patient. The patient (64 year-old, female)
showed 9.6 mg/dl of serum creatinine and 108.0 mg/dl
of urea nitrogen. The guanidino compounds were
calculated to 927.1 µg/dl of GSA, 25.6 µg/dl of GAA and
29.2 µg/dl of MG.

Many investigators have demonstrated that creatinine can be
metabolized by identifiable bacteria to compounds such as N-
methylhydantoin, methylguanidine, urea, methylamine, sarcosine and
ammonia. Cohen[15] suggested that the metabolic origin of methylguanidine in uremia is mainly creatinine and partially arginine.
Mikami et al.[16] reported that the production of methylguanidine
was increased by creatinine and creatine infusion in experimental
uremic rats, and that its marked increase was observed in the case
of creatinine. Guanidinosuccinic acid production was increased
only by urea infusion. Their results suggest that most methylguanidine is produced through creatinine, and that guanidinosuccinic acid may be closely related to urea.

In the present study we examined the extent of correlation
between the levels of various guanidine derivatives and the aliphatic monoamines in chronic renal failure. Table 2 reveals that

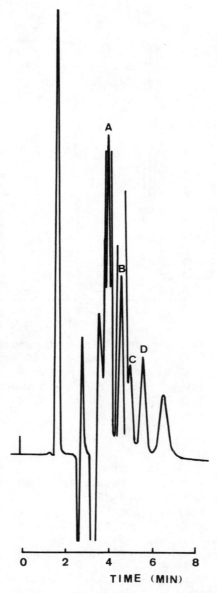

Fig. 5.   A chromatogram of 2,4-dinitrophenyl derivatives of
          amines in serum from a normal subject. Peaks are
          the same as in Fig. 2.

the methylguanidine values correlated more significantly with
creatinine with urea, whereas the reverse is true for guanidinosuccinic acid. This finding is in accordance with the suggestion
of Mikami et al.[16]. It is also clear that among the three aliphatic

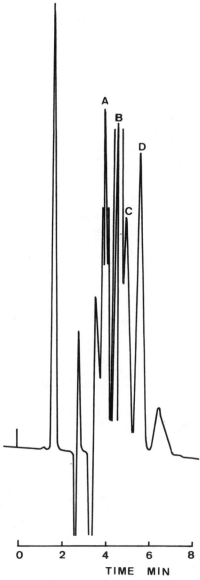

Fig. 6. A chromatogram of 2,4-dinitrophenyl derivatives of amines
         in serum from the uremic patient presented in Fig. 4.
         The conditions used are the same as in Fig. 5.

monoamines, dimethylamine had the highest correlation with methylguanidine (r=0.87), while the guanidinosuccinic acid concentration
was poorly correlated with any one of the aliphatic monoamines.
This finding suggests that the production of methylguanidine is

Table 1.   Serum levels of guanidino compounds and aliphatic mono-
           amines in 30 patients with chronic renal failure.

Compound		Uremic	Normal
urea nitrogen	(mg/dl)	70.23 ± 30.16	14.00 ± 3.00
creatinine	(mg/dl)	8.27 ± 4.14	0.90 ± 0.10
guanidino compounds	(μg/dl)		
guanidinosuccinic acid		314.56 ± 286.46	1.64
guanidinoacetic acid		29.86 ± 11.24	32.50
methylguanidine		19.34 ± 23.43	0.80
aliphatic monoamines	(μg/dl)		
ethanolamine		58.15 ± 50.57	72.37 ± 25.34
methylamine		16.73 ± 10.41	2.30 ± 0.40
dimethylamine		79.94 ± 41.74	14.17 ± 3.97

(Mean ± SD)

Table 2.   Zero order correlation coefficients between guanidino-
           succinic acid (GSA), guanidinoacetic acid (GAA),
           methylguanidine (MG), ethanolamine (EA), methylamine
           (MA), dimethylamine (DMA), creatinine (CRN) and urea
           nitrogen (UN) in serum of 30 patients with chronic
           renal failure.

Variable	guanidino compounds		aliphatic monoamines			CRN	UN
	GAA	MG	EA	MA	DMA		
GSA	-0.12	0.18	-0.12	-0.08	0.10	0.18	0.88*
GAA		0.53+	0.16	0.47#	0.49#	0.53+	0.01
MG			0.01	0.76*	0.87*	0.89*	0.27
EA				-0.15	0.04	-0.02	-0.17
MA					0.68*	0.64*	0.10
DMA						0.91*	0.21
CRN							0.32@

*;p<0.001,   +;0.005>P>0.001,   #;0.01>P>0.005,   @;0.1>P>0.05

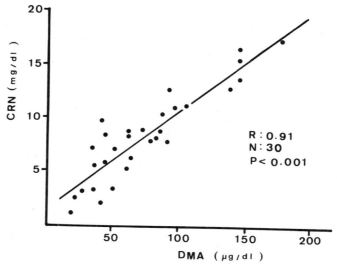

Fig. 7. Relationship between serum dimethylamine (DMA) and serum creatinine (CRN) in uremic patients.

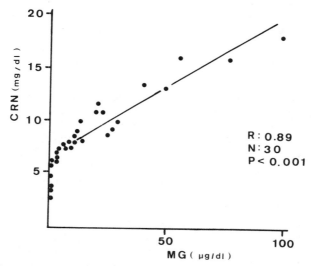

Fig. 8. Relationship between serum methylguanidine (MG) and serum creatinine (CRN) in uremic patients.

closely linked with that of the aliphatic monoamines, whereas the production of guanidinosuccinic acid is not related so.

In conclusion, our results add to the evidence that creatinine is a precursor of both methylguanidine and dimethylamine and

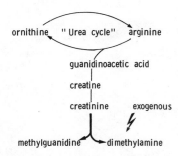

Fig. 9. Proposed metabolic pathway(s) for methylguanidine
         and dimethylamine.

also suggest that the metabolic pathway(s) for the production of
these two compounds from creatinine may be the same or closely
linked in chronic renal failure (Fig.9). Alternatively, the involvement of the serum type of gut flora in the conversion of
creatinine to methylguanidine or dimethylguanidine in chronic
renal failure is a matter of speculation and these points need
thorough investigation.

SUMMARY

     Serum levels of guanidino compounds and aliphatic monoamines
were measured in 30 patients with chronic renal failure of varying
etiology. Guanidino compounds in the serum were separated and
quantitated by high-performance liquid chromatography (HPLC) coupled with a Guanidino Compound Analyzer (Jasco). Separation and
determination of serum aliphatic monoamines by HPLC was achieved
after converting the aliphatic monoamines into their 2,4-dinitro-
phenyl derivatives followed by benzene extraction.

     Serum creatinine showed a more significant correlation with
methylguanidine (r=0.89) and dimethylamine (r=0.91) than with
guanidinosuccinic acid, guanidinoacetic acid, ethanolamine and
methylamine. Of all the aliphatic monoamines, dimethylamine
showed the highest correlation with methylguanidine (r=0.87) while
the guanidinosuccinic acid concentration showed poor correlation
with each individual aliphatic monoamine.

     Our data adds to the evidence that creatinine is a precursor
of both methylguanidine and dimethylamine and also suggests the
metabolic pathway(s) for the production of these two compounds
from creatinine may be the same or closely linked in chronic renal
failure.

ACKNOWLEDGMENTS

    B. K. Kishore is supported by a Research Fellowship from the
Ministry of Education, Science and Culture, Japan.

    We are indebeted to Ms Yamagiwa for expert technical
assistance.

REFERENCES

1. J. Bergstrom and E. E. Bittar, The basis of uremic toxicity,
    in:"The biological basis of medicine," E. E. Bittar and
    N. Bittar, eds., Academic press, New York (1969).
2. J. D. Kopple, Nitrogen metabolism, in:"Textbook of nephrol-
    ogy," S. G. Massry and R. J. Glassock, eds., Williams &
    Wilkins, Baltimore/London (1983).
3. G. Perez and R. Faluotico, Creatinine: A precursor of methyl-
    guanidine, Experientia, 29: 1473 (1973).
4. J. J. Pfiffner and C. V. Myers, On colorimetric estimation
    of guanidine base in blood, J. Biol. Chem., 87: 345 (1930).
5. M. Gonella, G. Barsotti, S. Lupetti, S. Giovannetti, V. Campa
    and G. Falcone, Role of the aerobic gut flora on the
    creatinine and methylguanidine metabolism, in:"Proc. 6th
    Int. Congr. Nephrol. Florence," S. Karger, Basel (1976).
6. D. J. Jones and P. C. Burnett, Creatinine metabolism and
    toxicity, Kidney Int., 7(Suppl. 3): 295 (1975).
7. M. L. Simenhoff, Metabolism and toxicity of aliphatic amines,
    Kidney Int., 7(Suppl. 3): 314 (1975).
8. M. L. Simenhoff, J. J. Saukkonen, J. F. Burke, R. W. Schaedler,
    W. H. Vogel, K. Bovee and N. Lasker, Importance of
    aliphatic amines in uremia, Kidney Int., 13 (Suppl. 8):
    16 (1978).
9. Y. Yamamoto, A. Saito, T. Manji, H. Nishi, K. Ito, K. Maeda,
    K. Ohta and K. Kobayashi, A new automated analytical
    method for guanidino compounds and their cerebrospinal
    fluid levels in uremia, Trans. Am. Soc. Artif. Intern.
    Organs, 24: 61 (1978).
10. A. Ando, T. Kikuchi, H. Mikami, M. Fujii, K. Yoshihara,
    Y. Orita and H. Abe, Quantitative determination of gua-
    nidino compounds: The excellent preparation of biological
    samples, in:"Urea cycle diseases," A. Lowenthal, A. Mori and
    B. Marescau, eds., Plenum Publishing Co., New York (1983).
11. S. Baba, Y. Watanabe, F. Gejo and M. Arakawa, High-performance
    liquid chromatographic determination of serum aliphatic
    amines in chronic renal failure, Clin. Chim. Acta,
    136: 49 (1984).
12. F. Gejyo, Studies on the guanidine derivatives in chronic
    renal failure. A study on the specific method for
    measurement of guanidine derivatives, especially methyl-

guanidine, and its clinical application, Jap. J. Nephrol.,
        16: 379 (1974).

13. S. R. Dunn, M. L. Simenhoff and L. G. Wesson, Gas chromato-
        graphic determination of free mono-, di-, and trimethyl-
        amine in biological fluids, Anal. Chem., 48: 41 (1976).

14. M. L. Simenhoff, H. E. Ginn and P. E. Teschan, Toxicity of
        aliphatic amines in uremia, Trans. Am. Soc. Artif. Intern.
        Organs, 23: 560 (1977).

15. B. D. Cohen, Uremic toxins, Bull. N. Y. Acad. Med., 51: 1228
        (1975).

16. H. Mikami, Y. Orita, A. Ando, M. Fujii, T. Kikuchi,
        K. Yoshihara, A. Okada and H. Abe, Metabolic pathway of
        guanidino compounds in chronic renal failure, in:"Urea
        cycle diseases," A. Lowenthal, A. Mori and B. Marescau,
        eds., Plenum Publishing Co., New York (1983).

GUANIDINOACETIC ACID (GAA) IN PATIENTS WITH

CHRONIC RENAL FAILURE (CRF) AND DIABETES MELLITUS (DM)

Y. Tsubakihara, N. Iida, S. Yuasa, T. Kawashima,
I. Nakanishi, M. Tomobuchi and T. Yokogawa

Kidney Disease Center, Osaka Prefectural Hospital
3-1-56 Bandaihigashi, Sumiyoshi-ku, Osaka, Japan

INTRODUCTION

Guanidinoacetic acid (GAA) is the precursor of creatine, an essential element in the energy metabolism of musle and nerve tissue[1]. GAA itself may have some physiological role such as the stimulation of insulin secretion[2]. As shown in Fig. 1, GAA is mainly produced in the kidney by the transamidination of glycine from arginine.

Recently it has been reported that the transamidinase activity in the perfused kidney of streptozotocin-induced diabetic rats was suppressed[3]. Therefore, production of GAA may be decreased leading to GAA deficiency in patients with kidney disease and DM. However, it has been reported that serum GAA concentration is increased in patients with chronic renal failure is accompanied by a decrease in urinary GAA excretion[4,5,6]. There is no data currently available concerning serum GAA concentration in patients with DM.

The present study makes use of an autoanalyzer developed in Japan whcih takes advantage of rapid high performance liquid chromatography and a sesitive fluorometric system for detection of guanidino compounds[7]. Many unknown guanidino compounds have been detected in uremic serum using these analysers[8,9]. It is possible that the presence of these newly discovered compounds was obscured using the less sensitive conventional column method[10].

In this study, we determined serum concentration and urinary excretion of GAA in patients with chronic glomerulonephritis (CGN) and diabetes mellitus (DM) by using this autoanalyzer.

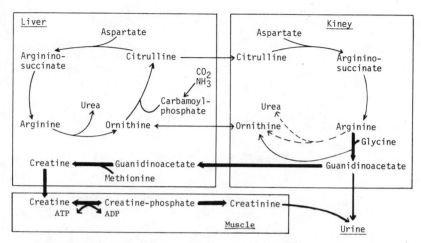

Fig. 1.   Metabolic pathway of guanidinoacetic acid.

MATERIALS AND METHODS

The subjects were 15 healthy adults (11 males), 104 patients
(67 males) with CGN and 33 non insulin-dependent diabetic patients
(18 males).   Twelve patients (7 males) with CGN and 6 patients
(4 males) with diabetic nephropathy were undergoing regular hemodialysis treatment.   One ml of serum and 1 ml of properly diluted
urine were deproteinized with 0.5 ml of 30% trichloroacetic acid.
This deproteinizing method was recommended for the determination
of guanidino compounds by Ando et al.[11] and Hoshino[12].   GAA was
determined with the JASCO G-520 guanidino autoanalyzer (Japan
Spectroscopic Co., Tokyo Japan).   The buffers and detection system
used have been described.

RESULTS

Fig. 2 shows the typical elution pattern of standard guanidino
compounds (Tau; taurocyamine, GSA; guanidinosuccinic acid, GAA; guanidinoacetic acid, GPA; guanidinopropioic acid, Cr; creatinine, GBA;
guanidinobutylic acid, GCA; guanidinocapronic acid, Arg; arginine, G;
guanidine and MG; methylguanidine) compared to normal and uremic
serum.   In uremic serum, guanidinosuccinic acid and methylguanidine
peaks were very high and many unknown peaks were seen.   The GAA
peak, however was lower than that seen in normal serum.

Table 1a and 1b show the serum concentration and urinary
excretion of GAA and creatinine (Cr) in patients with CGN and DM.
In the patients with CGN, the urinary excretion of GAA (U-GAA) was
significantly lower than that of normal subjects, and significantly

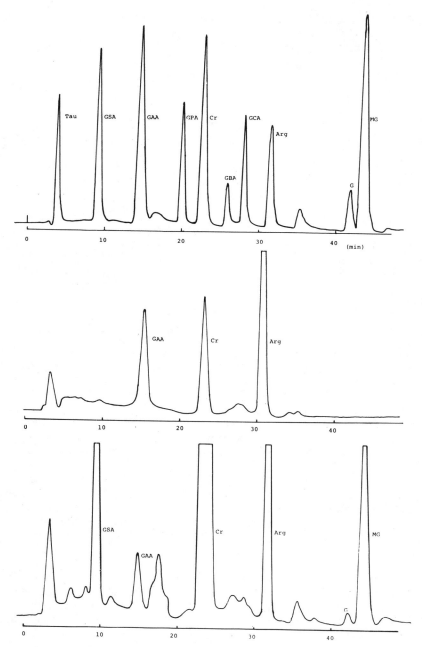

Fig. 2. Typical elution pattern of a solution of standard guanidino
         compounds (upper), a normal serum (middle) and a uremic
         serum (lower).

Table 1a. Serum concentration and urinary excretion of creatinine
in patients with chronic glomerulonephritis (CGN) and
diabetes mellitus (DM)

	N	S-Cr mg/dl	U-Cr mg/day	Ccr ml/min
Normal	15	0.9±0.1	1635±66	122±5
CGN (Ccr>100)	12	1.0±0.1	1742±73	122±1
CGN (30<Ccr<100)	24	1.5±0.1*	1385±50*	65±2*
CGN (Ccr<30)	56	7.1±0.5*	715±47*	11±1*
CGN under HD	12			
pre HD		11.7±0.9*		
post HD		4.3±0.3*		
DM (without proteinuria)	16	0.9±0.1	775±56*	60±3*
DM (Ccr<30)	11	5.0±0.7*	581±84*	10±2*
DM under HD	6			
pre HD		10.1±0.7*		
post HD		4.1±0.2*		

S; serum concentration, U; urinary excretion, HD; hemodialysis,
*; p < 0.05 (vs normal) ,   Values are expressed as mean ± S.E..

correlated with creatinine clearance (Ccr) as shown in Fig. 3.
However, as the Ccr decreased the U-GAA/Ccr ratio increased.
Urinary excretion of Cr, the end product of GAA metabolism, in
patients with a Ccr < 100 was significantly lower than that of
healthy adults.

The serum GAA concentration in these patients did not increase.
We found that there was a tendency for it to decrease especially in
patients with advanced azotemia (Ccr < 30), and it was significantly
lower than that of normal adults in those patients undergoing
regular hemodialysis treatment.

In the diabetic patients, the urinary excretion of Cr and GAA
were significantly lower than normal. The serum GAA concentration
of diabetic patients was significantly reduced irrespective of
renal function and, for those undergoing regular hemodialysis, it
was much lower than that of CGN patients on regular HD.

Fig. 4 shows the change in serum GAA concentration (S-GAA)
after nephrectomy in a patient with chronic pyelonephritis. S-GAA
decreased for 1 week and then increased gradually and stabilized
at half of the preoperative value.

Table 1b. Serum concentration and urinary excretion of GAA in
          patients with chronic glomerulonephritis (CGN) and
          diabetes mellitus (DM)

	N	S-GAA µg/dl	U-GAA mg/day	U-GAA/Ccr mg/day/ml/min
Normal	15	38.2±2.5	103.0±6.7	0.88±0.06
CGN (Ccr>100)	12	41.1±0.4	41.3±5.4*	0.34±0.04*
CGN (30<Ccr<100)	24	37.7±0.4	39.2±4.4*	0.59±0.06*
CGN (Ccr<30)	56	24.4±2.0*	5.9±0.9*	0.67±0.06*
CGN under HD	12			
pre HD		24.1±2.4*		
post HD		16.2±1.7*		
DM (without proteinuria)	16	27.8±1.8*	46.2±4.7*	0.81±0.08
DM (Ccr<30)	11	22.1±1.7*	2.3±0.5*	0.17±0.08*
DM under HD	6			
pre HD		16.5±3.5*		
post HD		13.4±1.7*		

S; serum concentration, U; urinary excretion, HD; hemodialysis,
*; $p < 0.05$ (vs normal),   Values are expressed as mean ± S.E..

DISCUSSION

The kidney is known to be the main organ producing GAA. It
is reasonable to expect that production of GAA will be suppressed
in patients with kidney desease. In fact, the urinary excretion
of Cr, which is the end product of GAA metabolism, in the patients
with CRF is demonstrated to be significantly lower than predicted
values. Jones et al. have termed this discrepancy "creatinine (Cr)
deficit"[13]. Since it has been reported that serum GAA in patients
with CRF is higher than normal, Jones et al. ascribed the "Cr
deficit" to the degradation of accumulated Cr caused by gut flora[13].
Meanwhile, we demonstrate in the present study that the serum GAA
concentration in patients with CRF is lower than that found in
normal subjects. Consequently, we suggest that GAA deficency is
one factor in the "Cr deficit".

The guanidino autoanalyzer, incidentally, detected many unknown guanidino compounds in uremic serum (as shown in Fig. 2)
which may be obscured using conventional column methods.

Funahashi et al. demonstrated that the transamidinase activity
in the perfused kidney of rats rendered diabetic by streptozotocin

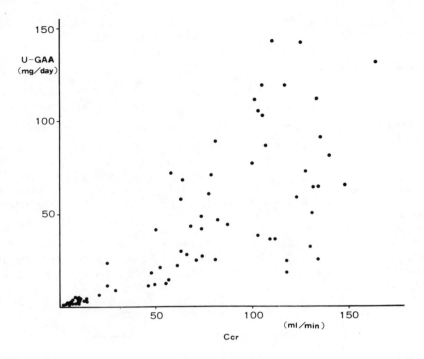

Fig. 3.   The relationship between urinary GAA excretion (U-GAA) and
          creatinine clearance (Ccr) in normal adults and patients
          with chronic glomerulonephritis.

was significantly lower than that in the perfused normal kidney[3].
However, there was no evidence suggesting GAA deficiency in
diabetic patients. We show that the urinary excretion of GAA and
Cr, and serum GAA concentration are significantly low in diabetic
patients as well (Table 1a and 1b).

      Since GAA is the precursor of creatine which is an essential
substrate for the energy metabolism of musle and nerve tissue, the
GAA deficiency demonstrated in CRF and DM patients may be responsible for some of the uremic and diabetic symptoms.

      Judging from the serum GAA concentration after nephrectomy,
extrarenal production of GAA in humans is confirmed. The pancreas
of dogs[14] and rats[15] have been shown to have transamidinase activity. However, nephrectomized rats have been shown to convert
only 6-20 % as much ^{14}C-labelled glycine to creatine as unoperated
controls[16]. One should be concerned, therefore, about producing a
GAA deficiency in nephrectomized patients.

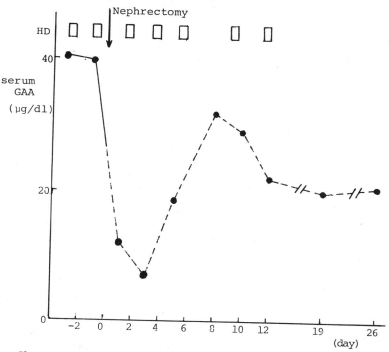

Fig. 4.   Changes in serum GAA concentration after nephrectomy in a
          patient with chronic pyelonephritis.

SUMMARY

     We examined the serum concentration (S) and the urinary ex-
cretion (U) of guanidinoacetic acid (GAA) in patients with chronic
glomerulonephritis (CGN) and diabetes mellitus (DM) by using an
autoguanidino analyser composed of a rapid high performance liquid
chromatographic system and a sesitive fluorometric system. In the
patients with chronic renal failure (CRF) due to CGN, U-creatinine
(Cr) and U-GAA were significantly lower than those of normal subjects. In contrast to previous reports, we found a significant
decrease in S-GAA in these patients. S-GAA in patients undergoing
regular hemodialysis treatment was also significantly low compared
to normal subjects. Among DM patients, S-GAA, U-GAA and U-Cr were
significantly lower than that seen in normal subjects and even
lower in patients with CRF due to DM nephropathy.

     These results call attention to a deficiency of GAA in the
patients with CRF and DM, especially in diabetic patients with
CRF, and some uremic and/or diabetic symptoms could be caused by
the disruption of nergy metabolism which results from a decrease
of creatine in muscle or nerve tissue.

REFERENCES

1. F. J. R. Hird, S. P. Davuluri and R. M. McLean, Evolutionary relationships between arginine and creatine in musle," in: "Urea Cycle Disease," A. Lowenthal, A. Mori and B. Marescau, eds., Plenum Press, New York, 401 (1982).

2. R. N. Alserver, R. H. Georg and K. E. Sussman, Stimulation of insulin secretion by guanidinoacetic acid and other guanidine derivatives, Endocrinology, 86:332 (1970).

3. M. Funahashi, H. Kato, S. Shiosaka and H. Nakagawa, Formation of arginine and guanidinoacetic acid in the kidney in vitro, J. Biochem., 89:1347 (1981).

4. B. D. Cohen, Guanidinosuccinic acid in uremia, Arch. Intern. Med., 126:846 (1970).

5. G. Perez, A. Rey, M. Michlus and I. Stein, Cation-exchange chromatography of guanidine derivatives in plasma of patients with chronic renal failure, Clin. Chem., 22:240 (1976).

6. C. H. Gold, V. Margaratha and D. V. A. Mzamane, The lack of effect of guanidino acetic acid on in vitro osmotic fragility of erythrocytes, Nephron, 31:17 (1982).

7. Y. Yamamoto, T. Manji, A. Saito, K. Maeda and K. Ohta, Ion-exchange chromatographic separation and fluorometric determination of guanidino compounds in physiological fluids, J. Chromato., 162:327 (1979).

8. M. Matsumoto, H. Kishikawa and A. Mori, Guanidino compounds in the sera of uremic patients and in the sera and brain of experimental uremic rabbits, Biochem. Med., 16:1 (1976).

9. R. Shainkin, Y. Berkenstadt, Y. Giatt and G. M. Berlyne, An automated technique for the analysis of plasma guanidino acids, and some findings in chronic renal disease, Clin. Chim. Acta., 60:45 (1975).

10. Y. Tsubakihara, N. Iida, Y. Orita, A. Ando, K. Nakata, H. Mikami, S. Miki, N. Maruta and H. Abe, Determination of uremic toxins, Jap. J. Clin. Chem., 7:208 (1979).

11. A. Ando, T. Kikuchi, H. Mikami, Y. Orita and H. Abe, An automated analytical method for guanidino compounds, Jap. J. Clin. Chem., 9:19 (1980).

12. T. Hoshino, Recommended deproteinizing methods for plasma guanidino compound analysis by liquid chromatography, in: "Urea Cycle Disease," A. Lowenthal, A. Mori and B. Marescau, eds., Plenum Press, New York, 391 (1982).

13. J. D. Jones and P. C. Burnett, Creatinine metabolism in humans with decreased renal function; creatinine deficit, Clin. Chem., 20:1204 (1974).

14. J. B. Walker, Role for pancreas in biosynthesis of creatine, Proc. Soc. Exp. Biol. and Med., 98:7 (1958).

15. W. H. Horner, Transamidination in nephrectomized rat, J. Biol. Chem., 234:2386 (1956).

16. R. Goldman and J. X. Moss, Synthesis in nephrectomized rats, Am. J. Physiol., 197: 865 (1959).

METABOLIC CHANGES OF GUANIDINO COMPOUNDS IN ACUTE RENAL
FAILURE COMPLICATED WITH HEPATIC DISEASE

Yōji Ochiai, Rikiya Matsuda, Kyōko Nishitani*,
Yoshinori Kōsogabe, Shinya Abe*, Yoshitarō Itano*,
Teruo Yamada* and Futami Kosaka*

Division of Intensive Care Unit, Okayama University
Hospital, and *Department of Anesthesiology, Okayama
University Medical school, Okayama 700, Japan

INTRODUCTION

Acute uremia may be associated with increased arginine (ARG)
utilization and increased synthesis of urea. Perez et al.[1]
demonstrated an increased synthesis of guanidinosuccinic acid
(GSA) by livers of acutely uremic rats. We found clinical evidence that in actual cases of acute renal failure (ARF) complicated with hepatic disease, concentrations of guanidino compounds
(GC) vary from those seen in ARF without hepatic disease[2]. The
present study was designed to examine the metabolic relationship
of liver to kidney and to investigate the effect of renal and
hepatic function on the metabolism of ARG and other urea cycle
intermediates. In this study, we compare GC concentrations in ARF
without hepatic disease with those in ARF with hepatic disease by
the use of guanidinograms, which are circle diagrams composed of
eight GC.

CASE REPORT

The patient was a 34 years old male with unremarkable past
history. He suffered from common cold symptoms and was diagnosed
as having acute hepatitis. After treatment for about two weeks,
glutamic oxalacetic transaminase (GOT) and glutamic pyruvic transaminase (GPT) increased markedly and he lapsed into a coma. He
was transferred to the Intensive Care Unit of Okayama University
Hospital. Liver function tests revealed a 514 IU/1 GOT, 735 IU/1
GPT, 29.1 mg/dl total bilirubin and a 238 µmol/1 ammonia (NH_3).

317

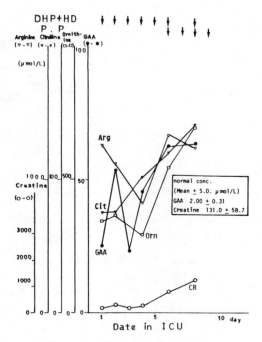

Fig. 1.   Plasma cincentrations of GAA, creatine, arginine,
          citrulline and ornithine.

Renal function tests showed a 32.7 ml/min creatinine clearance,
6.5 mg/dl blood urea nitrogen (BUN) and 1.03 mg/dl creatinine.   In
the ICU, hemodialysis, hemoperfusion and plasmapheresis were performed, but electroencephalographic slow waves became flat without
a decrease in NH_3, while the BUN and creatinine gradually increased
and the creatinine clearance fell to 3.0 ml/min by the tenth day
in the ICU.   Fig. 1 shows the rise and fall in the plasma concentration of guanidinoacetic acid (GAA), creatine (CR), arginine
(ARG), citrulline and ornithine.   CR and citrulline gradually increased while GAA, ARG and ornithine decreased initially and then
rose progressively.   Meanwhile, the control value of GAA was
2.0±0.31 µmol/l, while that of CR was 131.0±58.7 µmol/l.   GAA concentrations moved in parallel with ARG concentrations at the terminal stage.   In this case, guanidinosuccinic acid(GSA), guanidinopropionic acid (GPA), guanidine (G) and methylguanidine (MG) were
not detected at any time.   He died on the thirteenth day of ICU
before further plasmapheresis or hemodialysis could be undertaken.

MATERIALS AND METHODS

Patients

Details of patients with ARF and hepatic diseases treated at the ICU of Okayama University Hospital were as follows; two cases were of the viral type, eight cases showed postoperative hepatic failure, one case was of the drug-induced type, two cases were due to gramoxon poisoning, three cases were associated with hepato-cirrhosis and one case was of unknown cause. There were seventeen cases in all.

Analysis of guanidino compounds

Blood samples were obtained from each patient with ARF ran-domly. The plasma was deproteinized by trichloroacetic acid and centrifuged. The next in the pretreatment and separation by HPLC are previously described[2]. The plasma concentration of GC was calculated as the difference between the initial and oliguric stages in ARF, and as the difference between ARF with hepatic disease (Group-H) and ARF without hepatic diseases (Group-NH). The data is presented as the mean ± S.E.

RESULTS

In Group-H compared with Group-NH, in the initial stage

In Group-H compared with Group-NH, in the initial stage creatine (CR), guanidinopropionic acid (GPA), guanidine (G) and methylguanidine (MG) decreased and arginine (ARG) increased. In the oliguric stage, guanidinosuccinic acid (GSA), GPA, GBA, G and MG decreased and CR, GAA and ARG increased. In the end stage, CR, GAA and ARG markedly increased (Table 1).

Guanidinogram

The chart shows the variations seen by means of guanidinograms to more easily demonstrate the differences. A guanidinogram is a circle diagram which has eight components of GC. In cases in Group-H who died, there was a marked increase of CR, GAA and ARG. However, in Group-NH, CR and MG increased. GSA tended to decrease in production, and six cases out of seventeen showed no production of GSA. Fig. 2 shows the graphic patterns of GC in ARF with hepatic disease (left) and without hepatic disease (right).

The relationship of GAA to ARG

In the initial stage r=0.91 and in the oliguric stage r=0.66 (Fig. 3). This suggests that factors other than ARG are related

Table 1. Guanidino compounds in acute renal failure with hepatic diseases (group-H) and without hepatic diseases (group-NH).

	initial stage		oliguric stage		end stage	
	Groupe-H (n=11)	Groupe-NH (n=15)	Groupe-H (n=12)	Groupe-NH (n=15)	Groupe-H (n=6)	Groupe-NH (n=6)
GSA	10.6±4.2	10.0±2.2	4.8±3.2	24.7±4.8	8.2±5.7	29.8±4.8
CR	355.6±63.5	669.1±178.8	2293.7±1340.0	609.2±145.6	3067.9±160.0	1574.9±724.6
GAA	10.0±3.3	1.4±0.4	16.2±6.3	1.9±0.6	28.1±10.9	0.6±0.2
NAA	4.2±1.3	7.9±2.3	6.4±1.6	8.5±1.3	16.3±5.1	11.7±1.1
GPA	0.7±0.3	2.5±0.7	0.3±0.2	2.1±0.2	2.7±1.3	1.3±0.6
GBA	2.2±1.5	4.3±2.4	0.3±0.1	6.4±4.3	3.6±2.1	2.2±0.9
ARG	198.6±87.4	68.4±9.2	376.2±83.4	136.9±31.5	514.7±179.2	114.8±30.2
G	0.2±0.1	1.2±0.2	0.2±0.2	2.5±0.6	0.2±0.2	3.3±1.5
MG	0.2±0.1	0.7±0.2	0.02±0.02	3.4±0.9	0.07±0.07	8.6±4.3

(μmol/l, mean±S.E.)

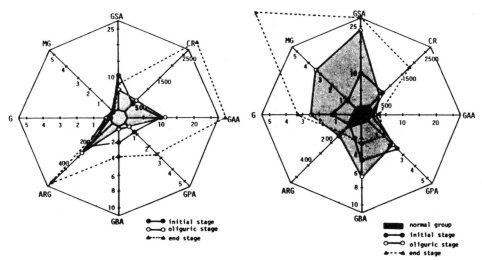

Fig. 2.   Guanidinograms: Circle diagrams of guanidino compounds in
          acute renal failure.   Left; with hepatic diseases,
          right; without hepatic diseases.

to GAA production.   In the initial stage when renal and hepatic
dysfunction occurred, the metabolic relationship between liver and
kidney are relatively constant despite the acute loss in renal and
hepatic function.

The relationship of the ARG/GAA ratio to the CR/GAA ratio

     The relationship of the ARG/GAA ratio to the CR/GAA ratio
were studied during the oliguric stage where r=0.62 and is not
significant (Fig. 4).   This shows that neither ARG nor CR are related to the production of GAA.   In the oliguric stage, the metabolic relationships between liver and kidney are not greater than
those in the initial stage despite the loss of the capacity of
these organs to compensate for one another.

DISCUSSION

     In experimental acute uremia, there is evidence that the
turnover rate of the urea cycle is accelerated[1,3].   Since arginine
(ARG) is a precursor of urea and other guanidino compounds (GC),
we have evaluated the metabolic relationships between liver and
kidney in actual cases of acute renal failure (ARF) with hepatic
disease.   In these cases marked increases in serum levels of gua-

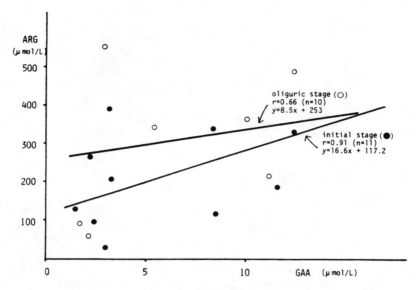

Fig. 3.   Relatioship between GAA and ARG in acute renal failure
          with hepatic diseases.

nidinoacetic acid (GAA) are seen, so that a relationship between
liver and kidney in ARG metabolism must be assumed. In this study,
the plasma concentrations of ARG and CR were very high in the
patients with ARF with hepatic disease. This means that vigorous
catabolism occurs and acute losses of renal and hepatic function
results in high concentrations of ARG and CR. ARG, a urea cycle
intermediate, has a pivotal role in protein metabolism not only
because of its role in protein synthesis but also as a precursor
of urea and GC[4]. Furthermore, Hopper-Seyler et al.[3] described a
significant increase in argininosuccinic acid synthetase and
carbamyl phosphate synthetase activities in the liver of starved
rats forty-eight hours after bilateral nephrectomy. We show that
in cases of acute renal and hepatic failure, GAA concentrations
are high but concentrations of guanidinosuccinic acid (GSA),
guanidine (G) and methylguanidine (MG) are generally reduced.
Cohen et al.[5] suggested that GSA can be formed in the liver from
transamidination of ARG to aspartic acid. Since plasma GSA levels
were increased in ARF without hepatic damage, but reduced in ARF
with hepatic damage where GAA levels increased, we suggest that
the metabolic production of GSA in the liver must decrease and GAA
production in the kidney increase in compensation.

     In the chart showing liver and kidney correlations (Fig. 5),
metabolism in the kidney is increased in response to hepatic
damage. Decreased production of GSA occurs in the presence of
acute hepatic damage, and increased production of GAA suggests an

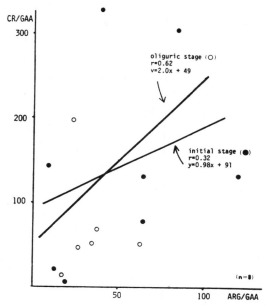

Fig. 4.   Relationship between ARG/GAA and CR/GAA in acute renal
          failure with hepatic deseases.

enhanced metabolic rate in the kidneys. Funahashi et al.[6] examined the physiological role of the kidney in the biosynthesis of ARG
and GAA using a more quantitative method. They suggested that the
liver releases citrulline into the blood stream, and that the
kidney takes it up to synthesize ARG. The kidney is suitable for
ARG formation because of its ability to take up citrulline from
the blood, although the kidney itself cannot form citrulline due
to a lack of ornithine carbamoyltransferase activity. The fact
that the liver is the only organ that has high activities of all
urea cycle enzymes suggests that citrulline is supplied by the
liver. ARG is converted to GAA and ornithine in the presence of
glycine by glycine amidinotransferase in the kidney. ARG formation in the kidney is regulated by a negative feedback mechanism,
but it seems unlikely that this regulatory mechanism works at the
level of enzymes associated with citrulline metabolism. In states
of hepatic and renal dysfunction, these enzymatic activities change
and normal regulatory machanisms are lost.

It is known that GAA is mainly methylated to form CR in the
liver, though the kidney may also participate in this reaction.
CR is transported in the blood to form phosphocreatine in muscle
and brain. In the case presented, we show that ARG and CR levels
are very high at the end stage. The problem of whether this incomplete urea cycle in the kidney is important in the removal of

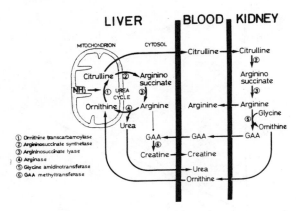

Fig. 5. Metabolic relationship between liver and kidney.

ammonium ions from the blood in liver desease is now under investigation. Natelson et al.[7] proposed the guanidine cycle as a
source of GSA. We think that other pathways possibly exist[5] and
more effective compensating mechanism act in hepatic and renal
dysfunction.

We conclude that the relationship between the liver and kidney
is important in the removal of ammonium ions, and that in hepatorenal failure the kidney compensates for the functional deficit in
the liver. Judging from the fact that the guanidinoacetic acid
concentration in the plasma decreases in acute renal failure but
increases in acute hepatic failure, the kidney is the main organ
for the synthesis of arginine from citrulline and for the synthesis
of guanidinoacetic acid from arginine.

REFERENCES

1. G. O. Perez, B. Rietberg, B. Owens and E. R. Schiff, Effects
       of acute uremia on arginine metabolism and urea and gua-
       nidino acid production by perfused rat liver, Pfugers Arch.,
       372:275 (1977).
2. Y. Ochiai, S. Abe, T. Yamada, K. Tada and F. Kosaka, Guanidino
       compounds and hemodialysis, in:"Urea cycle diseases,"
       A. Louenthal, A. Mori and B. Marescau, eds., Plenum press,
       New York and London (1982).
3. G. Hopper-Seyler, K. P. Maier, P. Schollmeyer, J. Frohlich,
       H. Talke and W. Gerok, Studies on urea cycle enzyme in rat
       liver during acute uremia, Eur. J. Clin. Invest., 5:15 (1975).

4. G. Perez, A. Rey and E. Schiff, Biosynthesis of guanidino-
     succinic acid by perfused rat liver, J. Clin. Invest.,
     57:807 (1976).
5. B. D. Cohen, I. M. Stein and J. E. Bonas, Guanidinosuccinic
     aciduria in uremia a possible alternate pathway for urea
     synthesis, Am. J. Med., 45:63 (1968).
6. M. Funahashi, H. Kato, S. Shiosaka and H. Nakagawa, Formation
     of arginine and guanidinoacetic acid in the kidney In Vivo.
     Their relations with the liver and their regulation, J.
     Biochem., 89:1347 (1981).
7. S. Natelson and J. E. Sherwin, Proposed mechanism for urea
     nitrogen re-utilization: Relationship between urea and pro-
     posed guanidine cycles, Clin. Chem., 25:1343 (1979).

URINARY GUANIDINOACETIC ACID EXCRETION

IN GENTAMICIN NEPHROTOXICITY

Hideo Itabashi, Hisaki Rinno and Hikaru Koide

Division of Nephrology, Department of Medicine
Juntendo University School of Medicine
Tokyo 133, Japan

INTRODUCTION

Bonas et al. reported in 1963 that four patients with kidney disease were shown to excrete markedly decreased amounts of guanidinoacetic acid (GAA) in their urine[1].

In recent years, more sensitive methods for the determination of guanidino compounds have been developed and many guanidino compounds have been measured simultaneously and rapidly with the use of high pressure liquid chromatography. Using this method, guanidino compounds have been studied in experimental and clinical uremia.

We studied the effect in rats of various doses of gentamicin on the urinary GAA excretion and enzyme activity. The results indicate that alteration in urinary GAA excretion might be a useful guide to the early diagnosis of gentamicin nephrotoxicity. We also studied the protective effect of CoQ_{10} on gentamicin nephrotoxicity.

MATERIALS AND METHODS

Male Fischer 344 rats weighing 190 - 210g were used in all studies. The animals were given a single intravenous injection of 10, 20, 30, or 50 mg of gentamicin per kilogram of body weight in 0.4 ml of 0.9 % saline. Physiological saline of 0.4 ml was used as a control. Twenty miligrams of CoQ_{10} were administered intraperitoneally to the rats for two days before gentamicin injection. After administration of gentamicin, the animals were housed singly

in metabolic cages and fasted but allowed free access to water.
After twenty-one hour urine specimens were collected, blood samples
were collected for the determination of serum urea nitrogen,
creatinine and serum CoQ_{10} concentration and the kidney was removed
for histological examination as well as tissue CoQ_{10} assay.   The
urine specimens were passed through 45 µm membrane filters and
urinary creatinine excretion in the 21 hour sample was measured by
the method of Jaffe[2].   A part of the urine obtained by filtration
was deproteinized by AMICON-GF25 at 1000 x g for 30 min.   The
supernatant was diluted twenty fold with 0.01 N HCl and submitted
for urinary GAA determination by the JASCO G-250 guanidine analyzer.
The amount of GAA excreted was expressed as µg per mg creatinine
per 21 hours.   The remaining urine specimens were dialyzed against
tap water for 2hr at 4°C, and analyzed for N-acetyl-glucosaminidase
(NAG), leucine aminopeptidase (LAP) and γ-glutamyltranspeptidase
(γ-GTP) activities.   The activities of NAG, LAP and γ-GTP were
determined using m-cresolsulfonphthaleinyl-N-acetyl-β-glucosaminide,
L-leucine p-nitroanilide and γ-glutamyl p-nitroanilide as substrate
and expressed in international units (IU) or mIU creatinine per
21-hours urine.   The coenzyme Q_9 and Coenzyme Q_{10} per mg in the
serum and the kidney were determined by high performance liquid
chromatography[3].

RESULTS

Urinary GAA excretion in gentamicin nephrotoxicity

    Fig. 1 shows the urinary activities of LAP and NAG after dif-
ferent doses of gentamicin.   As compared with controls, enzyme
activity in the groups of 30 or 50 mg/kg of gentamicin injection
were significantly increased.   Serum urea nitrogen and creatinine
levels were not changed significantly in any gentamicin-injected
group.   The urinary GAA excretion at different doses of gentamicin
is shown in Fig. 2.   Excretion is significantly decreased in the
20, 30 and 50 mg/kg dosage groups.

    Fig. 3 shows the coenzyme Q_9 and Coenzyme Q_{10} concentration in
the renal cortex at different doses of gentamicin.   The coenzyme Q_9
concentration was significantly decreased in the 10 to 50 mg/kg
dosage group and the decrease was dose-dependent.   The concentration of coenzyme Q_{10} significantly decreased in the 50 mg/kg group.

The effect of coenzyme Q_{10} on gentamicin nephrotoxicity

    The activities of urinary NAG were not different significantly
between the groups receiving gentamicin alone and the groups which
received 20 mg/kg of coenzyme Q_{10} for two days before the gentamicin
injection (Fig. 4).   In the 30 mg/kg group, there was no effect of
coenzyme Q_{10} pretreatment on the amount of GAA excretion.   However,

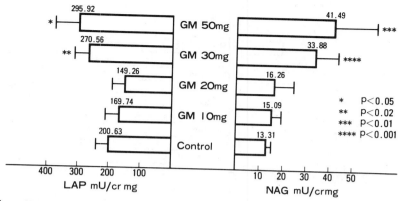

Fig. 1.   Urinary activities of LAP and NAG at different gentamicin
          doses.

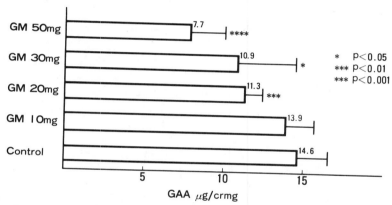

Fig. 2.   Urinary GAA excretion at different gentamicin doses.

the urinary excretion of GAA was significantly increased in the
group which was treated by coenzyme Q_{10} before 50 mg/kg of
gentamicin injection (Fig. 5).

    Electron microscopy shows degeneration of the proximal tubu-
lar cell after 50 mg/kg gentamicin. Following pretreatment with
coenzyme Q_{10} 20 mg/kg, these abnormalities were reduced in severity
(Fig. 6).

DISCUSSION

    Guanidinoacetic acid is formed from arginine in the presence
of glycine by glycine amidinotransferase in the kidney and the

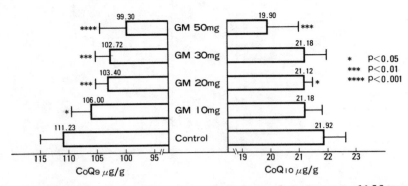

Fig. 3.   Kidney concentration of CoQ₉ and CoQ₁₀ at different
          gentamicin doses.

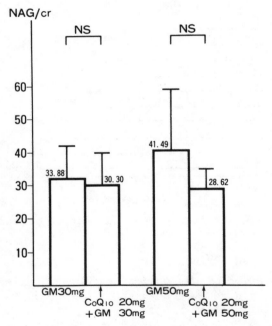

Fig. 4.   Effect of coenzyme Q₁₀ on the activities of urinary NAG
          against gentamicin nephrotoxicity.

pancreas of rats[4]. Funabashi et al.[5] demonstrated that arginine
and guanidinoacetic acid were formed in the kidney and that their
precursor, citrulline, was mainly transported from the liver to
the kidney. Guanidinoacetic acid is then methylated to form creatine in the liver. The urinary excretion and kidney cencentration
of GAA are shown to be markedly decreased in uremic patients[1,6].

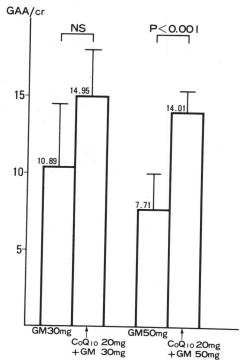

Fig. 5. Effect of coenzyme Q_{10} on the excretion of urinary GAA
following gentamicin nephrotoxicity.

Urinary enzymes are particularly helpful in detecting subtle
tubular dysfunction before renal damage has proceeded to the point
of producing a decrease in glomerular filtration rate[7]. The increased excretion of urinary enzyme has been noted after high-doses
of salicylates[8] and aminoglycosides[9,10], although its sensitivity
for detecting nephrotoxicity has not been demonstrated to be
superior to other indices[11]. Since aminoglycoside antibiotics are
known to be nephrotoxic in animals and man, renal dysfunction was
anticipated when gentamicin was first made availabl for clinical
use[12,13]. However, gentamicin renal toxicity has been reported to
be relatively infrequent[13,14], although the estimation of clinical
incidence varies in accordance with the circumstance in which the
drug is given, such as serious sepsis, and by the insensitivity of
parameters to monitor toxicity in patients[12,15,16].

The Fischer 344 rats used in this study are reported to be
quite sensitive to aminoglycoside induced nephrotoxicity[17]. Poly-
uria, proteinuria, enzymuria, cyturia and azotemia are the usual
symptoms of nephrotoxicity induced by aminoglycosides including

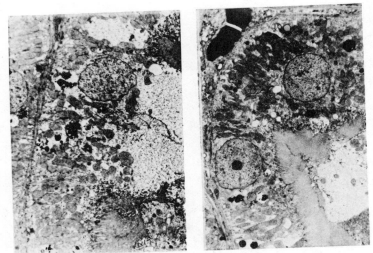

fig. 6.   Electron micrographs of proximal tubular cells from rats
          treated with gentamicin (50 mg/kg) alone or following co-
          enzyme Q_{10} (20 mg/kg) pretreatment.

gentamicin.   Although BUN and creatinine have been used extensively
to monitor nephrotoxicity, they indicate functional impairment
which is substantially late in appearance. They are, therefore,
less sensitive than enzymuria. Among the urinary parameters employed in this study, NAG and GAA showed the highest sensitivity.
The results of this study suggest that the determination of GAA in
urine would be useful for the early diagnosis of gentamicin
nephrotoxicity.

Coenzyme Q_{10} has been used in recent years not only in cardiac
disease but also in other pathological conditions such as ischemic
renal damage[18]. However, the exact mechanisms by which CoQ_{10} pro-
tects cells against cytotoxic agents are still not known. Since
CoQ_{10} is an essential component of the electron transport system in
mitochondria, it might promote such transport in the mitochondria
of renal tissue.

In these studies, urinary GAA excretion was higher in the
CoQ_{10} pretreated group receiving 50 mg/kg of gentamicin. Furthermore, improvement in the electron microscopic findings in the
kidney was also observed in the CoQ_{10} pretreated group.

CONCLUSION

The determination of urinary GAA excretion was useful for the early diagnosis of gentamicin nephrotoxicity. Coenzyme Q_{10} has a protective effect in gentamicin nephrotoxicity.

REFERENCES

1. J. E. Bonas, B. D. Cohen and S. Natelson, Separation and estimation of certain guanidino compound. Application to human urine, Microchem. J., 7:63-77 (1963).
2. M. Jaffe, Ueber den Niedershlag, welchen Pikrinsäure in normalem Harn erzeugt und über eine neue Reaction des kreatinins, Hoppe-Seyler Z, physiol. chem., 10:391 (1886).
3. S. Imabayashi et al., Determination of individual ubiqinone homologues by mass spectrometry and high performance liquid chromatography, Analytical chemistry., 51, 534-536 (1979).
4. J. B. Walker, Formanidine group transfer in extracts of human pancreas, liver and kidney, Biochim. Biophys. Acta., 73:241-247 (1963).
5. M. Funahashi et al., Formation of arginine and guanidino-acetic acid in the kidney, J. Biochem., 89:1347-1356 (1981).
6. K. Kano et al., Clinical significance of guanidino compounds in uremia, J. Clin. Path., 28:660-664 (1980).
7. A. W. Mondorf, M. Zegelman et al., Comparative studies on the action of aminoglycosides and cephalosporins on the proximal tubule of the human kidney, Proceedings of the 10th International Congress of Chemotherapy, Zurich, 1977, Abstract 270.
8. R. A. Proctor and C. M. Kunin, Salicylate-induced enzymuria, Am. J. Med., 65:987-993 (1978).
9. P. R. Beck, R. B. Thompson and A. K. R. Chadhuri, Aminoglycoside antibiotics and renal function : Changes in urinary -glutamyltransferase excretion, J. Clin. Pathol., 30:432-437 (1977).
10. J. M. Wellwood, D. Lovell, A. E. Thompson et al., Renal damage caused by gentamicin: A study of the effects on renal morphology and urinary enzyme excretion, J. Pathol., 118:171-182 (1976).
11. W. E. Stroo and B. J. Hook, Enzymes of renal origin in urine as indicators of nephrotoxicity, Toxicol. Appl. Pharmacol., 39:423-434 (1977).
12. F. G. Falco, H. M. Smith and G. M. Aricieri, Nephrotoxicity of aminoglycosides and gentamicin, J. Infect. Dis., 119:406-409 (1969).
13. G. G. Jack, Current therapeutics. CCXXXIV. Gentamicin, Practitioner, 198:855-866 (1967).
14. M. Abramowicz and C. M. Edelmann Jr., Nephrotoxicity of antiinfective drugs, Clin. Pediat., 7:389-390 (1968).

15. J. N. Wilfert et al., Renal insufficiency associated with
     gentamicin therapy, J. Infect. Dis., 124 (Suppl) S148-S115
     (1971).
16. N. Milman, Renal failure associated with gentamicin therapy,
     Acta. Med. Scand., 196:87-91 (1974).
17. J. C. Kosek, R. I. Mazze and M. J. Cousins, Nephrotoxicity of
     gentamicin, Lab. Invest., 30:48-57 (1974).
18. Y. Tatsukawa, Y. Dohi et al., The role of coenzyme Q_{10} for
     preservation of the rats kidney: A model experiment for
     kidney transplantation, Life Science, 24:1309 (1979).

THE IMPORTANCE OF ENERGY INTAKE FOR PREVENTING THE ACCUMULATION OF GUANIDINO COMPOUNDS IN PATIENTS WITH CHRONIC RENAL FAILURE

Seiji Fujiwara, Toshiyuki Nakano, Kazuo Isoda and Tadashi Miayahara

The Second Department of Internal Medicine
The Jikei University School of Medicine
3-25-8 Nishishibashi, Minato-ku, Tokyo, Japan

INTRODUCTION

It is now recognized that methylguanidine (MG) and guanidino-succinic acid (GSA) are the prototypes of low molecular weight uremic toxins[1,2]. There remain, however, many problems concerning the origin of these substances. This study is designed to evaluate the effect of dietary energy intake in preventing the accumulation of MG and GSA. We report that in the presence of a high energy intake the serum levels of MG, GSA, and urea nitrogen diminish and the nutritional status improves.

METHODS

Eleven patients with chronic renal failure were studied. They consisted of seven males and four females, aged 39 to 72 with an average of 53.3 ± 11.7 yrs. They received conservative management for renal failure in our university hospital. Their creatinine clearance (Ccr) values ranged from 3.5 to 25 ml/min with an average of 9.05 ± 6.3 ml/min. The clinical data and daily diet program of the patients are shown in Table 1.

All patients were studied at two levels of dietary energy intake: a low energy diet (Period I) and a high energy diet (Period II). The duration of each period was 7 to 14 days. The actural energy intake during Period I averaged 30.6 ± 9.7 kcal/kg body wt/day and during Period II averaged 37.7 ± 7.9 kcal/kg body wt/day. Three patients began the study on the high energy intake, and eight patients initially received the low energy diet. The

Table 1. Daily diet characteristics in 11 chronically uremic patients.

Patients	Age yr	Sex	Original lesion	Ccr ml/min	Energy Per kg body wt (kcal) Period I	Period II	Protein Per kg body wt (g) Period I	Period II
S.M.	43	F	CGN	5.5	43.1	50.0	0.65	0.60
H.E.	40	M	Polycystic kidney	3.5	26.2	32.5	0.40	0.40
A.I.	72	M	CGN	6.9	44.1	48.4	0.80	0.89
H.S.	40	F	CGN	25.0	37.5	38.2	0.69	0.71
H.U.	54	F	CGN	6.0	35.3	38.1	0.54	0.52
H.U.	63	M	CGN	5.8	35.8	44.5	0.70	0.70
K.K.	39	M	Hydronephrosis	4.1	25.0	31.3	0.62	0.63
T.H.	72	M	DM	15.7	24.1	25.1	0.53	0.57
T.I.	54	M	CGN	12.3	23.9	36.9	0.66	0.73
H.Y.	50	M	Renal Tbc	3.7	9.3	26.6	0.23	0.66
Y.K.	59	F	CGN	11.1	32.1	43.1	0.65	0.74
Mean	53.3			9.1	30.6	37.7*	0.60	0.65**
±S.D.	11.7			±6.3	±9.7	±7.9	±0.15	±0.12

Period I; low energy diet, Period II; high energy diet.
* Mean value significantly higher than Period I; $p < 0.01$.   **; Not significant.

Table 2.   Serum levels of GSA, MG and nutritional parameters
            during each period

		Priod I	Period II	P-Value
sGSA	μg/100ml	411.6±363.4	239.7±219.6	p<0.05
sMG	μg/100ml	108.0±132.1	49.6±60.8	p<0.05
SUN	mg/100ml	72.7±30.4	62.3±24.6	p<0.05
Albumin	g/100ml	4.0±0.2	4.1±0.3	N.S.
Tansferrin	mg/100ml	184.2±25.9	208.3±33.2	p<0.05
C$_3$	mg/100ml	53.5±11.7	63.5±15.3	p<0.01
N-bal	g	−1.82±1.85	−0.04±1.84	p<0.05
NPU	%	2.89±49.6	41.12±19.43	p<0.05
UNA	g/day	5.83±3.01	5.05±2.17	N.S.

Values are the means ± S.D.
N.S: Not significant.

protein intake of these patients was fixed in the range of 0.5 −
0.7 g/kg body wt/day. These intakes were precisely monitored by
the dietitian during the study from diet records. Vitamin supple-
ments and aluminum hydrochroride were given daily as needed. No
patient received anabolic steroids.

    Serum was obtained for measurement of guanidino compounds,
urea nitrogen (SUN), creatinine, albumin, transferrin and C$_3$ for
the last three days of each period. Guanidino compounds were
measured by the HPLC-fluorometric method. Transferrin and C$_3$ were
measured with radial immuno-diffusion plates. Nitrogen balances
(N-bal) were calculated as the difference between nitrogen intake
and the sum of fecal and urinary nitrogen output. Urea nitrogen
appearance (UNA), a measure of net urea production, was calculated
according to Kopple's equation[3]. Net protein utility (NPU) was
calculated by Taylor's method[4].

    All data are presented as the mean ± S.D; statistical analy-
sis was performed with the paired t-test.

RESULTS

    All patients were in negative N-bal during Period I (Table 2).
As they took more calories during Period II, N-bal and NPU significantly increased (p<0.05). Serum transferrin and C$_3$ significantly increased; transferrin, 184.2±25.9 mg/100 ml to 208.3±33.2
mg/100 ml (p<0.05); C$_3$ 53.5±11.7 mg/100 ml to 63.5±15.3 mg/100 ml

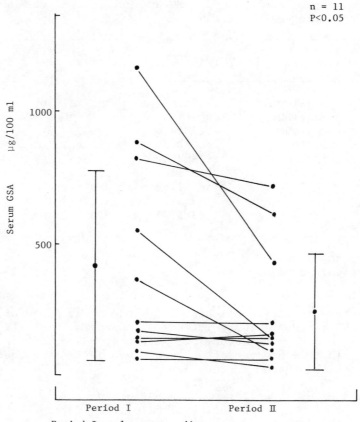

Fig. 1. Effect of dietary energy intake on serum levels of GSA.

(p<0.01), while SUN significantly decreased; 72.7±30.4 mg/100 ml
to 62.3±24.6 mg/100 ml (p<0.05), and UNA was slightly lower (5.05±
2.17 g/day) than that during Period I but the difference was not
significant. Serum levels of GSA significantly decreased (p<0.05);
411.6±363.4 µg/100 ml to 239.7±219.6 µg/100 ml (Fig. 1). Serum
levels of MG also significantly decreased (p<0.05); 108.0±132.1
µg/100 ml to 49.6±60.8 µg/100 ml (Fig. 2). Other guanidino compounds, those as, guanidinoacetic acid (GAA), guanidinopropionic
acid (GPA), guanidinobutyric acid (GBA) and guanidine (G) were
studied simultaneously during Period I and Period II (Table 3),
but there was no significant difference between Period I and
Period II.

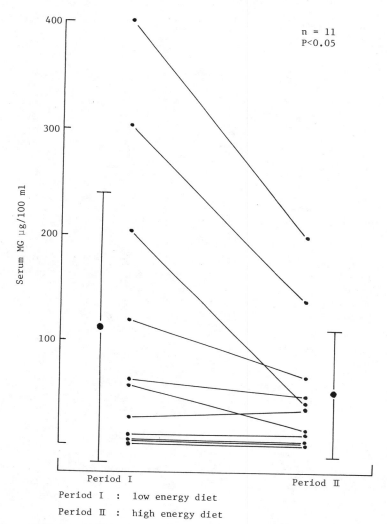

Fig. 2.   Effects of dietary energy intake on serum levels of MG.

DISCUSSION

Previous researchers have reported the effect of dietary treatment on guanidino compounds[2,5,6,7]. However, as a rule, nutritional studies in renal failure have measured the response to different quantities of protein rather than relative energy intake[8]. This study was carried out to evaluate the importance of energy intake for preventing the accumulation of guanidino compounds in patients with chronic renal failure.

Table 3.   Serum levels of other guanidino compounds during each
           period.

		Period I	Period II	P-Value
GAA	µg/100ml	32.2±12.4	26.7±17.9	N.S.
GPA	µg/100ml	n.d	n.d	N.S.
GBA	µg/100ml	1.33±4.2	1.42±3.02	N.S.
G	µg/100ml	1.18±3.74	1.52±3.36	N.S.

Values are the mean ± S.D , N=11
n.d : Not detected, N.S : Not significant

As all patients took in more total energy, N-bal improved and
NPU, transferrin and C_3 increased, suggesting that in the presence
of high energy intake protein sparing in uremic patients is more
efficient. During Period I, where a low energy intake was maintained, the serum levels of GSA, MG and SUN were higher than during
Period II.

Abitobol et al.[9], reported that the fractional rate of urea
synthesis increased markedly with the stress of inadequate energy
intake, suggesting a catabolic response at a critical level of
energy intake so that fractional urea synthesis is not related to
nitrogen intake but rather total energy intake.

Stein, Cohen and Kornhauser[10] reported that GSA is a product
of an alternative pathway for arginine metabolism when the normal
pathway is blocked by urea retention. Natelson et al.[11] proposed
the existence of a guanidine cycle linked with the urea cycle.
In either case the production of GSA could be stimulated by urea
loading.

SUMMARY

Our results demonstrated the importance of energy intake for
preventing the accumulation of GSA, MG and urea nitrogen. Our
study shows that the synthesis of GSA, MG and urea increase markedly
with the stress of inadequate energy intake while in the presence
of high energy intake GSA, MG and urea in serum decrease.

REFERENCES

1. P. L. Balestri, M. Biagini and P. Rindi, Uremic toxins, <u>Arch. Intern. Med.</u>, 126:843 (1970).

2. B. D. Cohen, I. M. Stein and J. E. Bonas, Guanidinosuccinic aciduria in uremia, <u>Am. J. Med.</u>, 45:63 (1968).

3. G. Grodstein, M. J. Blumenkrantz and J. D. Kopple, Nutritional and metabolic response to catabolic stress in uremia, <u>Am. J. Clin. Nutr.</u>, 33:1411 (1980).

4. Y. S. M. Taylor, N. S. Scrimshaw and V. R. Young, The relationship between serum urea levels and dietary nitrogen utilization in young men, <u>Br. J. Nutr.</u>, 32:407 (1974).

5. A. Ando, Y. Orita, K. Nakata, Y. Tsubakihara, Y. Takamitsu, N. Ueda, M. Yanase and H. Abe, Effect of low protein diet and surplus of essential amino acids on the serum concentration and the urinary excretion of methylguanidine and guanidino succinic acid in chronic renal failure, <u>Nephron.</u>, 24:161(1979).

6. S. Giovannetti, M. Biagini, P. L. Balestri, R. Navalesi, P. Giognoni, T. de Matteis, P. Ferro-Milone and C. Perfetti, Uraemia-like syndrome in dogs chronically intoxicated with methylguanidine and creatinine, <u>Clin. Sci.</u>, 36:445 (1969).

7. S. Giovannetti, P. L. Balestri, M. Biagini, C. G. Menichini, P. Rindi, Implications of dietary therapy, <u>Arch. Intern. Med.</u>, 126:900 (1970).

8. E. P. Cottini, D. L. Gallina and J. M. Dominguez, Urea excretion in adult humans with varing degrees of kidney malfunction fed milk, egg or amino acid mixture assessment of nitrogen balance, <u>J. Nutr.</u>, 103:11 (1973).

9. C. Abitobol, G. Jean and M. Broyer, Urea synthesis in moderate experimental uremia, <u>Kidney Int.</u>, 19:648 (1981).

10. I. M. Stein, B. D. Cohen and R. S. Kornhauser, Guanidinosuccinic acid in renal failure, experimental azotemia and in born errors of the urea cycle, <u>N. Engl. J. Med.</u>, 280:926 (1969).

11. S. Natelson and J. E. Sherwin, Proposed mechanism for urea nitrogen reutilization: relationship between urea and proposed guanidinie cycles, <u>Clin. Chem.</u>, 25:1343 (1979).

THE EFFECT OF LACTULOSE ON THE METABOLISM OF GUANIDINO COMPOUNDS IN CHRONIC RENAL FAILURE

Mitsuhiro Miyazaki, Kazumasa Aoyagi, Shoji Ohba,
Sohji Nagase, Mitsuharu Narita and Shizuo Tojo

Department of Internal Medicine, Institute of
Clinical Medicine, The University of Tsukuba
Sakura-mura, Ibaraki-ken, 305 Japan

INTRODUCTION

Serum concentrations of urea and creatinine increase in patients with decreased renal function. Urea has been demonstrated to undergo enterohepatic circulation[1,2,3,4]. Creatinine has also been thought to have an enterohepatic circulation in patients with decreased renal function[5,6,7]. Serum concentrations of some guanidino compounds, especailly guanidinosuccinic acid (GSA) and methylguanidine (MG) increase in uremic states[8,9,10,11]. We reported that urea stimulated the synthesis of GSA in isolated rat hepatocytes[12,13,14]. Cohen proposed that MG might be produced from creatinine[15].

In the present study, we report the effect of lactulose which inhibits the enterohepatic circulation of urea, on the metabolism of various nitrogen compounds in patients with chronic renal insufficiency.

MATERIALS AND METHODS

Materials

Twelve men and three women with chronic renal insufficiency, aged 25-74 years, were admitted for this study. The primary diseases which caused renal insufficiency were chronic glomerulonephritis in thirteen cases, diabetic nephropathy in one case and malignant hypertension in one case (Table 1).

Table 1. Patients and Clinical Diagnosis

No.	Name	Age	Sex	Body weight (kg)	Clinical Diagnosis
1	T.W.	71	F	42.0	CGN
2	K.N.	63	M	66.0	diabetic nephropathy
3	M.I.	64	F	40.5	CGN
4	Y.I.	55	M	46.5	CGN
5	K.I.	31	M	69.0	CGN
6	E.K.	28	F	69.0	CGN
7	T.S.	60	M	66.5	CGN
8	K.S.	48	M	45.5	malignant hypertension
9	K.H.	25	M	50.0	CGN
10	T.M.	28	M	45.5	CGN
11	N.H.	35	M	68.5	CGN
12	S.M.	30	M	70.0	CGN
13	T.K.	29	M	52.5	CGN
14	M.O.	55	M	50.5	CGN
15	F.F.	74	M	57.5	CGN

Schedule

The patients were placed on a standard diet containing 50 g of protein and 10 g of sodium chloride and were observed for two weeks. Then they were administered a 65 % lactulose (4-o-β-D-galactopyranosyl-D-α-fructose) solution orally after every meal, which is enough to cause loose bowels, and were observed for the following two weeks. Other drugs and therapies were not changed during this study. Clinical history, physical examination and serum concentration of guanidino compounds were checked at the end of the observation period and the administration period, and the values obtained were compared by paired t-test to determine the effect of lactulose.

Analysis of guanidino compounds

Guanidino compounds were determined by high pressure liquid chromatographic analysis using 9,10-phenantherenequinone for the labelling method according to Yamamoto et al.[16]. Guanidino compounds were separated on a stainless steel column, 4.6 x 125 mm, packed with Toyosoda IEX210SC cation-exchange resin with a mean particle size of 12 μm (Toyosoda Co. Ltd., Tokyo, Japan). They were eluted by Na-citrate buffers at pH 3.5, pH 5.5 and pH 10.0 with a flow rate of 1.0 ml/min. The column was washed with 1N NaOH at a flow rate of 1.0 ml/min for ten minutes and equilibrated with 0.4 N Na-citrate buffer pH 3.0 at a flow rate of 1.0 ml/min for 20 min.

The analysis was done automatically using a TRI-ROTOR III pump
(Japan Spectroscopic Co. Ltd., Tokyo, Japan) and an autosampler.
Collected plasma was deproteinized using an equal volume of 20 %
trichloroacetic acid. The supernatant was adjusted to pH 2.2 by
2.0 N NaOH and 200 μl of this pH adjusted extract was used for the
determination.

RESULTS

Clinical course and laboratory findings

      Loose bowel occurred in all patients following lactulose ad-
ministration. The dose of lactulose needed to cause loose bowels
was 30 ml/day in two patients, 45 ml/day in four patients and 60
ml/day in nine patients. There were no complaints such as abdominal
pain or severe diarrhea.

      In ten patients (No. 1-10), there was no remarkable change in
blood pressure or body weight. In the other five (No. 11-15),
general malaise and appetite loss was observed before lactulose
therapy and hemodialysis was started a few weeks after the study
because of increases in serum creatinine (S-creatinine), metabolic
acidosis, hemorrhagic diathesis or congestive heart failure.

      Hematological examination showed a significant dicrease in
RBC, Hb and Ht after lactulose therapy. There were no remarkable
changes in blood chemistry or acid-base balance after lactulose
therapy except for the ratio of BUN/S-creatinine (Table 2).
Guanidinobutyric acid (GBA) and guanidine (G) were not detected
in this study (Table 3).

Classification by serum creatinine concentration

      The serum concentration of creatinine during the observation
period correlated with the clinical course well enough to use it
as a parameter to classify the patients. According to the serum
creatinine concentration during the observation period, they fell
into two groups as follows: Group I contained ten patients whose
serum creatinine concentration ranged from 3.0 mg/dl to 7.0 mg/dl,
Group II was made of five patients whose serum creatinine concentrations were over 7.0 mg/dl. Further statistical analyses were
carried out in each group.

      The Hb and Ht decreased significantly after lactulose therapy
in group I and group II. Serum concentrations of calcium and
phosphate decreased significantly, but total protein, albumin and
cholesterol did not change after lactulose therapy in group I.
Serum concentration of phosphate increased slightly, but total protein, albumin, cholesterol and calcium did not change in group II.

Table 2.   Changes in Laboratory Values

		before		after		
RBC	$(\times 10^4/mm^3)$	317 ±15		281 ±26		#
Hb	(g/dl)	9.7 ± 0.5		9.2 ± 0.5		*
Ht	(%)	29.7 ± 1.5		28.0 ± 1.7		*
TP	(g/dl)	6.8 ± 0.2		6.7 ± 0.2		°
Albumin	(g/dl)	3.7 ± 0.1		3.6 ± 0.1		
Cholesterol	(mg/dl)	183 ±10		180 ±11		
BUN	(mg/dl)	80.4 ± 6.4		73.6 ± 8.3		°
Creatinine	(mg/dl)	6.5 ± 0.9		6.7 ± 1.1		
BUN/Creatinine		14.0 ± 1.3		12.4 ± 0.9		#
Na	(mEq/1)	139.0 ± 0.9		138.4 ± 1.3		
K	(mEq/1)	4.4 ± 0.1		4.4 ± 0.2		
Ca	(mEq/1)	4.5 ± 0.1		4.3 ± 0.1		
P	(mg/dl)	5.2 ± 0.3		5.2 ± 0.4		
PH		7.35± 0.01		7.35± 0.01		
HCO$_3$	(mEq/1)	19.0 ± 0.8		18.8 ± 0.9		
BE	(mEq/1)	-5.7 ± 0.8		-5.9 ± 1.5		

°;p<0.10, #;p<0.05, *;p<0.025, (Mean±SE).

Table 3.   Changes in serum levels of Guanidino Compounds

		before	after
GSA	(nmol/dl)	1330±210	1770±460
GAA	(nmol/dl)	154± 13	176± 29
GPA	(nmol/dl)	121± 16	138± 13
Creatinine	(μmol/dl)	49± 6	55± 9
Arg	(μmol/dl)	11± 1	10± 1
MG	(nmol/dl)	113± 23	126± 34

(Mean ± SE)

There were no remarkable changes in acid-bass balance in either group (Table 4).

The BUN decreased from 67.9 ml/dl to 56.9 mg/dl, the serum concentration of creatinine (S-creatinine) decreased from 4.6 mg/dl to 4.3 mg/dl and the ratio of BUN/S-creatinine decreased from 15.8 to 13.7 after lactulose therapy in group I. In group II, the BUN increased from 105.4 mg/dl to 107.0 mg/dl, the serum concentration

Table 4.   Changes in Laboratory Values in Groups I and II

		Group I (3.0≤creatinine<7.0 mg/dl)		Group II (creatinine≥7.0 mg/dl)	
		before	after	before	after
RBC	$(\times 10^4/mm^3)$	338±15	327±16 °	275±28	250±34 #
Hb	(g/dl)	10.3±0.4	9.9±0.5 #	8.6±0.9	7.8±1.0 #
Ht	(%)	31.4±1.5	30.0±1.6 #	26.4±3.1	24.0±3.3 *
TP	(g/dl)	6.9±0.1	6.8±0.1	6.8±0.4	6.7±0.4
Albumin	(g/dl)	3.6±0.1	3.6±0.1	3.7±0.3	3.7±0.3
Cholesterol	(mg/dl)	182±10	179±10	194±28	197±27
Na	(mEq/1)	138.1±1.0	140.3±1.5	139.4±1.8	135.0±1.3
K	(mEq/1)	4.4±0.2	4.5±0.3	4.5±0.1	4.1±0.2
Ca	(mEq/1)	4.5±0.1	4.3±0.1 #	4.3±0.1	4.2±0.1
P	(mg/dl)	4.8±0.3	4.4±0.2 #	6.1±0.4	6.6±0.6 °
PH		7.35±0.01	7.35±0.01	7.34±0.02	7.33±0.03
HCO_3	(mEq/1)	20.0±0.9	20.2±1.0	17.2±1.1	16.3±1.2
BE	(mEq/1)	−4.7±0.9	−4.7±1.0	−7.4±1.3	−8.1±1.5

°; $p<0.10$, #; $p<0.05$, *; $p<0.025$, (Mean±SE)

of creatinine increased from 10.4 mg/dl to 11.4 mg/dl and the ratio of BUN/S-creatinine decreased from 10.5 to 9.8, but there was no significant difference before and after lactulose (Fig. 1).

In group I, the plasma concentration of guanidinosuccinic acid (GSA) decreased from 850 nmol/dl to 690 nmol/dl and that of methylguanidine (MG) decreased from 52 nmol/dl to 41 nmol/dl after lactulose therapy. In group II, plasma concentration of GSA increased from 2300 nmol/dl to 3940 nmol/dl and that of MG increased from 212 nmol/dl to 264 nmol/dl after lactulose therapy (Fig. 2). Other guanidino compounds, for example guanidinoacetic acid (GAA), guanidinopropionic acid (GPA) and arginine (Arg) showed no remarkable change following lactulose therapy in either group.

DICUSSION

An enterohepatic circulation has been demonstrated for urea, which involves excretion into the gut, metabolism to ammonia by the gut flora, and absorption followed by synthesis to urea in the liver again[1][2][3][4]. Creatinine is also thought to have an enterohepatic circulation in patients with decreased renal function[5][6][7]

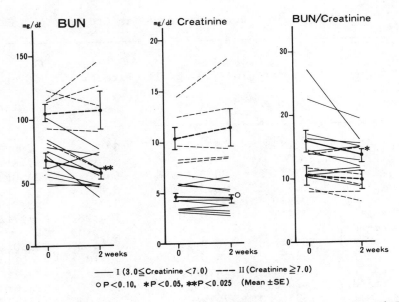

Fig. 1.   Changes in BUN and creatinine in Groups I and II.

We have already reported that urea stimulated the synthesis of GSA in isolated rat hepatocytes and in vivo[12,13,14]. On the other hand, creatinine may be one of the precursors of MG[15,17]. We applied lactulose in order to increase the extra renal excretion of nitrogen compounds and anticipated an effect on guanidino compounds such as GSA and MG, which are thought to be uremic toxins[8,9,10,11]. Lactulose alters the bacterial environment in the gut by lowering the intestinal pH and inhibits the production and absorption of ammonia[18,19,20,21]. It has been used in the therapy of hepatic coma[18,22].

    The BUN and the ratio of BUN/S-creatinine decreased signifi-cantly after lactulose therapy. Evidence of this was clearest in group I. There were highly significant decrease of, not only BUN and BUN/S-creatinine, but also GSA and MG after lactulose therapy in group I. These might be produced by the increase in extra-renal excretion of urea and creatinine due to the loose bowel caused by lactulose. However, the inhibition of the enterohepatic circulation of urea also accounts for the decrease of BUN and the synthesis of GSA might be suppressed by the decreased serum concentration of urea, and that of MG might be suppressed by the decreased serum concentration of creatinine. In addition, the absorption of MG in the gut could be decreased by the lowering of the intestinal pH. In group II, although the ratio of BUN/S-creatinine also decreased, the serum concentration of urea, creatinine, GSA and MG increased

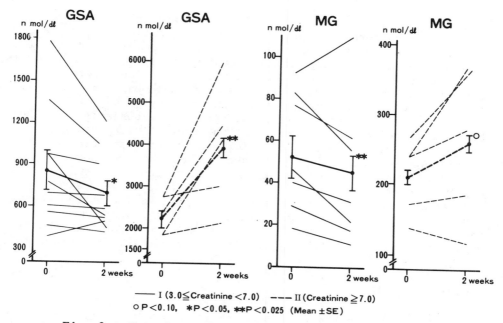

—— I (3.0≤Creatinine <7.0) ——— II (Creatinine ≥7.0)
○ P<0.10, *P<0.05, **P<0.025 (Mean ±SE)

Fig. 2.   Changes in GSA and MG in Groups I and II.

after lactulose therapy.   Levels of GSA and MG increased progressively.   Judging from the decreased BUN/S-creatinine ratio,
lactulose might increase the enteric excretion of nitrogen compounds, but not enough to decrease the serum concentration of
nitrogen compounds in this group.   Man and coworkers reported that
the excretion of urea in the stool increased by the administration
of lactulose in patients with decreased renal function[23].

Serum concentrations of calcium and phosphate also decreased
possibly as a result of increased enteric excretion.   Progression
of anemia was demonstrated in both groups.   It could result from
the frequent blood sampling in a short period in both groups coupled
with the progression of renal failure in group II.   There were no
remarkable adverse effects caused by lactulose.

These results indicate that lactulose decreases serum concentrations of urea, GSA, MG and phosphate, and suggests that lactulose
might be useful in the therapy of chronic renal insufficiency.

SUMMARY

The enteric excretion of nitrogen compounds has been used as a
therapy for chronic renal insufficiency.   To stimulate the enteric

excretion of nitrogen compounds, we administered lactulose to
patients with decreased renal function. The patients were placed
on a standard diet for two weeks. They were administered a 65 %
lactulose solution orally in sufficient quantity cause loose bowels,
and were observed for the following two weeks. There were no
remarkable changes in body weight or blood acid-base balance. No
adverse effects were demonstrated from the drug. The ratio of
BUN/S-creatinine decreased significantly in all subjects. In the
ten patients whose serum concentration of creatinine during the
observation period was 3.0-7.0 mg/dl, the serum concentration of
urea, GSA, MG and phosphate decreased significantly. However, in
the five patients who had clinical manifestations of uremia at the
beginning of this therapy with serum creatinine concentrations over
7.0 mg/dl, the serum concentration of GSA and MG increased progressively. These results indicate that lactulose lowers the serum
concentration of urea, GSA, MG and phosphate, and suggests that it
might be useful in the therapy of chronic renal insufficiency.

ACKNOWLEDGEMENTS

        This study was supported in part by a Research Grant from the
Intractable Disease Division, Public Health Bureau, Ministry of
Health and Welfare, Japan and University of Tsukiba Project
Research.

REFERENCES

1. M. Walser and L. J. Bodenlos, Urea metabolism in man,
        J. Clin. Invest., 38:1617 (1956).
2. M. Walser, Urea metabolism in chronic renal failure,
        J. Clin. Invest., 53:1385 (1974).
3. J. J. Brown, M. J. Hill and P. Richards, Bacterial ureases
        in uraemic man, Lancet, II:406 (1971).
4. E. A. Jones, R. A. Smallwood, A. Craigie and V. M. Rosenoer,
        The enterohepatic circulation of urea nitrogen, Clin. Sci.,
        37:825 (1969).
5. S. Giovannetti, P. 1. Balestri and G. Barsotti, Methyl-
        guanidine in uremia, Arch. Intern. Med., 131:709 (1973).
6. J. D. Jones and P. C. Burnett, Creatinine metabolism in
        humans with decreased renal function - creatinine deficit,
        Clin. Chem., 20:1204 (1974).
7. J. D. Jones and P. C. Burnett, Creatinine metabolism and tox-
        icity, Kidney International., 7:294 (1975).
8. J. E. Bonas, B. D. Cohen and S. Natelson, Separation and
        estimation of certain guanidino compounds - application to
        the human urine, Microchem. J., 7:63 (1963).
9. S. Natelson, I. Stein and J. E. Bonas, Improvement in the
        method of separation of guanidino organic acids by column

chromatography – isolation and identification of guanidino-succinic acid from human urine, Microchem. J., 8:371 (1964).

10. B. D. Cohen, Guanidinosuccinic acid in uremia, Arch. Intern. Med., 126:846 (1970).

11. Y. Tofuku, M. Kuroda, T. KIta, and R. Takeda, Studies on guanidino compounds in uremia – reevaluation of serum GSA in relation to acute symptoms of uremia, Jpn. J. Nephrol., 22:249 (1980).

12. S. Ohba, K. Aoyagi and S. Tojo, On the biosynthesis of guanidinosuccinic acid in rat liver – effect of urea, ornithine and D,L-norvaline, Jpn. J. Nephrol., 24:1147 (1982).

13. K. Aoyagi, S. Ohba, M. Narita and S. Tojo, Biosynthesis of guanidinosuccinic acid in isolated rat hepatocytes – II inhibition of its synthesis by urea cycle members and D,L-norvaline, Jpn. J. Nephrol., 24:1137 (1982).

14. K. Aoyagi, S. Ohba, M. Miyazaki, S. Iida, K. Watanabe, M. Narita and S. Tojo, Biosynthesis of guanidinosuccinic acid in isolated rat hepatocytes – 1 stimulation of its synthesis by urea, Jpn. J. Nephrol., 24:345 (1982).

15. B. D. Cohen, Uremic toxins, Bull. N. Y. Acad. Med., 51:1228 (1975).

16. Y. Yamamoto, T. Manji, A. Saito, K. Maeda and K. Ohta, Ion-exchange chromatographic separation and fluorometric determination of guanidino compounds in physiological fluids, J. Chromatogra., 162:327 (1979).

17. S. Nagase, K. Aoyagi and S. Tojo, On methylguanidine synthesizing organs – estimation from short term effect of creatinine, Jpn. J. Nephrol., (in press)

18. T. Aoyagi and W. H. J. Summerskill, Inhibition by aceto-hydroxamic acid of human mucosal and faecal urease-specific activity, Lancet, 1:296 (1966).

19. J. Bircher, U. P. Haemmerli and G. Scollo-Lavizzari, Treatment of chronic portal – systemic encephalopathy with lactulose, Am. J. Med., 51:148 (1971).

20. D. O. Castell and E. W. Moore, Ammonia absorption from the human colon, the role of nonionic diffusion, Gastroent., 60:33 (1971).

21. F. Simmons, H. Goldstein and J. D. Boyle, A controlled cinical trial of lactulose in hepatic encephalopathy, Gastroent., 59:827 (1970).

22. G. Rorsman and I. Sulg, Lactulose treatment of chronic hepatopotal encephalopathy, a clinical and electro-encephalographic study, Acta Med. Scand., 187:337 (1970).

23. N. K. Man, T. Drueke, J. Paris, C. Elizalde-Monteverde, M. Rondon-Nucete, J. Zingraff and P. Jungers, Increased nitrogen removal from the intestinal tract of uraemic patients, Proc. Euro. Dialysis Transplant Ass., 10:143 (1973).

EVALUATION OF THE EFFICACY OF ANTI-REJECTION THERAPY USING

THE QUANTITATIVE ANALYSIS OF GUANIDINOACETIC ACID (GAA) URINARY

EXCRETION AS A GUIDE

Makoto Ishizaki, Hiroshi Kitamura, Hisashi Takahashi,
Hisako Asano, Kazuaki Miura, Hajime Okazaki

Sendai Shakai-Hoken Hospital, Sendai, Japan

INTRODUCTION

Due to insufficient recognition of rejection crises in kidney transplants, rejection of grafts is a common complication. When it is difficult to make diagnoses using conventional biochemical parameters, histopathological studies of grafts have been undertaken to clarify the clinical events. Even in cases of successful biopsies, consecutive procedures are not possible. Therefore, since the urinary excretion of GAA reflects the metabolic functioning of the kidney, we examined the daily level of GAA in urine after transplantation and considered the feasibility of using this measure as a guide to the early detection of rejection and whether or not the amount of GAA urinary excretion could contribute to the evaluation of the efficacy as well as the sufficiency of anti-rejection therapy.

MATERIALS AND METHODS

Eight cases of renal transplants from living related donors and 4 cases from cadavers were included in this study. All patients receiving renal transplants from living related donors received donor specific buffy coat transfusions (DSBCT) preoperatively. The 4 cadaver kidneys from brain-dead donors were preserved for 7 to 54 hours by means of simple cold storage. However, acute tubular necrosis (ATN) was noted postoperatively in all these cases and required hemodialysis therapy for about 2 weeks. Pertinent data for the recipients are presented in Table 1.

Administration of 2 mg/kg azathioprine and 1 mg/kg predniso-

lone was initiated 2 days before surgery to the intended recipients
of renal transplants from living related donors. The same amount
of the two drugs was administered to the intended recipients of
renal transplants from cadavers starting on the operative day.
When a rejection crisis occurred, intravenous injections of a
large quantity of methylprednisolone were administered. When this
was not effective, the dosage of prednisolone was raised to 200
mg/day and then reduced gradually over a period of 4 weeks to a
maintenance level. Furthermore, in cases of severe rejection,
monoclonal antilymphoblast antibody was administered[1,2]. Rejec-
tions were diagnosed by the usual clinical and laboratory findings
consistent with deterioration in renal functioning and were frequently confirmed by histopathological evaluation of tissue obtained by transplant biopsy. Urine was continuously collected
each day in a vessel containing thymol crytals; part of this was
utilized for creatinine clearance (Ccr) measurement and guanidino
compound analysis. For purposes of GAA estimation, all the urine
samples were frozen at -20 °C until analysed. Guanidino compounds
were fluorometrically measured by HPLC using a cation exchange
resin column[3].

RESULTS

Rejection episodes

     Among the 12 cases studied, a total of 16 rejection episodes
were observed consisting of 3 incidences of accelerated rejection
and 13 incidences of acute rejection. GAA analysis was impossible
because of anuria in 3 rejection crises during the ATN period in
cases of cadaver transplants. In only one case receiving transplant from a living related donor was there no rejection among
those patients.

Accelerated rejection in case VI with DSBCT

     Fig. 1 shows the clinical course of case VI following kidney
transplantation from a living related donor. The operation was
performed on day 0. The top section represents the dosage of
prednisolone and azathioprine, the middle section indicates the
dosage of plasma creatinine, and the bottom section shows the
dosage of endogenous creatinine clearance (o—o) and the daily
urinary excretion of GAA (•—•). Solu-Medrol was intravenously
injected twice on days 2 and 4 at a dosage of 500 mg. GAA urinary
excretion started to gradually diminish on day 1, and deterioration in renal functioning was seen, although the increased dosage
of 200 mg/day prednisolone was started on day 3. Then, as a strong
anti-rejection therapy, monoclonal anti-lymphoblast antibody was
given 9 times daily which caused GAA excretion to rise. The amount
of GAA excretion showed the same tendency as that of Pcr and the

Table 1. Pertinent data for the recipients

Patient	Sex	Age	Donor	DST/RBT	Immunosuppression	Rejection
I	M	9	L	DST	Medrol, ALG, Solu-Medrol, Azathioprine	+
II	F	27	L	DST	Prednisolone, Solu-Medrol, Azathioprine,	+
III	M	32	L	DST	Prednisolone, Solu-Medrol, Azathioprine,	+
IV	M	25	L	DST	Prednisolone, Solu-Medrol, Azathioprine, ALG, Monoclonal Antibody	+
V	M	25	L	DST	Prednisolone, Solu-Medrol, Azathioprine, ALG, Monoclonal Antibody	+
VI	M	30	L	DST	Prednisolone, Solu-Medrol, Azathioprine, Monoclonal Antibody	+
VII	M	27	C*	RBT	Prednisolone, Solu-Medrol, Azathioprine, ALG	+
VIII	F	25	L	DST	Prednisolone, Solu-Medrol, Azathioprine,	+
IX	M	20	L	DST	Prednisolone, ATG, Azathioprine,	–
X	F	27	C*	RBT	Prednisolone, Solu-Medrol, Azathioprine, ALG, Monoclonal Antibody	+
XI	M	34	C**	DST	Prednisolone, Solu-Medrol, Azathioprine,	+
XII	F	35	C*	DST	Azathioprine, Solu-Medrol	+

L, living donor; C, cadaver donor;
*, acute tubular necrosis; **, slight acute tubular necrosis;
DST, donor specific transfusion; RBT, random blood transfusion.

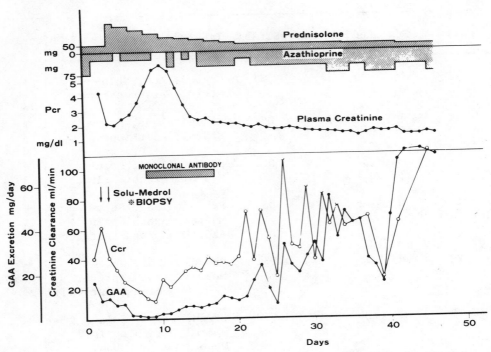

Fig. 1.   Clinical course of case VI in which accelerated rejection
          was observed.   Rejection onset on day 2.

reciprocal of Pcr throughout the rejection crisis.   However changes
in GAA excretion occurred earlier and were more markedly abnormal
than those of Pcr and Ccr.   Particularly in cases of accelerated
rejection, the amount of GAA excretion was initially at a minimal
normal value but then diminished daily.   A clinical rejection
crisis soon followed at which point anti-rejection therapy was
undertaken.   The kidney graft biopsy performed on day 6 during the
period of anti-rejection therapy showed the endothelial cells of
the vas afferens to be slightly swollen, the glomerulus to be
collapsed, a slight infiltration of mononuclear cells around the
glomeruli and small arteries, and interstitial edema.   Marked
hydropic changes in the tubular cells and distended tubular lumen
were also seen.   The above findings led to a histopathological
diagnosis of a humoral type of acute rejection (Fig. 2).

Acute rejection in case V of a renal transplant from a living
related donor

        Fig. 3 illustrates the clinical course in the case of a
patient who received a kidney from his 49-year-old mother.   DSBCT

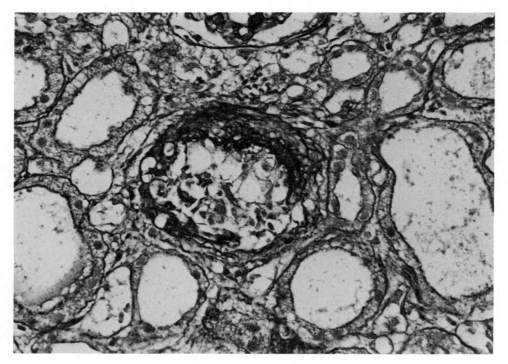

Fig. 2.   Graft biopsy of case VI on day 6.

was given preoperatively. Very favorable GAA excretion at a level between 30.1 mg/day and 76.6 mg/day was seen immediately after the operation. On day 8 the sudden onset of severe rejection occurred. Renal functioning markedly deteriorated in spite of the administration of Solu-Medrol in bolus therapy, increased dosage of prednisolone and the administration of ALG. Hemodialysis treatment was resumed on day 10, and then 7 treatments were required until diuresis was achieved. Although there was marked deterioration in renal function, GAA excretion started to rise on day 21 due to monoclonal antibody treatment and was followed by improvement in plasma creatinine. In cases of acute rejection of transplants from living related donors, a large quantity of GAA excretion was also observed immediately after the operation. However, it gradually or rapidly diminished prior to the onset of clinical rejection, at which point anti-rejection therapy was undertaken. The renal graft biopsy performed on day 15 showed swelling of the endothelial cells of the vas afferens, thickening of intima and collapse of the glomeruli. There was minor fibrosis of the interstitial tissue as well as an infiltration by mononuclear cells and hydropic changes in the tubular cells. The case was diagnosed as being an example of acute rejection.

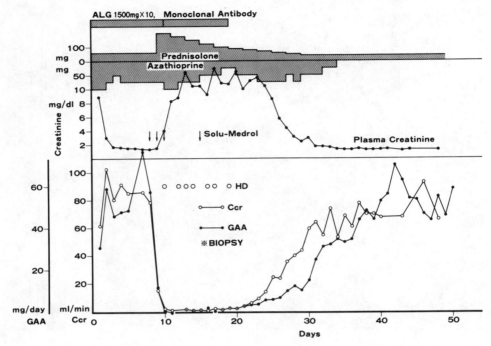

Fig. 3.   Clinical course of case V with acute rejection.
Rejection onset on day 8.

Acute rejection in case VII after renal transplantation from a cadaver donor

　　　Fig. 4 illustrates the clinical course following transplantation of a graft from a cadaver which was shipped from the U.S.A.
and revascularized after being preserved in cold Euro-Collins'
solution for 36 hours.   ATN was observed immediately after the
operation and lasted until day 16.   A total of 14 hemodialysis
treatments were performed.   On day 5 a clinical rejection crisis
occurred during ATN, and anti-rejection therapy consisting of Solu-
Medrol bolus therapy and an increased dosage of prednisolone was
undertaken.   This rejection crisis was confirmed by renal scan and
renal graft biopsy.   Starting on day 18 the amount of urine per
day rose to over 1500 ml and the recipient became completely free
of hemodialysis therapy.   On day 25 GAA excretion was 2.3 mg/day
and Ccr was 32.6 ml/min.   On day 28 the second clinical rejection
crisis commenced at which time GAA excretion was 0.8 mg/day and
Ccr was 35.7 ml/min.   In cases of grafts from cadavers, following
ATN, GAA excretion had a much lower value than that seen in cases
of transplantation from living related donors.   The post-ATN level
of GAA excretion in case VII tended to diminish to a much lower

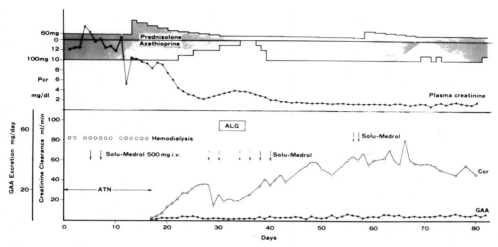

Fig. 4. Clinical course of case VII after renal transplantation from a cadaver donor. Rejection onset on day 5, 28 and 56.

level starting 3 days prior to the rejection crisis. However, the degree of Ccr in this case was increasing prior to the onset of the second rejection crisis. Ccr diminished during the rejection crisis, and GAA remained at a low level. This condition was improved by successful anti-rejection therapy, although the level of GAA excretion improved only slightly.

Comparison between the urinary GAA excretion and Ccr before the rejection crisis and at the onset of rejection

The highest level of GAA excretion in each case was observed from 1 to 8 days (mean 4.0 ± 4.8 days) before the onset of a rejection crisis. A value of 100 % was assigned to both the highest prerejection value of GAA and the corresponding value of Ccr at this point, and each of the two values was compared with the GAA and Ccr values at the onset of rejection in terms of the percentage change (Fig. 5).

GAA excretion decreased at the onset of the rejection crisis in all cases at a rate of 75.2 ± 17.3 %. On the other hand, the change in Ccr from 100 % ranged from 35.3 % to 148.2 % and specific deterioration of Ccr was not noted prior to the onset of rejection. The parameters of renal function indicated by the values of Pcr, Ccr and GAA were worst during rejection crises.

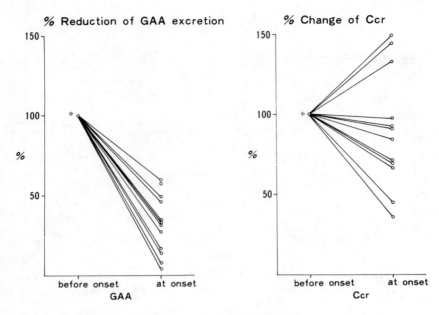

% Reduction of GAA excretion   % Change of Ccr

*This point shows highest GAA levels before onset of rejection
and Ccr at that time.

Fig. 5.   Comparison between GAA urinary excretion and the
           clearance of creatinine before a rejection crisis
           and at the onset of a rejection.

Variation of GAA excretion before, during and after anti-rejection therapy

Among cases of rejection of renal transplants from living
related donors, the GAA value was 36.5 ± 24.5 mg/day (mean ± SD)
before rejection, decreasing to a level of 1.68 ± 1.78 mg/day
during rejection and almost recovering to a level of 31.9 ± 16.9
mg/day after successful therapy (about 3 weeks after the onset of
rejection) which was as high as the level before rejection.   On
the other hand, in transplantation of donor kidneys from cadavers,
corresponding values were 3.7 ± 2.75 mg/day before rejection,
1.25 ± 1.58 mg/day during rejection and 9.5 ± 4.83 mg/day after
rejection.   In cases of cadaver donors in which the rejection
crisis was overcome, the GAA value was 2.6 times better than the
value before the rejection crisis (Fig. 6).

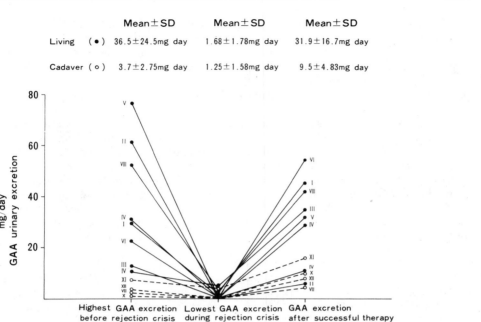

Fig. 6. Variation of GAA excretion before, during and after anti-rejection therapy.

DISCUSSION

Since Salvatierra reported that donor specific transfusion (DST) had a beneficial effect on graft survival of renal transplants from living related donors[4], preoperative DST has generally been employed. Nevertheless, rejection crises after surgery are still sometimes encountered and graft rupture and/or anuria due to postoperative humoral rejection are occasionally seen. In cases of severe rejection, postoperative hemodialysis treatment is required. Early detection of rejection crisis and sufficient anti-rejection therapy can prevent irreversible changes in renal transplants caused by acute rejection and can also promote graft survival. In 8 cases of renal transplants from living related donors in which DSBCT was administered, there was no rejection crisis in only one case, while in the remaining 7 cases, there were a total of 9 rejection crises. In all the cases with rejection crises, GAA excretion decreased prior to rejection, but adequate anti-rejection therapy elevated the GAA value. Therefore, the present finding that a deterioration in renal functioning and a decrease in GAA excretion occur concurrently agrees with previous findings which showed a correlation between renal dysfunction and the level of GAA excretion[5,6,7,8].

Regarding renal graft biopsy performed before and during

rejection in cases where GAA excretion tended to diminish, histopathological changes characteristic of rejection were always observed. This allowed better detection of rejection crisis after renal transplantation and the prevention of irreversible changes in renal grafts with sufficient anti-rejection therapy based on the observation of daily changes in GAA excretion.

In cases of cadaver transplantation in Japan, even kidneys from brain-dead donors need to be preserved for more than 40 hours and ATN is frequently observed. Regarding rejection during ATN, GAA doesn't function as a parameter of rejection because of anuria, and conventional renal scans or histopathological studies must be relied on. Once patients recover from ATN and the amount of urine exceeds 1000 ml/day, although GAA excretion remains very low, it tends to further diminish and effective therapy promotes a better GAA value than the value before rejection.

In conclusion, a daily quantitative analysis of GAA levels in excreted urine after renal transplantation can contribute not only to early detection of rejection crises but also to an evaluation of the effecacy of anti-rejection therapy which can facilitate efforts to prevent irreversible changes caused by renal transplant rejection.

REFERENCES

1. R. Billing, J. Wells, D. Zettel and P. I. Terasaki, Monoclonal and hetero-antibody reacting with different antigens common to human blast cells and monocytes, Hybridoma, 1:303 (1982).
2. R. Billing and S. Chatterjee, Prolongation of skin allograft survival in monkeys treated with anti-Ia and anti-blast/monocyte monoclonal antibody, Transplant Proc., 15:649 (1983).
3. K. Aoyama, C. Uno and M. Ishizaki, Fluorometrical Analysis of guanidino compounds, Nippon Denshi News, 18:18 (1978).
4. O. Jr. Salvatierra, F. Vincenti, W. Amend, D. Potter, Y. Iwaki, G. Opelz, P. Terasaki, R. Duca, K. Cochrum, D. Hans, R. J. Stoney and N. J. Feduska, Deliberate donor specific blood transfusions prior to living related transplantation: A new approach, Ann. Surg., 192:543 (1980).
5. J. E. Bonas, B. D. Cohen and S. Natelson, Separation and estimation of certain guanidino compounds: Application to human urine, Microchem. J., 7:63 (1963).
6. M. Sasaki, K. Takahara and S. Natelson, Urinary guanidinoacetate/guanidinosuccinate ratio: An indicator of kidney dysfunction, Clinical Chemistry, 19:315 (1973).
7. K. Kadono, M. Kokawa, A. Irie, H. Kushiro, J. Kodama, G. Kimura, M. Satani and C. Hayashi, Clinical significance of urinary excretion in guanidino compounds, Abstracts of the 3rd annual meeting of Japan guanidino compounds research association, 3:96 (1980).

8.   A. Okada, Y. Mikami, M. Fujii, E. Imai, K. Kuniba, A. Ando,
      Y. Orita and H. Abe, Clinical evaluation of guanidino
      compounds (GC) with special reference to GC (MG, GSA, GAA)
      and renal function, Abstracts of the 5th annual meeting of
      Japan guanidino compound research association, 5:91 (1982).

METABOLIC PROFILES OF GUANIDINO COMPOUNDS IN

VARIOUS TISSUES OF UREMIC RATS

Hikaru Koide and Chieko Azushima

Division of Nephrology, Department of Medicine
Juntendo University School of Medicine
Tokyo 113, Japan

INTRODUCTION

Several guanidino compounds are implicated as toxic metabolites responsible for the systemic effects of uremia. Serum concentrations of methylguanidine and guanidinosuccinic acid were reported increased in patients with end-stage renal failure[1,2].
Serum profiles of guanidino compounds in uremia, however, might
not reflect metabolic alterations of these compounds in tissues.
Therefore, we investigated the levels of guanidino compounds in
various tissue preparations obtained from rats made experimentally uremic by sub-total nephrectomy or bilateral ligation of the
ureters.

MATERIAL AND METHODS

Production of experimental renal failure

Male Sprague-Dawley rats, weighing 250 g were rendered chronically uremic by right nephrectomy and two-third resection of the
left kidney. Acute renal failure was attained by bilateral ligation of the ureters. After surgery, rats were given a standard
laboratory chow diet and water ad lib. Two weeks or two days
later, the control and uremic animals were sacrificed and blood
was collected by aortic puncture for determination of the blood
urea nitrogen (BUN) and the serum level of guanidino compounds.
The tissues were immediately removed and frozen in an acetone-dry
ice mixture. Only those rats with BUN levels above 45 mg/100ml
were considered uremic and were used in the experimental studies.

365

Table 1.   Plasma guanidino compounds in chronically uremic rats.

	Control (n=3)		Uremia (n=7)		P
T C	222	± 25	242	± 27	N. S.
GSA	2.2	± 0.5	15.9	± 6.1	<0.01
GAA	93	± 14	61	± 7	<0.001
A A	28	± 8	41	± 8	<0.05
GPA	1.3	± 0.2	1.8	± 0.7	N. S.
GBA	6.8	± 2.2	26.3	± 7.4	<0.01
G	2.6	± 0.5	5.4	± 1.4	<0.02
M G	0.31	± 0.07	1.52	± 0.28	<0.001

(μg/100ml)

Pretreatment of biological samples

Plasma was centrifuged at 1,000 x g for 30 minutes in a Centriflo membrane CF-25 (Amicon Corp.). The ultrafiltrate was adjusted to pH 2.2, and used for the determination of guanidino compounds. The tissues were deproteinized using ice cold trichloroacetic acid at a final concentration of 10 %. After centrifugation at 2,000 x g for 10 minutes, trichloroacetic acid was removed from the supernatant using water-saturated ethyl ether.

Guanidino compounds analysis

The concentration of guanidino compounds in plasma and tissues were determined by the Automated Guanidine Analyzer G-520 (Japan Spectroscopic Co., Tokyo, Japan). Using this high pressure liquid chromatography, the following guanidino compounds have been measured simultaneously: taurocyamine (TC), guanidinosuccinic acid (GSA), guanidinoacetic acid (GAA), N-acetylarginine (AA), γ-guanidinopropionic acid (GPA), creatinine (Cr), γ-guanidinobutyric acid (GBA), arginine (Arg), guanidine (G) and methylguanidine (MG).

RESULTS

Experimental chronic renal failure

In chronic uremic rats, plasma GSA, AA, GBA, G and MG were significantly increased, whereas plasma GAA was markedly decreased and plasma GPA levels remained unchanged (Table 1).

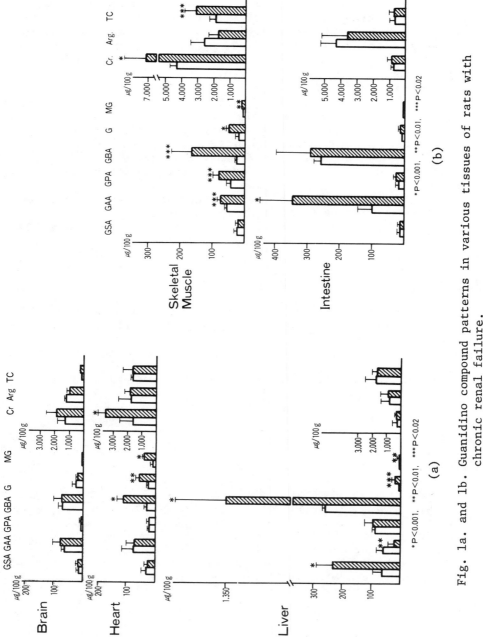

Fig. 1a. and 1b. Guanidino compound patterns in various tissues of rats with chronic renal failure.

The concentrations of guanidino compounds in various tissues
of the rats with chronic renal failure are shown in Fig. 1a and
Fig. 1b. GSA exists mainly in the liver with lower concentrations
in the heart and skeletal muscle. The concentration of GSA in the
liver is more than ten times higher than that in the plasma. The
concentration of GAA in the plasma is almost the same as that in
the tissues. Tissue G and MG levels were ten to twenty times
higher than plasma G and MG levels. In chronic renal failure, the
concentration of G and MG in the heart, the skeletal muscle and
the liver were increased.

The concentration of guanidino compounds in the brain did not
change at all. In the heart, GBA, G and MG levels were increased.
In the liver, the concentrations of GSA, GBA, G and MG were
elevated and the concentration of GAA was significantly decreased.
The metabolic rate of the alternative pathway by which GSA is produced from arginine might be accelerated at the expense of GAA
production from arginine. A marked enhancement of GBA production
would explain the elevation of plasma GBA levels. In skeletal
muscle, the concentration of GAA, GPA and GBA along with G and MG
were all increased. Only the GAA level was increased in the intestine.

Experimental acute renal failure

The guanidino compound pattern in various tissues of rats
with acute renal failure are shown in Fig. 2a and Fig. 2b. In the
brain, there was an increase of G and MG levels and a decrease of
arginine, which is different from the changes seen in chronic
renal failure. In the heart, there was an increase in Cr, MG and
GSA and a decrease in GAA and arginine. The changes in guanidino
compounds in the liver in acute renal failure were almost the same
as that seen in chronic renal failure. In the kidneys, the concentration of GAA was markedly decreased and those of GSA, GPA and
MG were significantly increased. The changes in guanidino compound
levels in skeletal muscle are similar to that in the heart. In
the intestine, the tissue levels of G, MG and GSA were increased,
whereas GBA and arginine were substantially decreased.

Fig. 3 shows the serial alterations of plasma guanidino
compounds in acute renal failure. The increase of plasma GSA and
G and the decrease of plasma GAA and arginine were detected at
the early stage of acute renal failure. On the contrary, the
concentrations of plasma Cr, MG, AA and GBA were increased in the
later stages of acute renal failure.

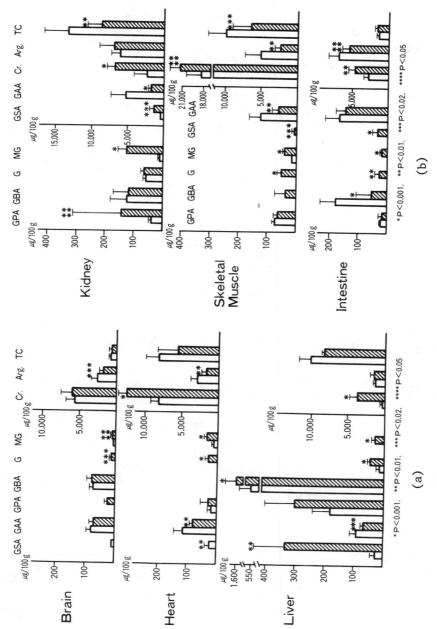

Fig. 2a. and 2b. Guanidino compound patterns in various tissues of rats with acute renal failure.

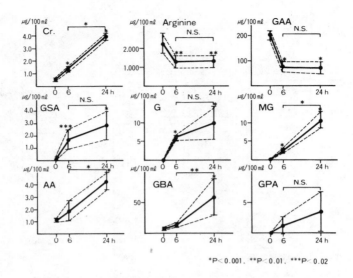

*P<0.001. **P<0.01. ***P<0.02

Fig. 3.   Serial determinations of plasma guanidino compounds in
acute renal failure.

Comparison of the metabolic profiles of guanidino compounds in chronic versus acute renal failure

Table 2 shows a comparison of the metabolic profiles of guanidino compounds in chronic versus acute renal failure. In both
cases, the decrease in GAA levels was always accompanied by an increase in GSA levels. On the other hand, elevations in GAA were not
accompanied by changes in GSA. The changes in MG concentration were
remarkable in tissues such as the heart and skeletal muscle which
contain large amounts of creatinine. Although the concentration of
arginine in the tissues was found to be unchanged in chronic renal
failure, its concentration was decreased in acute renal failure.

DISCUSSION

In 1968, Cohen[3] reported that serum GAA levels were elevated
and urinary GAA excretion was reduced in patients with renal failure. Our studies, however, show a reduction in plasma GAA in
chronically uremic rats. GAA is formed from arginine and glycine
in a reaction catalyzed by transaminidase of which the kidney is a
major site. Therefore, a decrease in plasma GAA may be due to an
impairment of enzyme activity in the kidney and a reduction in the
biosynthesis of GAA in the liver.

Table 2. Comparison of the changes in tissue guanidino compounds between chronic renal failure and acute renal failure in rats.

		GSA	GAA	GPA	GBA	G	MG	Cr.	Arg.	TC	AA
Brain	CRF	—	—	—	—	—	—	—	—	—	—
	ARF	—	—	—	—	↑	↑	—	↓	—	↓
Heart	CRF	—	—	—	↑	↑	↑	↑	—	—	—
	ARF	↑	↓	—	—	↑	↑	↑	↓	—	—
Liver	CRF	↑	↓	—	↑	↑	↑	—	—	—	—
	ARF	↑	↓	—	↑	↑	↑	↑	—	—	↑
Skeletal Muscle	CRF	—	↑	↑	↑	↑	↑	↑	—	↑	—
	ARF	↑	↓	—	—	↑	↑	↑	↓	↓	—
Intestine	CRF	—	↑	—	—	—	—	—	—	—	—
	ARF	↑	—	—	↓	↑	↑	↑	↓	—	—
Kidney	ARF	↑	↓	↑	—	—	↑	↑	—	↓	—
Plasma	CRF	↑	↓	—	↑	↑	↑	↑	—	—	—
	ARF	↑	—	↑	↑	—	↑	↑	—	—	↑
Erythrocyte	ARF	↑	—	—	—	↑	↑	↑	—	—	—

↑ : indreased,   ↓ : decreased,   — : not changed.

Although Shainkin[4] found an elevation in plasma GPA in subjects with chronic renal disease probably due to disorders of lipid metabolism, our results did not show any change in the plasma GPA levels in chronically uremic rats.

In the case of both chronic and acute renal failure in rats, the decrease in GAA levels was always accompanied by an increase in GSA levels in various tissues, suggesting an acceleration of the alternative pathway for arginine metabolism in uremia. Furthermore, in acute renal failure, an increase in plasma GSA and a decrease in plasma GAA and arginine were detected at an early stage and an increase of plasma Cr and MG were observed at a later stage. These observation suggest that the metabolic changes involving the arginine-GAA-GSA pathway occur at an early period and the metabolic changes involving Cr-MG pathway occur at a later period of acute renal failure.

SUMMARY

Using high pressure liquid chromatography, the profiles of guanidino compounds from plasma and various tissues were analyzed. Guanidino compounds were determined fluorometrically after reaction with 9,10-phenenthrenequinone in alkaline solution.

Chronic uremia was induced experimentally by the method of Platt and acute uremia was induced by the bilateral ligation of ureters in Sprague-Dawley rats.

In chronically uremic rats, there was a marked rise in the plasma concentrations of GSA, AA, GBA, G and MG. On the contrary, plasma levels of GAA were significantly reduced. The concentration of GSA, GBA, G and MG in the liver and that of GBA, G and MG in the heart were markedly increased in the rats with chronic uremia. All guanidino compounds except for GSA were increased in skeletal muscle, while all of the guanidino compounds remained within normal limits in the brain.

In acutely uremic rats, plasma levels of GSA, GBA and MG were markedly increased. The concentration of GAA was remarkably decreased in the kidney, liver, heart and skeletal muscle and the concentration of GSA was increased in all tissues except for the brain. In the brain, there was an increase in G and MG and a decrease in arginine. The decrease in GAA is always accompanied by an elevation of GSA, suggesting that the alternative pathway for arginine metabolism is accelerated in uremia.

REFERENCES

1. I. M. Stein, G. Perez, R. Johnson and N. B. Cummings, Serum levels and urinary excretion of methylguanidine in chronic renal failure, J. Lab. Clin. Med., 77:1020-1024 (1971).
2. I. M. Stein, B. D. Cohen and R. S. Kornhauser, Guanidino-succinic acid in renal failure, experimental azotemia and inborn errors of the urea cycle, New Engl. J. Med., 280:926-930 (1969).
3. B. D. Cohen, I. M. Stein and J. E. Bonas, Guanidinosuccinic aciduria in uremia. A possible altenate pathway for urea synthesis, Amer. J. Med., 45:63-68 (1968).
4. R. Shainkin, Y. Berkenstadt, Y. Giatt and G. M. Berlyne, An automated technique for the analysis of plasma guanidino acids and some findings in chronic renal disease, Clin. Chim. Acta, 60:45-50 (1975).

GUANIDINOACETIC ACID (GAA) DEFICIENCY AND SUPPLEMENTATION

IN RATS WITH CHRONIC RENAL FAILURE (CRF)

Y. Tsubakihara*, N. Iida*, S. Yuasa*, T. Kawashima*,
I. Nakanishi*, M. Tomobuchi*, T. Yokogawa*,
A. Ando**, Y. Orita**, H. Abe**, T. Kikuchi***, and
H. Okamoto***

*   Kidney Disease Center, Osaka Prefectural Hospital
    3-1-56 Bandaihigashi, Sumiyoshi-ku, Osaka, Japan
** The 1st Dept of Medicine, Osaka University
    Medical School
*** Morishita Pharm. Co., Ltd

INTRODUCTION

Guanidinoacetic acid (GAA) is the precursor of creatine which is essential in the energy metabolism of muscle and nerve tissue[1]. GAA is demonstrated to be produced mainly by the kidney[2]. Therefore, a decrease in GAA-producing tissue due to diseased kidney, could lead to a deficiency in creatine, the major matabolite of GAA.

In this study, we try to demonstrate such a deficiency in experimental CRF rats and correct it by supplementation with GAA.

MATERIALS AND METHODS

CRF was produced in 45 male Wistar rats weighing between 175 g and 180 g by the method of Platt et al.[3]. Approximately 2/3 - 3/4 of the left kidney was removed and 2 weeks later a contralateral nephrectomy was performed. After 2 weeks these CRF rats were divided into 3 groups having similar serum Cr concentrations. Ten sham-operated rats were used as controls.

Both normal and CRF rats were given standard chow (creatine free) ad libitum. One group with CRF received GAA 10 mg/day added to their diet and another 100 mg/day. Four weeks later muscle

373

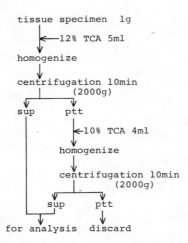

Figure 1.   Procedure for the extraction of GAA from tissues.

power was studied by measuring the sliding angle of the inclined
screen test, and physical strength was evaluated by measuring how
long they could survive in water (forced swimming method) .   After
these studies, the rats were sacrified and immediately the serum,
brain, lung, heart, liver, spleen, intestine, kidney, testis and
gastrocunemius muscle were collected and frozen.

     GAA, total creatine and creatinine (Cr) were extracted from
those organs using the method as shown in Fig. 1.   GAA was determined by an autoguanidino analyser JASCO G-520 (Japan Spectroscopic Co., Tokyo Japan).   Total creatine and Cr were measured by
the Folin-Wu method[5].

RESULTS

     Table 1a and 1b shows the GAA concentration of the serum and
organs and the urinary GAA excretion in the normal and CRF rats
without GAA administration.   The concentration in the serum and
organs (except the pancreas, brain and testis) in the CRF rats was
significantly lower than that seen in normal rats and the urinary
GAA excretion in CRF was also significantly reduced.   The GAA
concentration of the pancreas, which is the other organ capable of
producing GAA, is expected to be increased in the CRF rats in
compensation for the decrease in GAA production in the kidney.
However, there is no difference in pancreatic levels between the
normal and the CRF rats.

     The total creatine concentration in serum and organs and the
urinary excretion in normal and CRF rats with or without GAA ad-

Table 1a. Guanidinoacetic acid concentration in the serum, urine and organs of normal and CRF rats

	serum μg/dl	urine μg/day	kidney	pancreas	liver	testis
				μg/100g wet weight		
Normal (N=6)	78.1 ±8.9	666 ±98	4134 ±367	1995 ±288	123 ±8	4668 ±449
CRF (N=6) vs normal	5.8 ±2.1 **	19 ±5 **	553 ±128 **	1657 ±335 NS	41 ±5 **	5923 ±479 NS

Values are expressed as mean ± S.E..
NS; not significant, **; $p < 0.001$,

Table 1b. Guanidinoacetic acid concentration in the serum, urine and organs of normal and CRF rats

	heart	brain	lung	spleen	intestin	muscle
			μg/100g wet weight			
Normal (N=6)	159 ±15	195 ±24	240 ±19	280 ±19	167 ±9	186 ±14
CRF (N=6) vs normal	85 ±17 *	170 ±20 NS	122 ±42 *	166 ±15 **	72 ±18 **	105 ±16 **

Values are expressed as mean ± S.E..
NS; not significant, *; $p < 0.01$, **; $p < 0.001$,

ministration are shown in Table 2a and 2b. The total creatine content in the organs (except spleen) and the urinary creatine excretion in the CRF rats were significantly lower than those seen in the normal rats, but these factors increased significantly following GAA supplementation. There was no significant difference in total creatine content of the spleen among these four groups. There was no difference in the serum creatine concentration of the CRF and the normal rats, and it increased markedly in all animals fllowing the administration of 100 mg/day of GAA.

Table 2a. Total creatine concentration in the serum, urine and
           organs of normal and CRF rats with or without GAA
           administration.

	serum mg/dl	urine mg/day	muscle	heart	liver	brain
				mg/100g wet weight		
Normal	1.9	6.5	540.3	192.5	20.0	123.0
(N=6)	±0.2	±0.8	±22.9	±12.7	±1.6	±4.5
CRF(GAA 0)	1.5	3.8	359.3	124.8	13.8	108.3
(N=6)	±0.2	±0.3	±17.0	±10.3	±0.4	±2.6
vs normal	NS	*	***	***	***	*
CRF(GAA 10)	1.6	7.1	450.5	152.3	17.8	115.5
(N=6)	±0.3	±0.6	±15.5	±8.7	±1.4	±3.7
vs normal	NS	NS	**	*	NS	NS
vs CRF(GAA 0)	NS	**	***	NS	*	NS
CRF(GAA 100)	6.0	29.0	520.5	219.3	86.6	121.7
(N=6)	±1.0	±5.1	±16.6	±14.4	±14.3	±3.5
vs normal	***	***	NS	NS	***	NS
vs CRF(GAA 0)	***	***	***	***	***	***

GAA 0 ; without GAA administration,
GAA 10 ; with GAA administration of 10 mg/day,
GAA 100; with GAA administration of 100 mg/day,
NS; not significant, *; $p < 0.05$, **; $p < 0.01$, ***; $p < 0.001$,
(mean ± S.E.)

We found that muscle power and physical strength in the CRF
rats were significantly lower than those of the normal rats
(Table 3). They were significantly improved by GAA administration.
However, there was no significant difference between the two
groups of CRF rats.

DISCUSSION

Creatine (or creatine-phosphate) is an essential in energy
metabolism of muscle and nerve tissues[1]. GAA is the precursor of
creatine and produced by the kidney and pancreas[6-10]. Goldman et
al. demonstrated that nephrectomized rats converted only 6 - 20 %
as much ^{14}C-labelled glycine to creatine as unoperated controls[2].

Therefore, GAA deficiency is expected in kidney disease, es-

Table 2b. Total creatinine concentration in the serum, urine and organs of normal and CRF rats with or without GAA administration

	kidney	intestin	testis	lung	spleen	pancreas
			mg/100g wet weight			
Normal	16.2	46.6	294.3	17.3	21.3	20.7
(N=6)	±1.2	±4.0	±4.4	±1.4	±1.7	±1.6
CRF(GAA 0)	10.2	27.5	224.5	12.1	22.5	11.1
(N=6)	±1.2	±4.3	±11.9	±5.4	±2.5	±1.5
vs normal	**	**	***	**	NS	***
CRF(GAA 10)	15.9	39.9	285.0	16.5	24.8	18.2
(N=6)	±2.3	±2.8	±12.1	±1.1	±2.1	±3.5
vs normal	NS	NS	NS	NS	NS	NS
vs CRF(GAA 0)	NS	NS	***	***	NS	NS
CRF(GAA 100)	39.1	57.0	493.6	31.5	17.2	64.2
(N=6)	±6.8	±4.7	±19.8	±5.0	±3.9	±15.3
vs normal	**	NS	***	NS	NS	NS
vs CRF(GAA 0)	***	***	***	***	NS	**

GAA 0   ; without GAA administration,
GAA 10 ; with GAA administration of 10 mg/day,
GAA 100; with GAA administration of 100 mg/day,
NS; not significant, *; $p < 0.05$, **; $p < 0.01$, ***; $p < 0.001$,
Values are expressed as mean ± S.E..

pecially in chronic renal failure. To confirm this, we determined GAA and total creatine concentration in serum, urine and the major organs of normal and experimental CRF rats. We found a deficiency of GAA and creatine in the CRF rats (Table 1a, 1b, 2a and 2b). Some uremic manifestations may be caused by such a deficiency through disruption of energy metabolism in muscles, nerves and other organs.

Therefore, we studied the effect of GAA administration on muscle power measured by the inclined screen test and physical strength evaluated by the forced swimming method in CRF rats as a reflection of energy metabolism.

We found these functions to be conspicuously depressed in the CRF rats, and significantly improved with the increase in total creatine content in major organs which could be induced by GAA administration (Table 3).

Table 3. Muscle power and physical strength
in normal and CRF rats with or
without GAA administration

	muscle power (sliding angle) (°)	physical strength (swimming time) (min)
Normal	55.2 ± 0.9	52.0 ± 3.8
CRF(GAA 0) vs normal	42.3 ± 1.9 ***	21.9 ± 1.2 ***
CRF(GAA 10) vs normal vs CRF(GAA 0)	47.4 ± 1.3 *** *	25.5 ± 1.1 *** *
CRF(GAA 100) vs normal vs CRF(GAA 0)	48.8 ± 2.1 * *	29.0 ± 3.3 *** *

GAA 0   ; without GAA administration,
GAA 10 ; with GAA administration of 10 mg/day,
GAA 100; with GAA administration of 100 mg/day,
*; p <0.05, ***; p <0.001, (mean ± S.E.)

The GAA content in all the organs studied was significantly
higher than the serum concentration in the normal and CRF rats
(Table 1a, 1b). The GAA concentration of the testis was as high as
that of the kidney which is the organ producing GAA. These results
suggest that GAA may have some physiological role in these organs.

CONCLUSION

In this study, we demonstrate a deficiency of GAA and creatine in experimental chronic renal failure (CRF) in rats produced
by Platt's method. GAA (10 or 100 mg/day) was administered to
those rats p.o. for 8 weeks with standard chow.

GAA, total creatine and creatinine were determined in the
serum, urine, brain, heart, lung, liver, spleen, pancreas, kidney,
intestine, testis and gastrocunemius muscle of normal and CRF rats
with or without GAA administration. The result show that GAA and

creatine are deficient in organs and serum in CRF. Total creatine concentrations in these organs increased to normal following GAA administration.

Muscle power was measured by the inclined screen test and physical strength was evaluated by measuring how long they could survive in water. Both muscle power and physical strength were demonstrably low in the CRF rats compared to normals. We also show that these factors significantly recover fllowing GAA administration. These results suggest that there is a symptomatic deficiency of creatine due to a limited production of GAA in the kidney, and that the decrease in creatine content of major organs is a cause of uremic symptoms.

REFERENCES

1. F. J. R. Hird, S. P. Davuluri and R. M. McLean, Evolutionary relationships between arginine and creatine in muscle, in: "Urea Cycle Diseases," A. Lowenthal, A. Mori and B. Marescau, eds., Plenum Press, New York, 401 (1982).

2. R. Goldman and J. X. Moss, Synthesis of creatine in nephrectomized rats, Am. J. Physiol., 197:865 (1959).

3. R. Platt, M. H. Roscoe and F. W. Smith, Experimental renal failure, Clin. Sci., 11:217 (1952).

4. T. Tamura, S. Tsutsumi and K. Kizu, Study of the glucuronic acid matabolism, Jap. J. Pharm., 59:78 (1963).

5. R. W. Bonsnes and H. H. Taussky, On the colorimetric determination of creatinine by the Jaffe reaction, J. Biol. Chem., 158:581 (1945).

6. H. Borsook and J. W. Dubnoff, The formation of glycocyamine in animal tissues, J. Biol. Chem., 138:389 (1941).

7. J. B. Walker, Role of pancreas in biosynthesis of creatine, Proc. Soc. Exp. Biol. and Med., 98:7 (1958).

8. J. B. Walker and M. S. Walker, Formation of creatine from guanidinoacetate in pancreas, Proc. Soc. Exp. Biol. and Med., 101:807 (1959).

9. J. F. Van Pilsum, B. Olsen, D. Taylor, T. Rozycki, and J. C. Pierce, Transamidinase activities, in vitro, of tissues from various mammals and from rats fed protein-free, creatine-supplemented and normal diets, Arch. Biochem. Biophys., 100:520 (1963).

10. J. F. Van Pilsum, G. C. Stephens, and D. Taylor, Distribution of creatine, guanidinoacetate and the enzymes for their biosynthesis in the animal kingdom, Biochem. J., 126:325 (1972).

V: NATURAL GUANIDINO COMPOUNDS: A REVIEW

NATURAL GUANIDINO COMPOUNDS

Y. Robin and B. Marescau☆

Laboratoire du Métabolisme Minéral des Mammifères
(E.P.H.E.), Département de Physiologie, Faculté de
Pharmacie, 92290, Chatenay-Malabry, France, and ☆Labora-
tory of Neurochemistry, Born-Bunge Foundation, U.I.A.
B-2610 Wilrijk, Belgium

I. LINEAR GUANIDINO COMPOUNDS AND THEIR ANHYDRIZED FORMS
  A. Structure. Distribution. Biological Properties.
    1. Basic Mono- and Diguanidino Compounds
    2. ω-Guanidino-α-Amino Acids
    3. Octopine and the Guanidino "Opines"
    4. ω-Guanidino Acids
    5. Guanidinoethylphosphate Derivatives
    6. Phosphagens
    7. Guanidino Amides
  B. Physiological and Pharmacological Properties.
    1. Convulsivant Activity
    2. Hypoglycaemic Activity
    3. Antihypertensive Activity
    4. Antibacterial and Antitumor Activities

II. HETEROCYCLIC GUANIDINO COMPOUNDS
    1. Imidazole Derivatives
    2. Bromopyrrole Derivatives
    3. Pyrimidine Derivative
    4. Purine Derivatives
    5. Pteridine Derivatives
    6. Tetracyclopentazulene Pigments
    7. Cipridina Luciferin
    8. Guanidino Toxins
    9. Guanidino Antibiotics from Streptomyces

Guanidine was given this name because it was first isolated from the oxidation products of guanine (Strecker, 1861). This substance and many of the mono-, di- (symmetrical and asymmetrical) and tri-substituted derivatives constitute a large and important family of natural nitrogeneous compounds.

Interest in these substances arose from the discovery of creatine in meat extracts (Chevreul, 1835) and arginine in all animal and vegetal tissues (Schulze and Steiger, 1887; Hedin, 1895). The importance of their biological role was established later, when arginine was shown to participate in ureogenesis (Krebs and Henseleit, 1932) and creatine in muscular contraction (Eggleton and Eggleton, 1927). About the same time, more complex compounds were also noticed as a result of their high toxicity (e.g. the marine poison tetrodotoxin, Tawara, 1909) or their antibiotic activity (e.g. streptomycin, Schatz et al., 1944).

At the present time, more than one hundred naturally occurring guanidino compounds have been identified. Their common feature is the presence of a strongly basic guanidinium radical. The basicity of this radical has been explained by its specific thermodynamic properties (Cohn and Edsall, 1943; Pauling, 1960). The guanidino compounds differ considerably in their structure, distribution, metabolic functions and biological properties. Several well-documented reviews have appeared on these substances (Guggenheim, 1951: Thoai and Roche, 1960; Reinbothe and Mothes, 1962; Bell, 1964; Thoai, 1965; Thoai and Robin. 1969; Needham, 1970; Baker and Murphy, 1976; Mori, 1980, 1983; Chevolot, 1981).

The aim of the present review is to provide a general survey of the current knowledge of the structure, distribution and biological activities of natural guanidino compounds from various sources (vertebrates, invertebrates, plants and microorganisms).

I. LINEAR GUANIDINO COMPOUNDS AND THEIR ANHYDRIZED FORMS

Although these compounds are widely distributed in nature, many of them are restricted to a few groups of organisms and sometimes to a single species.

It is of interest to note that all biological guanidino compounds found in man belong to this class of substances. Therefore, they have been extensively studied with respect to their involvement in: (i) normal metabolic processes; (ii) pathological pathways; and (iii) physiological and pharmacological activities.

The first part of this section will be concerned with the structure, distribution and biological properties of linear guanidino compounds. The second part will deal with some aspects of their physiological and pharmacological properties.

A. STRUCTURE. DISTRIBUTION. BIOLOGICAL PROPERTIES

1. BASIC MONO- AND DIGUANIDINO COMPOUNDS

1.1. Guanidine and its derivatives

Guanidine (1) was first isolated from plant seeds (Schulze, 1891), and later found in several animal and vegetal tissues and in bacterial fermentation products (Guggenheim, 1951; Ackermann and Menssen, 1959; Baker and Murphy, 1976; Robin and Guillou, 1980). Guanidine is normally present at low concentration in human blood and urine (Van Pilsum et al., 1956). Higher levels have been reported in blood in cases of hypertension (Major and Weber, 1927; de Wesselow and Griffiths, 1932; Andes et al., 1937), essential epilepsy (Murray and Hoffmann, 1940) and migraine (Palmer et al., 1943) and in the urine of patients with chronic renal failure (Stein and Micklus, 1973). Monomethylguanidine (2) has been found in meat extracts, in muscle autolyzates, and in various tissues and biological fluids (Guggenheim, 1951). Its urinary excretion is increased in pathological conditions such as parathyroidectomy (Koch, 1913), newborn spasmophily (Burns and Sharpe, 1917) and adult tetany (Findlay and Sharpe, 1920). Elevated levels of methylguanidine and guanidine were also found in uremic sera (Carr and Schloerb, 1960; Giovannetti et al., 1973), and in the brain of mice with an epileptic disposition (Hiramatsu, 1980a). N,N-dimethylguanidine (3) is present in several animal tissues and in normal and pathological urine (Guggenheim, 1951).

$$NH_2 - C - NH_2 \qquad NH_2 - C - NH - CH_3 \qquad NH_2 - C - N - CH_3$$
$$\quad \; \| \qquad\qquad\qquad\quad\; \| \qquad\qquad\qquad\quad\; \| \; \; \; |$$
$$\quad NH \qquad\qquad\qquad\quad NH \qquad\qquad\qquad NH \; \; CH_3$$
$$\quad \textbf{1} \qquad\qquad\qquad\qquad\quad \textbf{2} \qquad\qquad\qquad\qquad \textbf{3}$$

Not much is known about the formation of guanidine and of its methyl derivatives in mammals. Reiter and Horner (1979) have recently reported the production of guanidine from L-arginine and L-canavanine in the rat. Creatinine was shown to be a precursor of monomethylguanidine in the rat (Perez et al., 1976). N,N-dimethylguanidine could be formed from creatine by decarboxylation.

Guanidine and its methyl derivatives are convulsive agents. Their toxicity increases with the degree of methylation. Similarities between the symptoms of guanidine intoxication and uremia have been reported (Stein et al., 1969). Normal dogs severely intoxicated with methylguanidine (Giovannetti et al., 1973) showed symptoms resembling those of uremia, thus confirming that this compound is an important uremic toxin. High levels of this compound in brain of mice with an epileptic disposition suggest that methylguanidine plays an important role in the seizure mechanism (Hiramatsu, 1980a). Guanidine and its derivatives display hypotensive and hypoglycaemic activities, but only at toxic doses (Guggenheim, 1951). Guanidine hydrochloride has been used in the treatment of infantile and juvenile atrophy (Angelini et al., 1980).

1.2. Agmatine and its derivatives

Agmatine (4) was first identified in hydrolyzates of herring sperm (Kössel, 1910) and then found in various animal and vegetal tissues (Guggenheim, 1951; Thoai and Roche, 1960; Smith and Richards, 1962; Baker and Murphy, 1976). Agmatine can be produced by bacteria, either by decarboxylation of arginine (Gale, 1946) or by hydrolysis of arcaine (Linneweh, 1931a). Production of agmatine by arginine decarboxylation has also been observed in marine invertebrates (Thoai, et al., 1953a). The same mechanism may be responsible for the formation of agmatine in higher plants, which was shown to be particularly active in potassium-deficient specimens (Smith and Richards, 1962). Agmatine can also be formed by transamidination of putrescine from arginine. This is the case in the leech, where agmatine constitutes an intermediary step in the biosynthesis of arcaine (Robin et al., 1967). Agmatine has weak hypotensive and hypoglycaemic activities (Kumagai et al., 1928). N-methylagmatine (5) has been reported in a medusa (Haurowitz and Waelsch, 1926) and in Octopus muscle (Iseki, 1931). However, its natural occurrence has been questioned (Baldwin, 1947). N-acetylagmatine (6) has been isolated from saké (Ohtaka and Uchida, 1959) and from sea-anemones of the genus Actinia (Guillou and Robin, 1979). Hydroxyagmatine (7), has been identified in a sea-anemone, Anthopleura japonica V. (Makisumi, 1961). Here, this compound could result from the decarboxylation of γ-hydroxyarginine. Agmatine lowers blood glucose after an initial rise (Kumagai et al., 1928).

$$NH_2 - C - NH - (CH_2)_4 - NH_2 \qquad\qquad NH_2 - C - NH - (CH_2)_4 - NH - CH_3$$
$$\quad\ \parallel \qquad\qquad\qquad\qquad\qquad\qquad\qquad\qquad\ \parallel$$
$$\quad\ NH \qquad\qquad\qquad\qquad\qquad\qquad\qquad\qquad NH$$

4   **5**

$$NH_2 - C - NH - (CH_2)_4 - NH - C - CH_3$$
$$\quad\ \parallel \qquad\qquad\qquad\qquad\qquad\quad \parallel$$
$$\quad\ NH \qquad\qquad\qquad\qquad\qquad\quad O$$

6

$$NH_2 - C - NH - CH_2 - CH - CH_2 - CH_2 - NH_2$$
$$\quad\ \parallel \qquad\qquad\qquad\quad | $$
$$\quad\ NH \qquad\qquad\qquad\ OH$$

7

1.3. Diguanidino compounds

The first diguanidino compound reported in nature, vitiatin (8),

$$NH_2 - C - NH - (CH_2)_2 - N - C - NH_2$$
$$\quad\ \parallel \qquad\qquad\qquad\quad | \quad \parallel$$
$$\quad\ NH \qquad\qquad\qquad CH_3 \ NH$$

8

was isolated from mammalian muscle and urine (Kutscher, 1907; Engeland, 1908). However, the proposed structure has not been unequivocally confirmed by synthesis (Schotte and Prieve, 1926). It is possible that this substance is an artefact.

Three other diguanidino compounds were found later in invertebrates, exclusively in molluscs and in worms. **Arcaine (9)** (diamidinoputrescine) was isolated from the marine mollusc Arca noae (Kutscher and Ackermann, 1931) and later found in another mollusc (Suzuki and Muraoka, 1954) and in annelids (Robin et al., 1957a; Roche et al., 1965). The homologous compound, **audouine (10)** (diamidinocadaverine) has been identified only in the marine annelid Audouinia tentaculata (Roche et al., 1965). **Hirudonine (11)** (diamidinospermidine) has been isolated from the leech, Hirudo medicinalis (Robin et al., 1957b; Robin and Thoai, 1961a) and further characterized in several terrestrial annelids (Robin and Roche, 1965). The biosynthesis of the three compounds in annelids has been established by in vivo isotopic studies. It was shown that they result from the successive transfer of two amidine groups of arginine to the corresponding diamines, with the intermediary formation of a monoamidino derivative (Robin and Audit, 1966; Audit et al., 1967; Robin et al., 1967).

$$NH_2 - C - NH - (CH_2)_4 - NH - C - NH_2$$
$$\parallel \qquad\qquad\qquad\qquad \parallel$$
$$NH \qquad\qquad\qquad\qquad\quad NH$$

9

$$NH_2 - C - NH - (CH_2)_5 - NH - C - NH_2$$
$$\parallel \qquad\qquad\qquad\qquad \parallel$$
$$NH \qquad\qquad\qquad\qquad\quad NH$$

10

$$NH_2 - C - NH - (CH_2)_3 - NH - (CH_2)_4 - NH - C - NH_2$$
$$\parallel \qquad\qquad\qquad\qquad\qquad\qquad \parallel$$
$$NH \qquad\qquad\qquad\qquad\qquad\qquad\quad NH$$

11

The biological role of the diguanidino compounds is unclear. They do not participate in the formation of a new phosphagen, even in the leech in which hirudonine is the principal guanidino compound present in the muscle, but in which, exceptionally, no phosphagen could be detected (Thoai and Robin, 1969). Arcaine, audouine and hirudonine were shown to promote in vitro the polymerization of actin as efficiently as the corresponding diamines (Oriol-Audit, 1978); they could therefore play a part in the control of growth processes. Their natural distribution, however, is very restricted.

The effect of arcaine on blood sugar is a decrease after an initial rise. The toxicity is about the same as that of guanidine (Linneweh, 1931b).

1.4. Guanidino alkaloids

Galegin (12) was isolated from the seeds of Galega officinalis
by Tanret (1914). Its structure was established later (Barger and
White, 1922; Späth and Procopp, 1924). More recently, another
guanidine has been isolated from the same material and identified
as 4-hydroxygalegin (13) (Pufahl and Schreiber, 1961; Thoai and
Desvages, 1963).

$$NH_2 - C - NH - CH_2 - CH = C - CH_3$$
$$\| \qquad\qquad\qquad\quad |$$
$$NH \qquad\qquad\qquad CH_3$$

12

$$NH_2 - C - NH - CH_2 - CH = C - CH_2 - OH$$
$$\| \qquad\qquad\qquad\quad |$$
$$NH \qquad\qquad\qquad CH_3$$

13

In vivo isotopic studies with labeled precursors have shown that
the amidine group of galegin is derived from arginine and that this
substance is not a nitrogen storage product in Galega (Reuter, 1964).
The biological role of galegin and hydroxygalegin is unknown.
Their levels in plant organs increase instead of decreasing. Since
they are formed from arginine, Thoai and Desvages (1963) have suggested that the intensive utilization of the amidine group of arginine
for their synthesis shoud save a part of this amino acid from proto-
genesis.

At toxic doses, galegin displays a strong hypotensive activity
(Kumagai et al., 1928). This substance produces paralysis of brain
and nervous centres in warm-blooded and cold-blooded animals.

Pterogyne (14) and pterogynidine (15) are two isomeric guanidine

$$NH_2 - C - N - CH_2 - CH = C - CH_3$$
$$\| \quad | \qquad\qquad\qquad |$$
$$NH \quad CH_2 \qquad\qquad CH_3$$
$$|$$
$$CH$$
$$\|$$
$$C - CH_3$$
$$|$$
$$CH_3$$

14

$$CH_3 - C = CH - CH_2 - NH - C - NH - CH_2 - CH = C - CH_3$$
$$| \qquad\qquad\qquad\qquad \| \qquad\qquad\qquad\qquad\quad |$$
$$CH_3 \qquad\qquad\qquad\quad NH \qquad\qquad\qquad CH_3$$

15

alkaloids isolated from the trunk-bark of Pterogene nitans by Corral et al. (1970, 1972). Their structure has been established and confirmed by synthesis by the same authors.

2. ω-GUANIDINO-α-AMINO ACIDS

2.1. Arginine and its derivatives

Arginine (16), first isolated from Cucurbita pepo and Lupinus luteus sprout extracts (Schulze and Steiger, 1887), was found to be a universal constituent of animal and plant proteins (Hedin, 1895). Free arginine is present in all living organisms (review by Guggenheim, 1951), but it is particularly abundant in plant vegetative organs and seeds, and in muscles of most invertebrates, who utilize phosphoarginine as phosphagen. Pathologically high blood levels of arginine have been found in hyperargininemia, a congenital disease of the urea cycle with arginase deficiency (Terheggen et al., 1972, 1975).

Various aspects of the biosynthesis of arginine and its biological role will not be emphasized here (for details see reviews by Thoai and Roche, 1960; Thoai, 1965; Meister, 1965; Reinbothe and Mothes, 1962; Marescau, 1981; Robin, 1982). Briefly, arginine is formed in ureotelic vertebrates via the urea cycle (Krebs and Henseleit, 1932). There is now evidence that all or most reactions of this cycle also exist in microorganisms, invertebrates and plants. Many organisms which do not have a functional urea cycle obtain arginine in their diet. The pathways leading to the formation of the ornithine chain and to the building of the amidine moiety (the most extensively studied part of the molecule) are highly diversified, specially in microorganisms. Arginine serves a primordial role in living organisms. In particular, it participates in: (i) nitrogen fixation, transfer, storage and detoxication; (ii) the energetic metabolism of muscle in most invertebrates; and (iii) the formation of most of the other guanidino compounds, either by an alteration of the ornithine chain, or by a transfer of the amidine group to appropriate amines.

Several hydroxylated, acetylated and methylated derivatives of arginine have been found in nature, but are rather uncommon.

$$NH_2 - \underset{\underset{NH}{\parallel}}{C} - NH - (CH_2)_3 - \underset{\underset{NH_2}{\mid}}{CH} - COOH$$

16

$$NH_2 - \underset{\underset{NH}{\parallel}}{C} - NH - CH_2 - \underset{\underset{OH}{\mid}}{CH} - CH_2 - \underset{\underset{NH_2}{\mid}}{CH} - COOH$$

17

γ-Hydroxyarginine (17) was discovered in two marine invertebrates,

the sea cucumber Polycheira rufescens (Fujita, 1959, 1960) and the
sea anemone Anthopleura japonica (Makisumi, 1961). This substance
was characterized in large amounts in the seeds of various Papilionoideae of the genus Vicia and isolated as its lactone from the seeds
of V. sativa (Bell and Tirimanna, 1963, 1964). The presence of γ-
hydroxyarginine was also reported in brain (Hosotani, 1973). The accumulation of γ-hydroxyarginine in Vicia seeds probably constitutes
a reserve of nitrogen for the developing plant. γ-Hydroxyarginine
lactone (18) is readily formed from γ-hydroxyarginine in acid solution
and reverts to a free acid in alkaline solution. The lactone can be
detected in concentrated extracts from the seeds of γ-hydroxyarginine-
containing Vicia, but lactonization may occur during or after extraction (Bell, 1965). α-N-acetylarginine (19) has been isolated from
cattle brain (Ohkusu and Mori, 1969) and found in normal human urine
(Terheggen et al., 1972) and serum (Matsumoto et al., 1976a). In
hyperargininemia, elevated values occur in urine (Terheggen et al.,
1972) and serum (Marescau et al., 1982). α-N-acetyl-γ-hydroxy-L-arginine (20) has been isolated and identified from human placenta (Mori
et al., 1969). δ-N-hydroxyarginine (21), studied by Fisher et al.,
(1973), was isolated from culture filtrates of Ninnizzia gypsea, a
dermatophyte belonging to the class Ascomycetes and its structure
has been elucidated mainly by spectral methods. This substance has
antibiotic activity against several bacteria and fungi.

$$NH_2 - C - NH - CH_2 - \overset{\displaystyle O}{\underset{NH_2}{\diagdown}} = O$$
$$\underset{NH}{\overset{\|}{}}$$

18

$$NH_2 - C - NH - (CH_2)_3 - CH - COOH$$
$$\underset{NH}{\overset{\|}{}} \qquad\qquad \underset{\substack{NH \\ | \\ C = O \\ | \\ CH_3}}{|}$$

19

$$NH_2 - C - NH - CH_2 - CH - CH_2 - CH - COOH$$
$$\underset{NH}{\overset{\|}{}} \qquad\quad \underset{OH}{|} \qquad\quad \underset{\substack{NH \\ | \\ C = O \\ | \\ CH_3}}{|}$$

20

$$NH_2 - C - N - (CH_2)_3 - CH - COOH$$
$$\underset{NH}{\overset{\|}{}}\ \underset{OH}{|} \qquad\qquad \underset{NH_2}{|}$$

21

N^G,N^G-dimethylarginine (22) and N^G,N'^G-dimethylarginine (23) have been isolated and identified from human urine and characterized in human serum, where they are apparently released by hydrolysis of methylated proteins (Kakimoto and Akazawa, 1970). Methylation of the arginine residues of these proteins was shown to occur subsequent to the synthesis of peptide bonds (Paik and Kim, 1975). The presence of roughly equal concentrations of N^G,N^G- and N^G,N'^G-dimethylarginines in urine has been confirmed, but mono- and trimethylated arginines could not be detected under the same conditions (Patthy et al., 1977).

$$CH_3 - N - C - NH - (CH_2)_3 - CH - COOH$$

with CH3 and NH below the N-C, and NH2 below the CH

22

$$CH_3 - NH - C - NH - (CH_2)_3 - CH - COOH$$

with N—CH3 below the C, and NH2 below the CH

23

$$HOOC - CH_2 - CH - NH - C - NH - (CH_2)_3 - CH - COOH$$

with COOH below the CH, NH below the C, and NH2 below the last CH

24

Argininosuccinic acid (24), an intermediate in the biosynthesis of arginine via the urea cycle , is a metabolite which normally does not accumulate in tissues and body fluids of ureotelic vertebrates (Ratner et al., 1953). However, high levels of this substance have been found in human urine in argininosuccinic aciduria, a hereditary disorder of the urea cycle caused by the deficiency or absence of argininosuccinase (Allan et al., 1958; Cusworth and Dent, 1960; Formstecher, 1978). Increases of argininosuccinic acid levels have also been observed in serum and cerebrospinal fluid of patients with this disease (Sakiyama et al., 1982). Argininosuccinic acid has been found in vegetal seeds, in which its level increases during germination (Davison and Eliott, 1952; Barber and Boulter, 1963).

The biosynthesis of argininosuccinic from arginine and fumarate has been demonstrated in various organisms (Davison and Eliott, 1952; Walker, 1952; Walker and Myers, 1953; Barber and Boulter, 1963).

The accumulation of argininosuccinic acid in body fluids and tissues may be responsible for the neurological symptoms of patients with argininosuccinic aciduria.

2.2. Homoarginine and its derivatives

Homoarginine (25) and γ-hydroxyhomoarginine (26) were originally detected in significant concentration in the seeds of many Lathyrus

species (Bell, 1962a,b; 1963). Homoarginine was isolated from the
seeds of L. cicera (Bell, 1962b) and L. sativus (Rao et al., 1963).
The hydroxy compound was obtained from the seeds of L. tingitanus
(Bell, 1964). The comparative distribution of guanidino compounds
in the seeds of two genus of the Papilionoideae, Lathyrus and Vicia,
exhibits interesting features (Bell and Tirimanna, 1964). For example,
the genus Lathyrus stores predominantly the C_7 amino acids homoarginine, γ-hydroxyhomoarginine and the related compound lathyrine, while
the genus Vicia stores relatively high concentrations of canavanine
and of the C_6 amino acids arginine and γ-hydroxyarginine. Apart from
plants, homoarginine was found in small amounts in human body fluids,
urine (Armstrong, 1967), serum (Matsumoto et al., 1976a) and cerebrospinal fluid (Mori et al., 1981b) and in mammalian brain (Mori et al.,
1978, 1979). Increased urinary excretion was reported in hyperlysinemia (Armstrong and Robinow, 1967) and citrullinemia (Scott-Emuakpor
et al., 1972).

Homoarginine may be formed in mammals from lysine, by a homologous urea cycle in which arginine is replaced by lysine (Ryan and
Wells, 1964; Scott-Emuakpor et al., 1972). The excretion of homocitrulline and homoarginine in urine following administration of
lysine to children supports this theory (Buergi et al., 1966). The
biosynthesis of homoarginine has not been examined in plants, but
γ-hydroxyhomoarginine is synthesized from homoarginine in Lathyrus
seeds (Bell, 1964). Subsequently both compounds have been identified
as the precursors of the cyclic guanidino derivative, lathyrine, also
present in Lathyrus seeds (Bell and Przybylska, 1965). Homoarginine
and γ-hydroxyhomoarginine probably serve as nitrogen reserves in
Lathyrus seeds.

Inhibition of growth by homoarginine has been reported in various
microorganisms (Walker, 1955; Rao et al., 1963).

$$NH_2 - C - NH - (CH_2)_4 - CH - COOH$$
$$\parallel \qquad\qquad\qquad\qquad \mid$$
$$NH \qquad\qquad\qquad\qquad NH_2$$

25

$$NH_2 - C - NH - (CH_2)_2 - CH - CH_2 - CH - COOH$$
$$\parallel \qquad\qquad\qquad \mid \qquad\qquad \mid$$
$$NH \qquad\qquad\qquad OH \qquad\quad NH_2$$

26

$$NH_2 - C - NH - (CH_2)_2 -$$
$$\parallel$$
$$NH$$

27

γ-Hydroxyhomoarginine lactone (27) isolated from the seeds of
L. tingitanus (Bell, 1964) probably results from the cyclization of
the open chain compound under acidic conditions.

2.3. Canavanine and its derivatives

Canavanine (28) was originally discovered in the seeds of Cana-valia ensiformis and C. obtusifolia (Kitagawa and Tomiyana, 1929). Subsequently, it was identified in the seeds of a great number of Leguminosae (Fearon and Bell, 1955; Bell, 1958, 1960). Its distribution in leguminous plants and some of its biological features have been reviewed (Reinbothe and Mothes, 1962; Bell, 1965). Canavanine is not restricted to seeds, but also occurs in vegetative organs, especially in roots, in which it is subject to an annual rhythm: in autumn and winter canavanine accumulates, while in spring this compound decreases. In seeds, canavanine disappears during germination. Thus it is probable that canavanine plays a role in the storage and transfer of nitrogen in the plant, and that it serves as a source of nitrogen and carbon for seed germination.

$$NH_2 - C - NH - O - (CH_2)_2 - CH - COOH$$

with $\|$ (NH) below the C, and $|$ (NH$_2$) below the CH

28

The biosynthesis of canavanine in plants is unclear. Possibly, reactions analogous to those involved in arginine biosynthesis occur, e.g. the intermediary formation of canavaninosuccinic acid (Walker, 1953). However, further experimental evidence is not available. An alternative hypothesis is the formation of canavanine by transamidination of canaline from arginine.

On the basis of their structural similarity, canavanine is a substrate for many of the enzymatic reactions that involve arginine (Kalyankar et al., 1958). Canavanine can interfer with arginine in protein synthesis by: (i) competitively inhibiting the incorporation of arginine into sRNA; and (ii) being incorporated into polypeptide chains in some of the positions normally occupied by arginine, which would possibly create proteins with altered properties (Allende and Allende, 1964 and cited references). The inhibition of growth by canavanine in plants (Reinbothe and Mothes, 1962) and microorganisms (Horowitz and Srb, 1948; Volcani and Snell, 1948: Walker, 1955; Schwartz and Maas, 1960; Nichols et al., 1980) may be explained by these metabolic changes.

Canavanine is an intermediary product in the canaline cycle, an alternative cycle proposed for urea detoxication in higher vertebrates (Natelson and Sherwin, 1979). However, canavanine has not been reported in animal tissues.

Canavaninosuccinic acid (29) has been obtained by in vitro

$$HOOC - CH_2 - CH - NH - C - NH - O - (CH_2)_2 - CH - COOH$$

with $|$ (COOH) below the first CH, $\|$ (NH) below the C, and $|$ (NH$_2$) below the last CH

29

incubation of canavanine with fumaric acid in the presence of a preparation of jack bean argininosuccinase (Walker, 1953). In vertebrates, this compound is supposed to be an intermediary metabolite
in the canaline cycle (Natelson and Sherwin, 1979) but the natural
occurrence of this substance has never been reported.

3. OCTOPINE AND THE GUANIDINO "OPINES"

All compounds of this group are imino acids resulting from the
condensation of an ω-guanidino-α-amino acid with an α-keto-acid,
followed by reduction.

Octopine (30) (derived from arginine and pyruvic acid) was first
isolated from Octopus muscle (Morizawa, 1927) and also found in a
large number of molluscs (Mayeda, 1936; Ackermann and Mohr, 1937;
Moore and Wilson, 1937; Irvin and Wilson, 1939; Regnouf and Thoai,
1970; Gäde and Zebe, 1973), in yeast cultures (Obata and Iimori,
1952), in sipunculid and nemertean worms (Thoai and Robin, 1969;
Robin, 1964a), in sea anemones (Zamit and Newsholme, 1976; Ellington,
1980), and in crown gall tumors induced by Agrobacterium tumefaciens
(Ménagé and Morel, 1964). Homooctopine (31) (derived from homoarginine and pyruvic acid) was isolated from crown gall tumor tissues
cultured on media containing homoarginine (Petit and Morel, 1966).
Canavanooctopine (32) (derived from canavanine and pyruvic acid)
was found in crown gall tumors induced by A. tumefaciens strain B6
on Canavalia ensiformis (Petit et al., 1968). Nopaline (33) (derived
from arginine and α-ketoglutaric acid) was identified in crown gall
tumors induced by A. tumefaciens strain T37 on Opuntia vulgaris
(Nopal in French) (Goldmann et al., 1969).

$NH_2 - C - NH - (CH_2)_3 - CH - COOH$ $\qquad$ $NH_2 - C - NH - (CH_2)_4 - CH - COOH$

$\qquad \| \qquad\qquad\qquad\quad |$ $\qquad\qquad\qquad\qquad\qquad \| \qquad\qquad\qquad\qquad |$

$\qquad NH \qquad\qquad\qquad\quad NH$ $\qquad\qquad\qquad\qquad\qquad NH \qquad\qquad\qquad\qquad NH$

$\qquad\qquad\qquad\qquad\qquad\quad |$ $\qquad\qquad\qquad\qquad\qquad\qquad\qquad\qquad\qquad\quad |$

$\qquad\qquad\qquad\qquad CH_3 - CH - COOH$ $\qquad\qquad\qquad\qquad\qquad\qquad CH_3 - CH - COOH$

30 $\qquad\qquad\qquad\qquad\qquad\qquad\qquad\qquad\qquad\qquad$ **31**

$NH_2 - C - NH - O - (CH_2)_2 - CH - COOH$

$\qquad \| \qquad\qquad\qquad\qquad\qquad |$

$\qquad NH \qquad\qquad\qquad\qquad\quad NH$

$\qquad\qquad\qquad\qquad\qquad\qquad\quad |$

$\qquad\qquad\qquad\qquad\qquad CH_3 - CH - COOH$

32

$NH_2 - C - NH - (CH_2)_3 - CH - COOH$

$\qquad \| \qquad\qquad\qquad\qquad |$

$\qquad NH \qquad\qquad\qquad\quad NH$

$\qquad\qquad\qquad\qquad\qquad\quad |$

$\qquad\qquad\qquad\qquad HOOC - CH - (CH_2)_2 - COOH$

33

The generic name "opines" refers to a class of substances with the type of structure described here and found in crown gall tumors (Guyon et al., 1980). Two well documented reviews on the opines have appeared recently (Tempé, 1982; Tempé and Goldmann, 1982). Only the guanidino opines are considered in the present review.

Octopine is biosynthesized in marine invertebrates by the reductive condensation of arginine and pyruvic acid catalyzed by a NAD^+-dependent dehydrogenase, octopine dehydrogenase and the reaction is reversible (Thoai and Robin, 1959a, 1961; Robin and Thoai, 1961b). The invertebrate enzyme has been purified and extensively studied (Monneuse-Doublet et al., 1980, and cited references). The same mode of formation has been found for the opines. The plant tumor dehydrogenases, however, utilize $NADP^+$ as coenzyme (Goldmann, 1977), and tumoral octopine dehydrogenase is immunologically different from the animal enzyme (Fort et al., 1982). Furthermore, their specificity for the ketonic acid substrate is controlled by the genome (Ti plasmid genes) of the bacterial strain which has caused the tumor, independent on the genome of the host plant (Goldmann et al., 1968; Petit et al., 1970). Thus, with respect to the guanidino opines, "octopine-strains" of A. tumefaciens induce the synthesis of octopine, homooctopine and canavanooctopine (formed from pyruvic acid) and "nopaline-strains" of A. tumefaciens induce that of nopaline (formed from α-ketoglutaric acid). The opine production trait is acquired during tumorous transformation. After transformation, cultures of tumor cells can synthesize opines, even in the absence of the Agrobacterium (Petit and Tempé, 1978, and cited references).

The biological role of octopine in invertebrates has been extensively studied. Octopine first was shown to be a part of an alternative anaerobic glycolytic pathway in a group of invertebrates whose lacticodehydrogenase activity is usually very low (Thoai and Robin, 1959a,b; Robin and Thoai, 1961b). Moreover, the reversibility of the reaction and the comparatively low levels of octopine in resting muscles suggested that octopine could serve as a sink for pyruvic acid and arginine formed during muscular work (Thoai and Robin, 1959b; Robin and Thoai, 1961b). Further studies of molluscs under various physiological conditions have confirmed the relationship between octopine formation, phosphagen depletion and NAD^+ regeneration and emphasized the biological role of octopine in muscular and non-muscular tissues, particularly in the maintenance of the redox potential (reviews by Baldwin and Opie, 1978; Gade, 1980).

In plant tumors, a striking correlation has been established between the opine production by the tumor and opine utilization by the inciting bacteria (Goldmann et al., 1968; Petit et al., 1970; Bomhoff et al., 1976). Both of these are controlled by the same type of Ti plasmid genes. The specific use of the induced opines by A. tumefaciens makes it probable that the opines serve as nutrient for the inciting bacteria. The opines may play an additional role as specific

inducers of the conjugational transfer of Ti plasmid genes to other
Agrobacteria (Guyon et al., 1980, and cited references). These substances are considered to be mediators of the host-parasite relationship (Tepfer and Tempé, 1981).

4. ω-GUANIDINO ACIDS

4.1. Carboxylic acids

Guanidinoacetic acid (34), commonly called glycocyamine, was
first isolated from mammalian urine (Weber, 1935) and subsequently
identified in most tissues and biological fluids of warm-blooded
animals (Guggenheim, 1951). In mammals organs, the higher levels are
found in the kidney (Dubnoff and Borsook, 1941; Mori et al., 1975).
Elevated glycocyamine levels have been observed in brain of mice
strains with an epileptic disposition (Hiramatsu, 1980a). In contrast,
lower values were found in brain of experimental uremic rabbits
(Matsumoto et al., 1976a). The urinary excretion of glycocyamine is
strongly increased in hyperargininemia (Terheggen et al., 1972).
Apart from mammals, glycocyamine has also been reported in various
marine invertebrates (Thoai and Robin, 1954a); this compound is
particularly abundant in the muscles of certain marine worms which
utilize phosphoglycocyamine as the muscular phosphagen (Thoai et al.,
1953b; Thoai and Robin, 1969).

Glycocyamine is biosynthesized in vertebrates by enzymatic
transfer of the amidine group of arginine to the amino group of glycine (Bloch and Schönheimer, 1941; Borsook and Dubnoff, 1941). This
transamidination reaction occurs in the kidney. The formation of
labeled glycocyamine from amidino-labeled arginine has been observed
in vivo in marine worms (Robin and Audit, 1966) but, curiously, the
reaction could not be demonstrated in vitro in many invertebrates
preparations (Robin and Audit, 1966; Van Pilsum et al., 1972).

The principal biological role of glycocyamine is to participate
in the biosynthesis of creatine in all vertebrates and, possibly, in
a limited number of invertebrates. In several marine polychaetes,
glycocyamine is a phosphagen precursor. Structural similarity with
γ-aminobutyric acid suggests that glycocyamine could be implicated
in some nervous processes in mammals.

$$NH_2 - C - NH - CH_2 - COOH$$
$$\overset{\|}{NH}$$

34

35

Glycocyamidine (35), the anhydrization product of glycocyamine,
has been observed as traces in normal and pathological human urine
(Kostir and Pristoupil, 1953).

Creatine (36) (N-methylglycocyamine) was isolated from meat extracts by Chevreul (1835) who considered it to be a derivative of urea. Its structure was elucidated several years later (review by Hunter, 1928). This compound was rapidly found as a normal constituent of all tissues and biological fluids of vertebrates (Guggenheim, 1951); it is particularly abundant in muscle, where it occurs in its free and phosphorylated forms (Eggleton and Eggleton, 1927; Fiske and Subbarow, 1929). In human urine, the concentration of creatine is normally low compared with that of creatinine. However, it increases in muscular dystrophy as a result of the inability of the muscle to retain creatine (Menne and Beckmann, 1955; Hurley and Williams, 1955). Creatine was further characterized in a number of protochordates and of invertebrates (Meyerhof, 1928; Greenwald, 1946; Roche et al., 1957; Thoai and Robin, 1969). In these animals, creatine is present either as a permanent constituent of muscle, or as a seasonal constituent, e.g. in mature spermotozoa, and is generally accompanied by its phosphoryl derivative.

$$NH_2 - C - N - CH_2 - COOH$$
$$\begin{array}{cc} \| & | \\ NH & CH_3 \end{array}$$

36

37

Creatine is biosynthesized in vertebrates by methylation of glycocyamine in the liver (Borsook and Dubnoff, 1940; Bloch and Schönheimer, 1941); the methylating agent is adenosylmethionine (Cantoni and Vignos, 1954). An alternative pathway for creatine formation has been suggested for vertebrates through the canaline cycle (Natelson and Sherwin, 1979). The ability of invertebrates to synthesize creatine is a controversial issue. Labelled creatine has been obtained by _in vivo_ administration of amidine-labelled arginine to worms; glycocyamine, its normal precursor, was also labelled (Robin and Audit, 1966). These findings suggest that the biosynthetic pathway could be the same in invertebrates as in vertebrates. On the other hand, the activity of the creatine-synthesizing enzymes could not be demonstrated _in vitro_, even in the invertebrates which contain creatine in their tissues (Robin and Audit, 1966; Van Pilsum et al., 1972). It has been suggested that creatine might be provided in their diet (Van Pilsum et al., 1972). However, the high seasonal raises observed in the genital tract of various invertebrates during the spawning period point to an endogenous origin.

The essential role of creatine is to participate in the formation of phosphocreatine, the muscular phosphagen for all vertebrates and a number of invertebrates.

Creatinine (37), the anhydrized form of creatine, was first identified in mammalian urine (Pettenkoffer, 1844) and later found in most tissues and biological fluids of vertebrates (Guggenheim,

in most tissues and biological fluids of vertebrates (Guggenheim, 1951). The concentration of creatinine in human blood and urine is normally very stable; N elimination as creatinine in urine represents about 5% that of urea. Increased levels of creatinine were found in serum in renal insufficiency (Adams et al., 1964). Levels below normal have been reported in muscular dystrophy (Menne and Beckmann, 1955; Hurley and Williams, 1955) and in schizophrenia (Subrahmanyan et al., 1962). The presence of creatinine has also been reported in the mollusc Abalone (Albrecht, 1921) and in several plants (Sullivan, 1911).

Creatinine is formed from creatine, probably by nonenzymatic anhydrization. This compound is valuable for glomerular filtration of the kidney. The relative constancy of its excretion in urine makes it a good index for determining the concentration of other urinary constituents.

β-Guanidinopropionic acid (38) has been found in rat liver (Rosenberg, 1959) and cattle brain (McLennan, 1959), and trace amounts were detected in normal and uremic human and rabbit brain and serum (Matsumoto et al., 1976a). Strong increases of this substance have been, however, reported in plasma of uremic patients (Shaikin et al., 1975). Among invertebrates, guanidinopropionic acid has been found in the sea-anemone Anthopleura japonica (Makisumi, 1961) and in the sipunculid worm Phascolion strombi (Robin and Guillou, unpublished data). The latter also contains the parent guanidinoamide, phascoline (Guillou and Robin, 1973). However, it has not been established whether guanidinopropionic acid is the precursor or the hydrolysis product of phascoline. Guanidinopropionic acid may be one of the factors responsible for the higher hemolysis of red blood cells in uremia (Shaikin et al., 1975).

β-Guanidinoisobutyric acid (39) has been isolated from the sipunculid worm Phascolosoma vulgare and characterized in several other marine worms (Robin, 1964b). It is the guanidine moiety of phascolosomine (Guillou and Robin, 1973). Both compounds derive from thymine for their β-aminoisobutyric chain in P. vulgare, but their metabolic interrelationship is unclear (Robin and Guillou, in preparation).

γ-Guanidinobutyric acid (40) was first identified as a product of the oxidative catabolism of arginine in several groups of marine invertebrates (Roche et al., 1952a) and then found in other inverte-

$$NH_2 - C - NH - (CH_2)_2 - COOH$$
$$\quad\quad \| $$
$$\quad\quad NH$$

38

$$NH_2 - C - NH - CH_2 - CH - COOH$$
$$\quad\quad \|\quad\quad\quad\quad\quad | $$
$$\quad\quad NH\quad\quad\quad\quad\quad CH_3$$

39

$$NH_2 - C - NH - (CH_2)_3 - COOH$$
$$\quad\quad \| $$
$$\quad\quad NH$$

40

brates (Lissitzki et al., 1954; Garcia et al., 1956; Irreverre et al., 1957; Makisumi, 1961), in plants (Mourgue et al., 1953; Mourgue and Dokhan, 1954; Irreverre et al., 1957; Morel and Duranton, 1958; Ito et al., 1967) and in mammals (Irreverre et al., 1957; Pisano et al., 1963). Its presence in mammalian brain has been confirmed (Irreverre and Evans, 1959; Blass, 1960; Mori et al., 1974). This substance has been characterized in human urine (Thoai et al., 1956c), in which its level increases in hyperargininemia (Terheggen et al., 1972).

γ-Guanidinobutyric acid is formed in invertebrates by non-enzymatic oxidative decarboxylation of α-keto-δ-guanidinovaleric acid (Thoai et al., 1953a; Garcia et al., 1956; Robin and Thoai, 1957). The same pathway is operative in plants (Barnes, 1962; Reinbothe, 1963) and in uricotelic vertebrates (Robin, 1954; Boulanger and Osteux, 1956). In mammals, the synthesis of γ-guanidinobutyric acid was obtained in vitro by transfer of the amidine group of arginine to γ-aminobutyric acid in kidney, pancreas and brain preparations (Pisano et al., 1963). According to its structural similarity with γ-aminobutyric acid, γ-guanidinobutyric acid may serve a role in some nervous processes (Irreverre et al., 1957). Intracisternal injection of this substance into the rabbit induces convulsions (Jinnai et al., 1966).

β-Hydroxy-γ-guanidinobutyric acid (41) has been found in the sea-anemone Anthopleura japonica (Makisumi, 1961). This substance could result from the oxidative catabolism of γ-hydroxyarginine, by a mechanism similar to that observed for the formation of γ-guanidino-butyric acid from arginine in a large number of marine invertebrates. β-hydroxy-γ-guanidinobutyric acid has also been identified in trace amounts in mammalian brain and serum (Matsumoto et al., 1976a).

$$NH_2 - C - NH - CH_2 - CH - CH_2 - COOH$$
$$\|\qquad\qquad\qquad |$$
$$NH\qquad\qquad\qquad OH$$

41

$$NH_2 - C - NH - (CH_2)_4 - COOH$$
$$\|$$
$$NH$$

42

δ-Guanidinovaleric acid (42) has been identified in human and rat urine (Thoai and Lacombe, 1958).

Studies of the rat have shown that labelled δ-guanidinovaleric acid is found in urine after administration of [^{14}C-U]-L-arginine (Boulanger and Osteux, 1960). Nevertheless, the proposed mechanism (Boulanger and Osteux, 1960), involving formation of α-keto-δ-guanidinovaleric acid followed by reduction of the keto group, has not been confirmed (Lacombe et al., 1964), and the metabolic pathway from arginine remains unclear.

α-Hydroxy-δ-guanidinovaleric acid (43), commonly named **argininic acid**, was first identified in plant tumoral tissues (Morel and Duranton, 1958). This substance has been found in normal and pathological human urine (Wiechert et al., 1976). Particularly high levels of argininic acid have been observed in the urine of hyperargininemic patients (Terheggen et al., 1972).

The biosynthesis of argininic acid has been studied in the rabbit. In vivo administration of α-keto-δ-guanidinovaleric acid results in a significant increase of argininic acid in the serum (Marescau et al., in press). The hydroxy compound perhaps results from an enzymatic hydrogenation of the keto analogue. This mode of formation could explain the high levels of argininic acid found in the urine of the hyperargininemic patients, which contains very large amounts of α-keto-δ-guanidinovaleric acid.

Argininic acid may provide an alternative pathway for the catabolism of excess arginine in hyperargininemic patients.

$$NH_2 - C - NH - (CH_2)_3 - CH - COOH \qquad NH_2 - C - NH - (CH_2)_3 - C - COOH$$
$$\parallel \qquad\qquad\qquad\quad | \qquad\qquad\qquad\qquad \parallel \qquad\qquad\qquad\qquad \parallel$$
$$NH \qquad\qquad\qquad OH \qquad\qquad\qquad\quad NH \qquad\qquad\qquad O$$

$$\textbf{43} \qquad\qquad\qquad\qquad\qquad\qquad\qquad \textbf{44}$$

α-Keto-δ-guanidinovaleric acid (44) was first found as a product of the oxidative catabolism of L-arginine in several classes of marine invertebrates (echinoderms, molluscs and crustaceans) (Roche et al., 1952a). This substance was subsequently identified in a fresh water mollusc (Robin and Thoai, 1957), insects (Garcia et al., 1956), plants (Brander and Virtanen, 1964; Durzan and Richardson, 1966; Miersch and Reinbothe, 1966) and uricotelic vertebrates (Robin, 1954; Boulanger and Osteux, 1956). α-Keto-δ-guanidinovaleric acid has been recently isolated from the urine of hyperargininemic patients, where it occurs at extremely high levels while only trace amounts are detected in normal human urine (Marescau et al., 1981).

α-Keto-δ-guanidinovaleric acid can be formed by enzymatic oxidative deamination of L-arginine. This pathway has been demonstrated in many invertebrates (Thoai et al., 1953a; Robin and Thoai, 1957; Garcia et al., 1956) and in uricotelic vertebrates (Robin, 1954; Boulanger and Osteux, 1956), which possess a L-amino acid oxidase particularly active on basic amino acids (Roche et al., 1952b; Boulanger and Osteux, 1956). The same pathway seems to be operative in plants (Durzan and Richardson, 1966) and in bacteria (Stumpf and Green, 1944). Another biosynthetic pathway exists in mammals, where the L-amino acid oxidase is not active on L-arginine (Meister and Wellner, 1963). Transamidinations between L-arginine and various keto acids, yielding α-keto-δ-guanidinovaleric acid, have been demonstrated in vitro in these animals, using various tissue preparations (Cammarata and Cohen, 1950; Quastel and Witty, 1951).

Biologically, α-keto-δ-guanidinovaleric acid seems to constitute
a secondary catabolic pathway of arginine in normal mammals. This
pathway is predominant in hyperargininemia, probably to compensate
for the urea cycle deficiency (Marescau et al., 1981). In several
groups of invertebrates, the oxidative deamination to α-keto-δ-
guanidinovaleric acid appears to be an active catabolic pathway of
arginine. This reaction, perhaps in association with transamination
reactions with non-basic amino acids, could play an important role
in ammonia detoxication in certain groups of marine invertebrates
(Robin, 1954).

Cyclic α-keto-δ-guanidinovaleric acid (45) has been found in
human urine in equilibrium with the open chain compound (Marescau et
al., 1981). The latter is probably the native product. The existence
of an equilibrium between the cyclic and open chain forms of α-keto-
δ-guanidinovaleric acid has been described previously (Cooper and
Meister, 1978).

Guanidinosuccinic acid (46) was first isolated from uremic
urine and identified by Natelson et al. (1964). Low level of this
substance is present in normal human urine, serum and cerebrospinal
fluid (Natelson et al., 1964; Stein et al., 1969; Mori et al., 1981b).
Significant variations are observed in pathological states. Increases
in guanidinosuccinic acid were found in urine, serum and cerebrospinal fluid of uremic patients (Natelson et al., 1964; Cohen et al.,
1968; Stein et al., 1969). On the other hand, lower values were reported in various inborn defects of urea synthesis (Stein et al., 1969;
Böhles et al., 1982), even in hyperargininemia (Lowenthal and
Marescau, 1981). Guanidinosuccinic acid has also been found in brain
(Mori et al., 1979).

$$NH_2 - C - NH - CH - CH_2 - COOH$$
$$\qquad \| \qquad\qquad |$$
$$\qquad NH \qquad\quad COOH$$

45   **46**

Two pathways have been proposed for guanidinosuccinic acid biosynthesis. The transamidination of arginine to aspartic acid yielding
guanidinosuccinic acid as suggested by Cohen (1970),has been established by in vivo experiments and hepatic perfusion studies of the
rat (Perez et al., 1976) and has been demonstrated in the rabbit
(Suwaki, 1978). The second pathway is via the canaline cycle, an
alternate cycle proposed for the conversion of urea nitrogen to
creatine, with the formation of guanidinosuccinic acid as an inert
overflow product (Natelson and Sherwin, 1979). Other studies (Cohen
and Patel, 1982) have confirmed that urea serves as a source of

guanidinosuccinic acid. In this case, guanidinosuccinic acid may be useful as a marker measuring the activity of the canaline cycle, a new metabolic pathway which can serve a practical function in the management of uremia.

Guanidinosuccinic acid has been implicated in the platelet defect thought to be responsible for uremic bleeding (Stein et al., 1969; Horowitz et al., 1970).

α-Guanidinoglutaric acid (47), normally present in small quantities in normal cerebral cortex (Mori et al., 1979), was found to increase markedly in the cobalt-induced epileptogenic focus of cat cerebral cortex, concomitant with the appearance of paroxysmal discharges (Mori et al., 1980, 1982b).

$$NH_2 - C - NH - CH - (CH_2)_2 - COOH$$
$$\quad\quad \| \quad\quad\quad |$$
$$\quad\quad NH \quad\quad COOH$$

47

α-Guanidinoglutaric acid can be produced _in vitro_ from α-ketoglutaric acid and arginine in brain homogenates (Mori et al., 1982b). However, the reaction mechanism is still unexplained. Paroxysmal discharges in electroencephalogram recordings from cats can be induced following topical application of this substance to the sensory motor cortex (Mori et al., 1982b).

4.2. Sulfonic and sulfinic acids

Asterubine (48) (N,N-dimethyl-2-guanidinoethanesulfonic acid) was the first sulfur guanidino compound discovered in nature. This substance was isolated from the starfish, _Masthasterias glacialis_ and _Asterias rubens_ (Ackermann, 1935a). Its structure was confirmed by synthesis (Ackermann, 1935b; Ackermann and Müller, 1935). Asterubine has no effect on blood pressure; the effect on blood glucose is a short initial hyperglycaemia, followed by a return to normal levels (Ackermann and Heinsen, 1935).

Taurocyamine (49) (2-guanidinoethanesulfonic acid) has been

$$CH_3 - N - C - NH - (CH_2)_2 - SO_3H \quad\quad\quad NH_2 - C - NH - (CH_2)_2 - SO_3H$$
$$\quad\quad |\quad \| \quad\quad\quad\quad\quad\quad\quad\quad\quad\quad\quad\quad \|$$
$$\quad\quad CH_3\ NH \quad\quad\quad\quad\quad\quad\quad\quad\quad\quad\quad\quad\quad\ NH$$

48   **49**

first isolated from the marine polychaete _Arenicola marina_ (Thoai and Robin, 1954a) and subsequently characterized in other marine worms (Thoai and Robin, 1969), sponges (Robin and Roche, 1954; Ackermann and Pant, 1961; Bergquist and Hartmann, 1969) and a sea anemone (Makisumi, 1961). Taurocyamine was also found in various biological fluids and tissues of mammals, e.g. in urine (Thoai et al., 1954), serum (Matsumoto et al., 1976a) and brain (Blass, 1960;

Mori et al., 1974). Pathological increases were reported in serum and brain in uremia (Matsumoto et al., 1976a) and in cerebrospinal fluid in epilepsia (Mori et al., 1982a). Higher brain levels were found in strains of mice with an epileptic disposition (Hiramatsu, 1980a). Significant variations were observed in mouse brain in relation to convulsions (Hiramatsu, 1980b).

It appears that the biosynthetic pathway for taurocyamine is not the same in vertebrates and invertebrates. In the former, tauro-cyamine results from the transamidination of taurine from arginine, as was shown by in vivo (Thoai et al., 1956c; Mori et al., 1981a) and in vitro (Shindo and Mori, 1980) studies in the rat. On the other hand, in vivo and in vitro studies have established that taurocyamine is formed in marine worms by oxidation of the sulfinic analogue, hypotaurocyamine (Thoai et al., 1963b).

Taurocyamine is a phosphagen precursor in various marine worms (Thoai et al., 1953b; Thoai and Robin, 1969); its biological role in other invertebrates is unknown. This substance induces convulsions in experimental animals when injected intracisternally (Mizuno et al., 1975). It could be one probable trigger for convulsions in epileptic subjects (Mori et al., 1982a). Taurocyamine has no effect on blood pressure. The effect on blood glucose is an initial hyperglycaemia, followed by hypoglycaemia (Ackermann and Heinsen, 1935).

Hypotaurocyamine (50) (2-guanidinoethanesulfinic acid) has been isolated from the sipunculid worm Phascolosoma vulgare (Robin and Thoai, 1962) and identified in various marine worms (Thoai and Robin, 1969). Hypotaurocyamine is formed in worms by transamidination from arginine; subsequent oxidation yields taurocyamine (Thoai et al., 1963b). The transamidination reaction seems to take place in prefe-rence in the gut and the oxidation of the sulfinic to sulfonic group in the muscle. Apart from being the precursor of taurocyamine in worms, hypotaurocyamine is a phosphagen precursor in a number of marine annelids and sipunculids (Thoai and Robin, 1969).

$$NH_2 - C - NH - (CH_2)_2 - SO_2H$$
$$\parallel$$
$$NH$$

50

5. GUANIDINOETHYLPHOSPHATE DERIVATIVES

Four guanidinoethylphosphate derivatives have been identified in living organisms. They only differ in the radical esterified on the phosphoryl residue (methyl for opheline, seryl for lombricine, N,N-dimethylseryl for thalassemine and aspartylseryl for bonellidine). Their natural occurrence seems to be restricted to worms (various annelids and sipunculids).

Lombricine (51) was first isolated from the earthworm <u>Lombricus</u> <u>terrestris</u> and identified mainly by degradation studies (Thoai and Robin, 1954b). The structure was confirmed by synthesis (Beatty and Magrath, 1959) and the serine residue was attributed the D-configuration (Beatty et al., 1959). Lombricine was further found in all terrestrial oligochaetes studied and in several marine polychaetes and echiurids (Thoai and Robin, 1969; Thoai et al., 1972; Robin and Guillou, 1980). In contrast with serine from earthworm lombricine, serine isolated from lombricine present in the echiurid <u>Urechis</u> <u>caupo</u> was found to have the L-configuration (Robin, 1964a). Opheline (52) was isolated from the marine polychaete <u>Ophelia neglecta</u> (Thoai et al., 1963a) and **thalassemine (53)** from the echiurid <u>Thalassema</u> <u>neptuni</u> (Thoai et al., 1972); their structure was established by identification of the hydrolysis products and confirmed by synthesis. **Bonellidine (54)** was isolated from the echiurid <u>Bonellia viridis</u> and identified by degradation studies (Thoai et al., 1967). Of two possible positions for the aspartyl residue, structure (54) was

$$
\begin{array}{c}
\text{O} \\
\text{\textbardbl} \\
\text{NH}_2-\text{C}-\text{NH}-(\text{CH}_2)_2-\text{O}-\text{P}-\text{O}-\text{CH}_2-\text{CH}-\text{COOH} \\
\;\;\;\;\;\;\text{\textbardbl}\;\text{\textbar}\;\;\;\;\;\;\;\;\;\;\;\;\;\;\;\;\;\;\text{\textbar} \\
\;\;\;\;\;\;\text{NH}\;\text{OH}\;\;\;\;\;\;\;\;\;\;\;\;\;\;\;\;\text{NH}_2
\end{array}
$$

51

$$
\begin{array}{c}
\text{O} \\
\text{\textbardbl} \\
\text{NH}_2-\text{C}-\text{NH}-(\text{CH}_2)_2-\text{O}-\text{P}-\text{O}-\text{CH}_3 \\
\;\;\;\;\;\;\text{\textbardbl}\;\text{\textbar} \\
\;\;\;\;\;\;\text{NH}\;\text{OH}
\end{array}
$$

52

$$
\begin{array}{c}
\text{O} \\
\text{\textbardbl} \\
\text{NH}_2-\text{C}-\text{NH}-(\text{CH}_2)_2-\text{O}-\text{P}-\text{O}-\text{CH}_2-\text{CH}-\text{COOH} \\
\;\;\;\;\;\;\text{\textbardbl}\;\text{\textbar}\;\;\;\;\;\;\;\;\;\;\;\;\;\;\;\;\;\;\text{\textbar} \\
\;\;\;\;\;\;\text{NH}\;\text{OH}\;\;\;\;\;\;\;\;\;\;\;\;\;\;\;\;\text{N}-\text{CH}_3 \\
\;\text{\textbar} \\
\;\text{CH}_3
\end{array}
$$

53

$$
\begin{array}{c}
\text{O}\;\text{O} \\
\text{\textbardbl}\;\text{\textbardbl} \\
\text{NH}_2-\text{C}-\text{NH}-(\text{CH}_2)_2-\text{O}-\text{P}-\text{O}-\text{CH}_2-\text{CH}-\text{NH}-\text{C}-\text{CH}_2-\text{CH}-\text{COOH} \\
\;\;\;\;\;\text{\textbardbl}\;\text{\textbar}\;\;\;\;\;\;\;\;\;\;\;\;\;\;\;\;\text{\textbar}\;\text{\textbar} \\
\;\;\;\;\;\text{NH}\;\text{OH}\;\;\;\;\;\;\;\;\;\;\;\text{COOH}\;\;\;\;\;\;\;\;\;\;\;\;\;\;\;\;\;\;\text{NH}_2
\end{array}
$$

54

choosen on the base of colour reactions (Thoai et al., 1967). A general study of the optical configuration of the serine residue in the lombricine, thalassemine and bonellidine molecules has shown a certain systematization in function of the animal origin; D-serine occurs in the compounds isolated from annelids and L-serine in those isolated from echiurids (Thoai et al., 1972; Robin and Guillou, 1980).

The biosynthesis of lombricine was studied by administrating various labelled precursors to earthworms (Rossiter et al., 1960); it was thus established that aminoethylphosphate is the direct precursor of lombricine and that the amidino group is derived from arginine. By similar methods, methionine was shown to be the methyl donor in the biosynthesis of opheline in Ophelia neglecta and of thalassemine in Thalassema neptuni (Thoai et al., 1966; Thoai et al., 1972).

Lombricine (Thoai and Robin, 1954b), opheline (Thoai et al., 1963a) and thalassemine (Thoai et al., 1972) were found to be phosphagen precursors in worms. Phosphobonellidine could not be characterized in Bonellia muscle. However, in Bonellia trunk muscle, the characterization of a phosphagen kinase active in vitro on bonellidine, the sole guanidino compound found in this tissue, suggests that bonellidine, like the other guanidinoethylphosphate derivatives, may contribute to the formation of a new muscular phosphagen in that echiurid (Robin and Guillou, unpublished data).

6. PHOSPHAGENS

The biological importance and the particular distribution of the phosphagens (N'-phosphorylguanidines) have been reviewed previously (Ennor and Morrison, 1958; Thoai and Roche, 1964; Thoai and Robin, 1969; Robin, 1974, 1980; Chevolot, 1981).

$$H_2O_3P - NH - C - NH - (CH_2)_3 - CH - COOH$$

with $\parallel$ below the second C (NH) and $\mid$ below CH (NH$_2$)

55

$$H_2O_3P - NH - C - N - CH_2 - COOH$$

with $\parallel$ below C (NH) and $\mid$ below N (CH$_3$)

56

Eight phosphagens have been so far identified in living organisms: **phosphoarginine (55)** (Meyerhof and Lohmann, 1928), **phosphocreatine (56)** (Eggleton and Eggleton, 1927; Fiske and Subbarow, 1929), **phosphoglycocyamine (57)** (Thoai et al., 1953b), **phospholombricine (58)** (Thoai and Robin, 1954b), **phosphoopheline (59)** (Thoai et al., 1963a), **phosphotaurocyamine (60)** (Thoai et al., 1953b), **phosphohypotaurocyamine (61)** (Robin and Thoai, 1962) and **phosphothalassemine (62)** (Thoai et al., 1972). Phosphocreatine and phosphoarginine

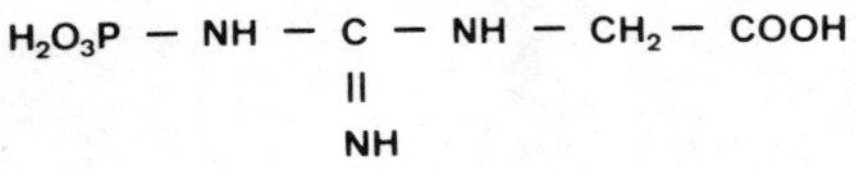

$$H_2O_3P - NH - C - NH - CH_2 - COOH$$
$$\|$$
$$NH$$

57

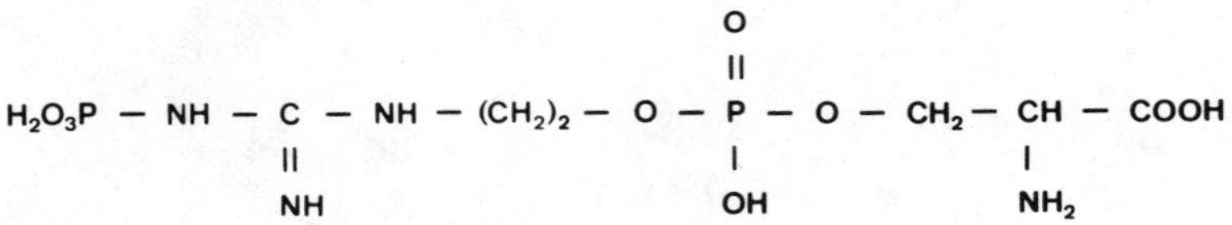

$$
\begin{array}{c}
O \\
\| \\
H_2O_3P - NH - C - NH - (CH_2)_2 - O - P - O - CH_2 - CH - COOH \\
\| \qquad\qquad\qquad\qquad\quad | \qquad\qquad\quad | \\
NH \qquad\qquad\qquad\qquad\qquad OH \qquad\qquad NH_2
\end{array}
$$

58

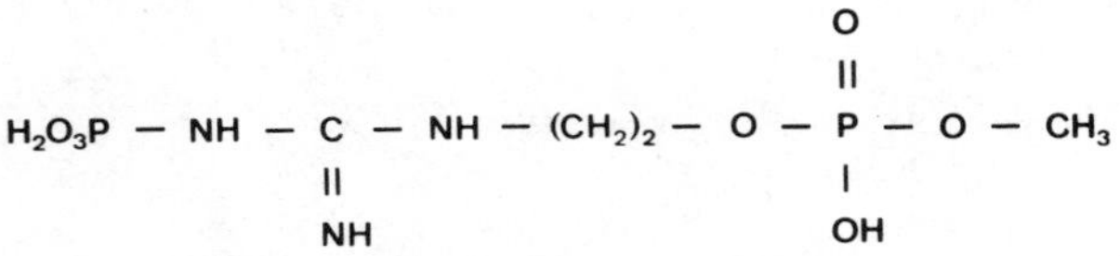

$$
\begin{array}{c}
O \\
\| \\
H_2O_3P - NH - C - NH - (CH_2)_2 - O - P - O - CH_3 \\
\| \qquad\qquad\qquad\qquad\quad | \\
NH \qquad\qquad\qquad\qquad\quad OH
\end{array}
$$

59

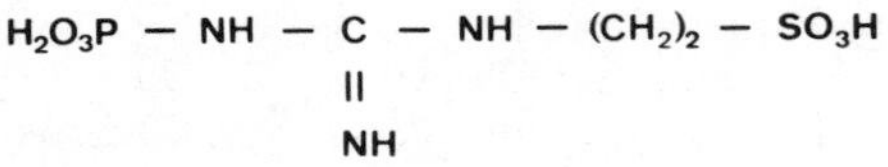

$$H_2O_3P - NH - C - NH - (CH_2)_2 - SO_3H$$
$$\|$$
$$NH$$

60

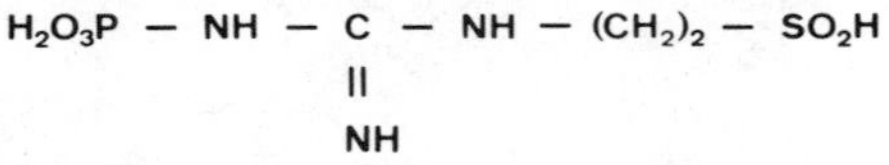

$$H_2O_3P - NH - C - NH - (CH_2)_2 - SO_2H$$
$$\|$$
$$NH$$

61

$$
\begin{array}{c}
O \\
\| \\
H_2O_3P - NH - C - NH - (CH_2)_2 - O - P - O - CH_2 - CH - COOH \\
\| \qquad\qquad\qquad\qquad\quad | \qquad\qquad\quad | \\
NH \qquad\qquad\qquad\qquad\qquad OH \qquad\qquad N - CH_3 \\
\qquad\qquad\qquad\qquad\qquad\qquad\qquad\qquad\qquad | \\
\qquad\qquad\qquad\qquad\qquad\qquad\qquad\qquad\qquad CH_3
\end{array}
$$

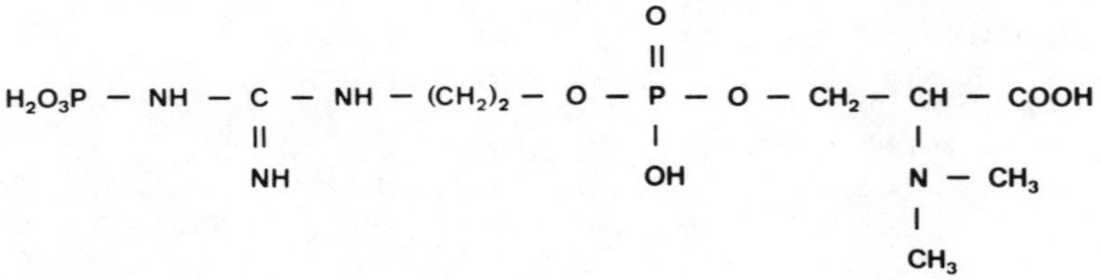

62

are widely distributed . Phosphocreatine has been found in all vertebrates (Eggleton and Eggleton, 1927) and in various protochordates
and invertebrates (Meyerhof, 1928; Ennor and Morrison, 1958; Thoai
and Robin, 1969). Phosphoarginine is present in most invertebrates
(Meyerhof, 1928; Meyerhof and Lohmann, 1928; Ennor and Morrison,
1958; Thoai and Robin, 1969) and in microorganisms (Robin and Viala,
1966; Di Jeso, 1967). In contrast , the other six phosphagens have
been found only in worms (Annelida and related phyla), where their
distribution is highly diversified (Thoai and Robin, 1969). The
phosphagens constitute a pool of labile phosphate freely available
for muscular contraction. Their high energetic potential results

from the lability of the phosphoryl bound, which is caused by the opposition between the resonances of the phosphoryl and the guanidine groups (Kalckar, 1941; Oesper, 1950). The enzymatic hydrolysis of the phosphagens provides a phosphoryl group for the resynthesis of ATP from ADP during the muscular work. The reaction is: phosphagen + ADP ⇌ guanidine derivative + ATP, and is catalyzed by specific phosphagen kinases. Since the reaction is reversible, the phosphagens can be restaured during muscular relaxation from other ATP sources, (e.g. glycolysis and oxidative phosphorylations). The advantages of the phosphagen system over an hypothetical use of high ATP concentrations in the tissues are (i) avoidance of the phosphate loss; (ii) low concentration of counter anions; (iii) lower production of ADP (a regulator of several enzymatic reactions) and (iv) no acidification by hydrolysis of ATP (Watts, D.C., oral communication, GABIM Meeting, Brest, 1977). As a result of their function, the phosphagens are present at higher concentration in muscle, particularly where intensive work is required (Beis and Newsholmes, 1975). They have been found at lower levels in the other tissues of vertebrates (Ennor and Morrison, 1958) and in the germinative cells of various invertebrates (Thoai and Robin, 1969). Here the phosphagen is often different from that of muscle and the greatest variety is observed in worms.

An attempt has been made to correlate the nature of the muscular phosphagen with the degree of evolution of the organism in terms of the present results (Robin, 1974). Curiously, in spite of inexplicable discrepancies such as those observed in worms, there is some evidence for a concomitant biochemical and phylogenic transition from phospho-arginine (bacteria, most invertebrates) to phosphocreatine (all vertebrates). This observation is strengthened by comparative studies of the phosphagen kinases at the genetic level (Watts and Watts, 1968). From an evolutionary point of view, one major advantage of replacing phosphoarginine with phosphocreatine is the isolation of the phosphagen from the rest of the amino acid metabolism.

Aside from their important function in muscular contraction, the phosphagens may play a role in the regulation of anaerobic glycolysis. This regulation apparently can be under the direct control of the phosphagen concentration in muscle (Storey, 1976). Alternatively, the acceptance of phosphate by arginine in invertebrate muscle and by creatine in vertebrate muscle following exercise, permits the resting muscle to remote inorganic phosphate to rate-limiting levels in order to control glycogenolysis and glycolysis (Davuluri et al., 1981).

7. GUANIDINO AMIDES

γ-guanidinobutyramide (63) is produced by enzymatic decarboxy-oxidation of arginine by <u>Streptomyces griseus</u> (Thoai et al., 1956a). The reaction is catalyzed by a decarboxyoxidase, an adaptive enzyme

induced in the microorganism when it is cultured in the presence
of arginine (Thoai et al., 1956b) and which is highly specific
towards L-arginine (Thoai and Olomucki, 1962). Unlike L-arginine,
γ-guanidinobutyramide cannot be utilized as the sole source of carbon
and nitrogen by S. griseus, but it is an indispensable intermediate
in the utilization of arginine by the microorganism (Thoai et al.,
1962). γ-Guanidinobutyramide was also characterized as a product of
the oxidative catabolism of arginine in the fresh-water mollusc
Limnaea stagnalis (Thoai et al., 1957). This animal also utilizes
the oxidative deamination pathway to α-keto-δ-guanidinovaleric
acid (Robin and Thoai, 1957).

   Phascoline (64) and **phascolosomine (65)** are two similar compounds
isolated from sipunculid worms, Phascolion strombi and Phascolosoma
sp., respectively (Guillou and Robin, 1973). These substances are present at very high concentration in the worms viscera. They consist of
an ω-guanidino acid (β-guanidinopropionic- or β-guanidinoisobutyric
acid, respectively) in amide linkage with a long chain linear amine
(2-hydroxy- or 2-methoxy-n-heptylamine, respectively).

$$NH_2 - C - NH - (CH_2)_3 - CONH_2$$
$$\| $$
$$NH$$

63

$$NH_2 - C - NH - (CH_2)_2 - C - NH - CH_2 - CH - (CH_2)_4 - CH_3$$
$$\|\qquad\qquad\qquad\|\qquad\qquad\quad|$$
$$NH\qquad\qquad\qquad O\qquad\qquad\quad OH$$

64

$$NH_2 - C - NH - CH_2 - CH - C - NH - CH_2 - CH - (CH_2)_4 - CH_3$$
$$\|\qquad\qquad\quad|\quad\|\qquad\qquad\quad|$$
$$NH\qquad\qquad CH_3\ O\qquad\qquad OCH_3$$

65

   The biological role of phascoline and phascolosomine is unknown.
Biosynthetic studies have established that the amidine group of the
two compounds is provided by arginine, and that the aminopropionic
and aminoisobutyric chains are derived from uracil and thymine, respectively (Robin and Guillou, in preparation). But the stage at which
the amidination occurs and the origin of the amine moiety could not be
established . The unusually high concentration of these substances in
the worms viscera and their rapid disappearance upon starving suggest
they serve as reversible storage products, for the guanidine moiety
in pyrimidine metabolism, and for the amine moiety in aminolipid metabolism.

   Phascoline and phascolosomine possess a negative chronotrop
activity on cultured heart cells and the activity seems to be located
in the amine moiety (Auclair et al., 1976). These substances have a
moderate inhibitory effect on acetylcholine esterase of rabbit brain

(Matsumoto et al., 1977). In both cases, phascolosomine was slightly
more efficient than phascoline.

Recently, two series of related long chain guanidino compounds
derived from lipids, the arcanidines and the polyandrocarpidines, have
been identified in marine animals (review by Chevolot, 1981). **Arcanidines a, b, and c (66)** were isolated from the sponge <u>Acarnus erithacus</u>
by Carter and Rinehart (1978), who established their structure by
intensive spectral and chemical studies. These compounds have in

$$NH_2 - C - NH - (CH_2)_5 - N - (CH_2)_3 - NH - C - CH = C - CH_3$$

with the following vertical substituents: $\|$ NH below the first C, $|$ R below the N, $\|$ O below the third C, and $|$ CH_3 below the C.

66

a $R = CO - (CH_2)_{10} - CH_3$

b $R = CO - (CH_2)_3 - CH = CH - (CH_2)_5 - CH_3$

c $R = CO - (CH_2)_2 - CH = CH - (CH_2)_2 - CH = CH - CH_2 - CH = CH - CH_2 - CH_3$

common a substituted homospermidine skeleton but differ in the fatty
acid substituents which consist of C_{11} to C_{13} saturated or unsaturated
fatty acids (Carter and Rinehart, 1978; Rinehart, personal communication, in Chevolot, 1981). **Polyandrocarpidines I (67, n=5)** and **II
(67, n=4)** were found in a colonial tunicate, <u>Polyandrocarpa sp</u>.

$$NH_2 - C - NH - (CH_2)_n - NH - C \;\triangle\; (CH = CH)_2 - (CH_2)_3 - CH_3$$

with vertical substituents: $\|$ NH below the first C, $\|$ O below the second C.

67

(Cheng and Rinehart, 1978). Their structure was established mainly
from mass spectral studies of various pyrimidyl derivatives and
that of polyandrocarpidine II was indirectly confirmed by synthesis.
An interesting feature of the polyandrocarpidines is the presence
of a cyclopropene ring, a structural group which is rarely encountered in nature (Cheng and Rinehart, 1978). The arcanidines and the
polyandrocarpidines are pharmacologically active substances. They
display antibacterial activity against various microorganisms and
antiviral activity against <u>Herpes simplex</u> virus. The polyandrocarpidines are cytotoxic agents.

Gongrine (68) and **gigartinine (69)** are two parent compounds
isolated from the red alga, <u>Gymnogongrus flabelliformis</u>, (Ito and
Hashimoto, 1965; 1966a). These compounds were identified by degradation studies and their structure was confirmed by synthesis
(Ito and Hashimoto, 1969). They are usually described as guanylureido
derivatives, but they may also be considered as amides of an hypo-

$$NH_2 - C - NH - C - NH - (CH_2)_3 - COOH$$
$$\qquad\; \| \qquad\qquad \|$$
$$\qquad\; NH \qquad\quad O$$

68

$$NH_2 - C - NH - C - NH - (CH_2)_3 - CH - COOH$$
$$\qquad\; \| \qquad\qquad \| \qquad\qquad\qquad |$$
$$\qquad\; NH \qquad\quad O \qquad\qquad\qquad NH_2$$

69

thetical guanidinoformic acid with γ-aminobutyric acid and ornithine, respectively. Both substances have been found in about 50% of the red algae tested (Ito and Hashimoto, 1966b); but they have not been detected in brown or green algae. The biosynthesis of gongrine and gigartinine has not been investigated. It can be considered that gigartinine is built from ornithine. Gongrine is likely to proceed from gigartinine through enzymatic oxydative deamination followed by oxidative decarboxylation, as γ-guanidinobutyric acid proceeds from arginine.

B. PHYSIOLOGICAL AND PHARMACOLOGICAL PROPERTIES

1. CONVULSIVANT ACTIVITY

Soon after their discovery, guanidine and guanidine-like substances were recognized as powerful convulsive agents, highly toxic to warm- and cold-blooded animals. Fühner (1923) was the first to point out that the "guanidinium ion" was responsible for their physiological activities. The convulsivant properties of guanidino compounds and their implication in neurological disorders have been reviewed (Guggenheim, 1951; Mori, 1983).

Recent studies have established that γ-guanidinobutyric acid, taurocyamine, glycocyamine, creatine, creatinine, phosphocreatine, N-acetylarginine and methylguanidine induced convulsions in the rabbit when they were injected intracisternally (Mori et al., 1982a, and cited references). Although these substances are normal constituents of brain and biological fluids of mammals, excess concentrations caused by metabolic disorders may induce convulsions (Mori et al., 1982a). Elevated values of taurocyamine, glycocyamine and methylguanidine in brain of mice with an epileptic disposition (Hiramatsu, 1980a) and variations of the taurocyamine level of mouse brain during convulsions (Hiramatsu, 1980b) have been reported. Increased levels of taurocyamine in the cerebrospinal fluid of epileptic patients (Mori et al., 1981b) and an increase of guanidinoglutaric acid in the cobalt-induced epileptogenic focus of cats concomitant with the appearance of paroxysmal discharges (Mori et al., 1980, 1982) also suggests a relationship with the seizure mechanism. Similarly, the accumulation of methylguanidine in uremic patients may trigger convulsions observed during uremia (Matsumoto et al., 1976a).

However, the mechanism by which guanidino compounds induce convulsions is still unclear. Recent works (Matsumoto and Mori, 1976; Matsumoto et al., 1977; Mori et al., 1981a; Watanabe et al., 1983) suggest that guanidine derivatives may directly affect neurotransmitters in the central nervous system (Mori, 1983).

Guanidinoacetic acid, β-guanidinopropionic acid and γ-guanidino-butyric acid were compared with γ-aminobutyric acid (GABA) for their blocking effect on single nerve cells (Edwards and Kuffler, 1959) and for their depressing effect on cortical neurons (Brnjevic and Phillips, 1963). In guanidinoacetic acid and β-guanidinopropionic acid, the distance between the terminal amine and carboxylic groups are almost the same as in GABA (3 carbon atoms) and they have activities very similar to GABA. In contrast, γ-guanidinobutyric acid was appreciably less active. Studies of the giant neurons of Achatina fulica have shown that, out of 20 guanidino compounds assayed, only guanidinoacetic acid at high concentration had an inhibitory effect on the tonically active neuron (Matsumoto et al., 1976c). Competitive binding studies on the postsynaptic membrane of the crayfish muscle provides good evidence that guanidinoacetic acid and, to a lesser extent β-guanidinopropionic acid and γ-guanidinobutyric acid, combine with the same receptors as GABA, although significant differences are observed in the effects on membrane conductance (Takeuchi and Takeuchi, 1975a,b).

Another observation that taurine can antagonize taurocyamine-induced seizures suggests that these two compounds play mutually antagonistic roles in excitation and inhibition in the central nervous system (Mori et al., 1981a).

2. HYPOGLYCAEMIC ACTIVITY

The decrease in blood glucose level in the presence of guanidine was first reported by Watanabe (1918), but this effect was only obtained at toxic doses. A search for drugs effective in the oral treatment of diabetus mellitus resulted in a systematic investigation of the hypoglycaemic properties of other natural guanidino compounds (review by Guggenheim, 1951). These compounds were too toxic for medical use, but they served as models for the preparation of synthetic analogues with increased hypoglycaemic activity and decreased toxicity. Synthetic long chain diguanidino compounds (Synthalins A and B), and shortly thereafter biguanides were used as oral antidiabetics (review by Schäfer, 1980). The interest of these compounds (some of which are still in use today) has now declined because of fatal accidents caused by lactate acidosis following biguanide treatment.

The reason for lactate acidosis, common to all natural and synthetic guanidino compounds, is increased anaerobic glycolysis caused by inhibition of cellular respiration (Schäfer, 1980, and cited references). Indeed, the mechanism of the hypoglycaemic effect of guanidine derivatives is fundamentally different from that of insuline. Insuline lowers the blood glucose level by increasing glycogen storage and glucose combustion in the tissues. In contrast, guanidino compounds (i) deplete glycogen storage by inhibiting hepatic gluconeogenesis and (ii) increase anaerobic glycolysis with production of lactic acid as a result of an inhibition of cellular oxidations. If respiratory inhibition induces an increase in glycolysis and peripheral lactate production while the removal of lactate and H^+ equivalents by hepatic gluconeogenesis is also inhibited, lactate acidosis will follow.

3. ANTIHYPERTENSIVE ACTIVITY

About 1930, natural guanidino compounds effects on blood pressure were investigated. Several compounds (e.g. guanidine, methylguanidine, arcaine and galegine) have hypo- or hypertensive effects. However, they cannot be used for medical purposes due to their toxicity (review by Guggenheim, 1951). Synthetic analogues, which are all cyclic compounds, have been developed and many of these are now utilized in the treatment of hypertension (Touitou and Perlemuter, 1976).

The complex mechanism of the hypotensive activity of guanidino compounds has been clarified by studies of these synthetic substances. These compounds are adrenergic neuron blocking agents. They all act by preventing noradrenaline (norepinephrine) release from the post-ganglionic nerve ending, either through displacement of stored noradrenaline from the granules or by interfering with neuronal depolarisation. The effect is to reduce the amount of transmitter substance available to the postsynaptic adrenergic receptor sites, thereby reducing sympathetic tone to the end organ vessel (Stafford and Fann, 1977).

4. ANTIBACTERIAL AND ANTITUMOR ACTIVITIES

The antibiotic and antimitotic properties of linear guanidino compounds have not been systematically investigated. Several derivatives recently isolated from microorganisms or from marine animals (e.g. δ-N-hydroxyarginine, the arcanidines and the polyandrocarpidines) possess antimicrobial, antifungal, antiviral and/or cytotoxic activities. However, most guanidino antibacterial and antitumor agents known at the moment are heterocyclic derivatives.

II. HETEROCYCLIC GUANIDINO COMPOUNDS

Several carbocyclic rings may incorporate the guanidine radical. Other more or less complex systems contain guanidine residue(s) linked in different ways.

Most of these interesting compounds have been found during the last two decades, often in marine invertebrates and/or in microorganisms. Many of them display antibacterial and antitumor activities.

1. IMIDAZOLE DERIVATIVES

2-Aminoimidazole (70) is the basic unit for this class of compounds. It was first obtained from cultured medium of Streptomyces eurocidus (Seki et al., 1970) and later from the sponge Reneira cratera (Cimino et al., 1974). The production of 2-aminoimidazole by S. eurocidus is stimulated by arginine, suggesting a possible role in the biosynthetic process (Seki et al., 1970).

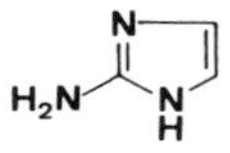

70

2-Aminoimidazole is the probable precursor of the antibiotic 2-nitroimidazole (azomycine), also produced by S. eurocidus. This is suggested by in vivo conversion of the amine to 2-nitroimidazole by washed cell-suspensions (Seki et al., 1970).

Aplysinopsine (71a) was isolated almost simultaneously from the sponge Thorecta sp. collected in Australia (Kazlaukas et al., 1977) and from another sponge, Verongia spengelii, collected in Florida (Hollenbeak and Schmitz, 1977). Its structure has been confirmed by synthesis. It is made up of an indole ring and a substituted 2-amino-imidazole guanidine moiety. Arginine may be responsible for the biogenesis of the guanidine unit of aplysinopsine. This substance

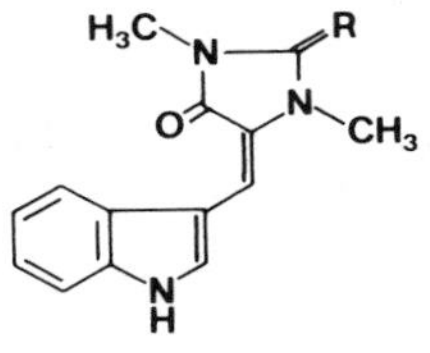

a R = NH

b R = N — CH$_3$

71

displays antitumor activity. Methylaplysinopsine (71b), also isolated from a marine sponge, exerts an antidepressant activity by competitively inhibiting monoamine oxidase (Taylor et al., 1981).

2. BROMOPYRROLE DERIVATIVES

These interesting compounds, isolated from sponges, have a common bromopyrrole-2-carboxamide unit. The guanidine radical is not incorporated into the pyrrole ring. In several of these derivatives, it is enclosed in a 2-aminoimidazole ring, as in the above imidazole derivatives.

N-amidino-4-bromopyrrole-2-carboxamide (72) is the simplest of these compounds. It was isolated from a mediterranean sponge of the genus Agelas and the structure determined by synthesis (Stempien et al., 1972). This substance possesses bactericidal activity.

Oroidin (73), isolated from the sponge Agelas oroides (Forenza et al., 1971) and found in two sponges of the genus Axinella (Cimino et al., 1975), is a somewhat more complex compound. The structure initially proposed was based on chemical and physical data (Forenza et al., 1971) and later corrected following synthesis of the compound (Garcia et al., 1973).

a $R_1 =$ H $R_2 =$ H

b $R_1 =$ Br $R_2 =$ H

4-Bromophakellin (74a) and 4,5-dibromophakellin (74b) were obtained from the australian sponge Phakellia flabellata (Sharma et al., 1970; Sharma and Burkholder, 1971). They were assigned a tetracyclic structure on the basis of NMR spectra, X-ray diffraction analysis and chemical studies (Sharma and Burkholder, 1971; Sharma and Magdoff-Fairchild, 1977). However, a tricyclic structure would also explain the NMR spectroscopic data (Baker, 1976b). An unusual feature of the phakellins is the low basicity of the guanidine residue (pKa<8) compared with pKa values > 13.4 of most other guanidines (Sharma and Burkholder, 1971). This low basicity is attributed to an inhibition of the resonance of the guanidinium cation imposed by the twisted spatial configuration of the imidazoline ring (Sharma and Magdoff-Fairchild, 1977).

With respect to the biosynthesis of the phakellins and the structurally related oroidins, it has been suggested that the pyrrole unit
may be derived from proline and the 2-aminoimidazole guanidine moiety
from arginine, as is the origin of 2-aminoimidazole in the sponge
<u>Reniera cratera</u> (Cimino et al., 1974).

The phakellins have a very mild antibacterial action against
<u>B. subtilis</u> and <u>E. coli</u> (Sharma and Magdoff-Fairchild, 1977).

3. PYRIMIDINE DERIVATIVE

Lathyrine (75) or β-(2-amino-pyrimidin-4-yl)alanine, has been
detected in numerous species of <u>Lathyrus</u> and isolated from the seeds
of <u>L. tingitanus</u> (Bell, 1961; Bell and Foster, 1962). The chemical

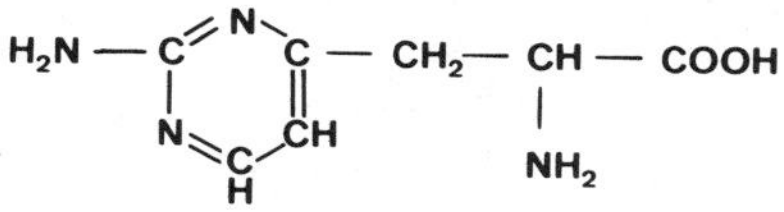

75

analogy between homoarginine, hydroxyhomoarginine and lathyrine, and
their natural occurrence in the same <u>Lathyrus</u> species suggested that
they might be metabolically related. The first assays for demonstrating the formation of lathyrine from labelled arginine in the stems
and fruit of <u>L. tingitanus</u> were unsuccessful (Bell, 1964). But this
reaction was later established in the roots, with the intermediate
formation of γ-hydroxyhomoarginine (Bell and Przybylska, 1965). The
mechanism of the cyclization and dehydrogenation has not been elucidated. Lathyrine is thought to be a product of nitrogen storage
in lathyrine-containing <u>Lathyrus</u>. The substance is toxic to man
and animals.

4. PURINE DERIVATIVES

Several simple purine derivatives have been isolated from marine
animals, most of them from sponges.

Guanine (76) (2-amino-6-oxypurine) has been found in the soft co-
ral, <u>Alcyonum digitatum</u> (Ackermann and Menssen, 1959). Guanosine (77)

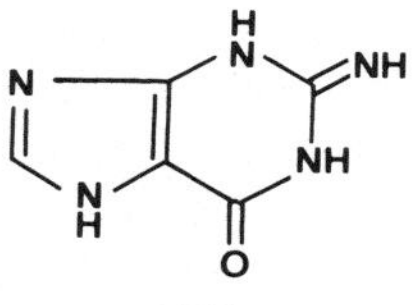

76

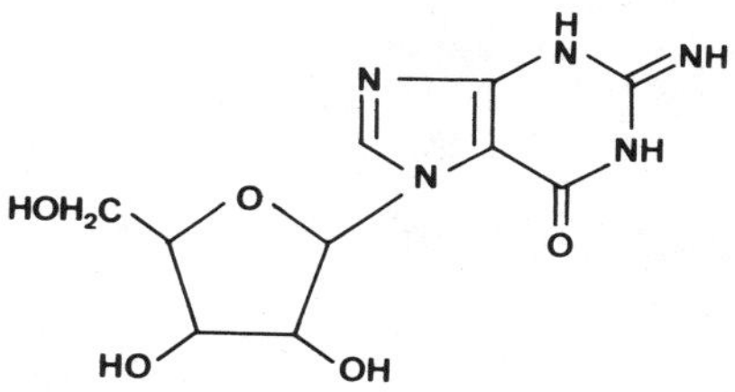

77

is present in the sponge, <u>Cryptotethia crypta</u> (Cohen, 1963). **2-Amino-6,8-dioxypurine (78)** was found in the ascidian, <u>Microcosmus poly-morphus</u> (Karrer et al., 1948). The guanidinium betaine, **herbipoline (79)** (derived from guanine) was isolated from the sponge, <u>Geodia gigas</u> (Ackermann and List, 1957, 1960).

78   **79**

5. PTERIDINE DERIVATIVES

Pterin (2-amino-4-hydroxypteridine) **(80a)**, xanthopterin (2-amino-4,6-dihydroxypteridine) **(80b)**, isoxanthopterin (2-amino-4,7-dihydroxy-pteridine) **(80c)** and related pigments constitute a large group of compounds with a common substituted pteridine skeleton. The first pteridines were isolated from the wings of butterflies by Hopkins (1889), who erroneously concluded that these pigments were derivatives of uric acid. Their structure as pteridines was established a long time after (review by Schöpf, 1964). These substances are frequently

a $R_1 =$ H	$R_2 =$ H
b $R_1 =$ OH	$R_2 =$ H
c $R_1 =$ H	$R_2 =$ OH

80

found in the epidermis and eyes of insects, in the skin of amphibians and in various marine organisms (reviews by Corrigan, 1970; Balinski, 1970; Chevolot, 1981). They mostly differ in the nature of the substituents in the 6-(radical R_1) and 7-(radical R_2) positions of the pteridine ring. The formation of the pteridines and their interconversion is the subject of much speculation. They probably originate from guanosine or a phosphorylated derivative of guanosine.

Since these compounds contain the same pterin unit as folic acid, they function in many metabolic processes (e.g. electron transport, hydroxylation and C_1 unit transfer). They are also pigments, and may have additional roles in marine animals as attractive substances and chemical communication agents.

6. TETRACYCLOPENTAZULENE PIGMENTS

Two series of yellow pigments which are highly fluorescent under ordinary light and possess interesting biological activities, have been recently isolated from colonial anthozoans of the order <u>Zoanthidea</u> (see reviews by Prota, 1980; Chevolot, 1981). These compounds

are highly methylated heterocyclic bases containing two guanidine
units. They have been divided into two groups according to the structure of their tricyclic skeleton: 1,3,5,7-tetracyclopent (f) azulene
for the zoanthoxanthins and 1,3,7,9-tetracyclopent (e) azulene for
the pseudozoanthoxanthins.

   Zoanthoxanthin (81), a representative of the first group, has
been isolated and identified from <u>Parazoanthus axinellae</u> (Cariello
et al., 1973; 1974a). Several closely related compounds (parazoantho-
xanthins A,B,C,D,E,F and G, epizoanthoxanthins A and B and palyzoantho-
xanthins A,B and C) have been identified from the same or from similar
species (Cariello et al., 1979 and cited references). These substances
essentially differ in their methylation pattern. The zoanthoxanthin
structure was fully determined by X-ray diffraction studies.

   Pseudozoanthoxanthin (82), a member of the second group, was
also isolated and identified from <u>Parazoanthus axinellae</u> (Cariello
et al., 1974b). This anthozoan also contains the demethylated derivative, 3-norpseudoanthoxanthin. A similar compound, paragracine, was
found in <u>Parazoanthus gracilis</u> (Komoda et al., 1975). The presence
of several additional related pigments has also been reported (Komoda,
personal communication, in Chevolot, 1981).

81   **82**

   Little is known about the biogenesis of these compounds. They
may be formed by duplication of two guanidine units derived from
arginine (Cariello et al., 1979).

   Paragracine has papaverine-like activity (Komoda et al., 1975).
Zoanthoxanthin and its demethylated derivative 3-norzoanthoxanthin
are DNA intercalating agents which selectively inhibit DNA synthesis
(Quadrifoglio et al., 1975).

7. CIPRIDINA LUCIFERIN

   <u>Cipridina</u> luciferin (83) was isolated by Shimomura et al. (1957)
from the marine ostracod crustacean <u>Cypridina</u> found along the coasts
of Japan. Its structure was established by chemical degradation studies (Kishi et al., 1966a,b) and confirmed by synthesis (Kishi et al.,
1966c).C. luciferin is probably formed by the condensation of three
amino acids, tryptophane, arginine and isoleucine (Scheuer, 1973).
It differs from other luciferins only in the nature of the constitu-

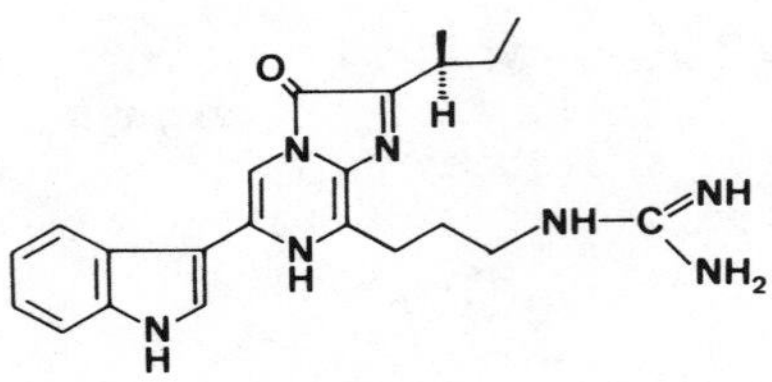

83

tive amino acids. C. luciferin is responsible for the bioluminescence
of the planktonic copepod Cipridina. Bioluminescence of marine organisms has recently been reviewed by Goto (1980).

8. GUANIDINO TOXINS

8.1. Tetrodotoxin. Saxitoxin

These compounds are among the most toxic non-protein substances
known. They have been recently reviewed (Scheuer, 1973, 1975, 1977;
Chevolot, 1981).

Tetrodotoxin (84), pufferfish (fugu) poison, is found in certain
species of the Tetrodontidae, one of the culinary delicacies of Japan.
The toxin occasionally appears in the fishes during the prespawning
period, and can be fatal. It is particularly concentrated in the
liver and ovaries. Tetrodotoxin was first isolated from pufferfish
ovaries in an impure form (Tawara, 1909), and later crystallized
(Tsuda and Kawamura, 1952). This substance was found to be identical
with tarichatoxin isolated from the salamander Taricha torosa
(Buchwald et al., 1964). It also occurs in the unrelated fish Gobius
criniger (Noguchi and Hashimoto, 1973), the octopus Hapalochlaena maculosa (Scheumack et al., 1978), and the frog Atelopus chiriquiensis
(Pavelka et al., 1977) which also contains a similar toxin, chirico-
toxin. Following extensive studies, the structure of tetrodotoxin was
elucidated in 1964 (review by Scheuer, 1973) and confirmed by total
synthesis (Kishi et al., 1972; Tanino et al., 1974). It contains an
unique hemilactal function and a guanidino group. The weak basicity
of the latter has been discussed on the basis of structural conside-
rations (Scheuer, 1973). Saxitoxin (85), resonsible for the "paralytic.

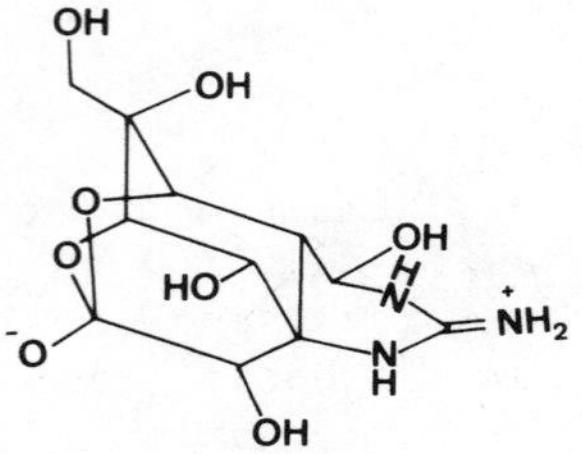

84

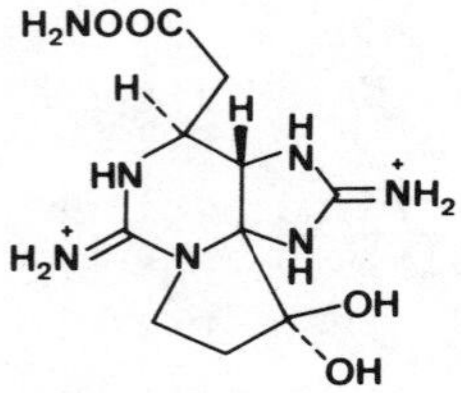

85

shellfish poisoning", was first isolated from toxic clams and mussels (Schantz et al., 1957). In fact, this compound is produced by various dinoflagellates of the genus Gonyaulax which become concentrated in the shellfish during massive blooms of these organisms ("red tides"). The connection between the dinoflagellate and the shellfish toxins was established by Schantz et al. (1966). The etiologic agents are G. catanella in the Pacific Ocean and G. tamarensis in the Atlantic Ocean. The latter produces in addition the related toxins neosaxi-toxin and gonyautoxins II and III. Saxitoxin has also been found in a blue-green alga (Moore, 1977) and a crab (Noguchi et al., 1969). Studies on the complex structure of saxitoxin have been summarized by Shimizu (1978). The molecule contains two guanidino groups. The total synthesis was carried out by Tanino et al. (1977). The guanidine group of these toxins can be derived from arginine but their biolo-gical origin is unclear. The maximum toxicity of pufferfishes during spawning months would support an endogenous origin, but does not explain why only half of the fishes are toxic during this period.

Tetrodotoxin and saxitoxin are nervous poisons which block nerve conduction by inhibiting the passage of sodium through electrically excitable membranes. They also have hypotensive properties. The me-chanism of their toxic action has been extensively studied. They are used as channel probes for the study of the molecular properties of ion channels in biomembranes. These toxins have been often reviewed (Russel, 1967; Kao, 1972; Narahashi; 1972; Reed and Trzos, 1979; Ritchie; 1980; Taylor et al., 1981).

8.2. Bufotoxin

Bufotoxin (86) is the principal toxin of the venom secreted in the parotid gland of the common european toad, Bufo vulgaris (Wieland

86

and Alles, 1922; Wieland et al., 1936; Meyer, 1949). This substance is an ester of suberylarginine and of a steroid glycoside, bufotaline, through the 3-hydroxyl of the steroid and the free carboxyl group of suberic acid. Bufotoxin is a cardiac poison and has similar effects on heart muscle to those of digitalin.

9. GUANIDINO ANTIBIOTICS FROM STREPTOMYCES

9.1. Streptomycin group

Streptomycin (87), obtained as a metabolite of Streptomyces griseus Waksman (Schatz et al., 1944) was the first antibiotic utilized in the treatment of tuberculosis. It belongs to the group of aminoglycoside antibiotics containing aminocyclitols. The guanidine moiety is streptidine, an inositol substituted with two guanidino groups. Streptomycin is the collective name for a family of compounds with similar structure. Streptomycin A (87a) is the basic unit. Streptomycin B (87b), streptomycin C (87c), and streptomycin D (87d) are mannosido- hydroxy- and dihydrostreptomycins, respectively. These substances are produced by specific strains of Streptomyces (reviews by Horner, 1967; Schlessinger and Medoff, 1975).

87

	R_1	R_2	R_3
a	CHO	H	CH_3
c	CHO	H	CH_2OH
d	CH_2OH	H	CH_3
b	CHO	(see structure)	CH_3

The biosynthesis of streptomycin has been the subject of several recent reviews (Pearce and Rinehart, 1981, and cited references) and will not be detailed here. D-glucose is the precursor for the carbon skeleton of the three units (streptidine, L-streptose and L-methyl-L-glucosamine). The two amidine groups of streptidine are provided by arginine and transferred to myoinositol in two similar consecutive series of five enzymatic steps involving oxidation, amination, phosphorylation, amidination and dephosphorylation. However, the dephosphorylation of streptidine phosphate after the introduction of the second amidine group probably occurs after it is incorporated into the streptomycin molecule.

Streptomycin is a broad spectrum antibiotic active against a range of gram-positive bacteria, gram-negative bacteria and myco-bacteria, particularly Mycobacterium tuberculosis. Dihydro- and hy-droxystreptomycin have virtually the same antibiotic spectrum as streptomycin. Mannosidostreptomycin is less active.

Streptomycin inhibits protein biosynthesis. It tindly binds to the ribosomes, affecting peptide chain initiation by blocking the 30S subunit initiation complex formation, by inducing breakdown of the 70S initiation complex and/or by inhibiting ribosomal dissocia-tion. It also interferes with peptide elongation process by causing misreading of the genetic code (reviews by Jacoby and Gorini, 1967; Schlessinger and Medoff, 1975; Tanaka, 1975).

Streptomycin has low acute toxicity and high chronic toxicity.

Bluensomycin (88), produced by Streptomyces bluensis, is another member of the streptomycin class of antibiotics. It differs from

88

streptomycin only in the structure of the guanidine-containing unit. One of the guanidine groups in the streptamine nucleus of streptomycin is replaced by $-OCONH_2$ in bluensomycin (Bannister and Argoudelis, 1963; Barlow and Anderson, 1972). Buensomycin is active against a range of bacteria and fungi. This substance causes misreading in vitro and in vivo (reviews by Jacoby and Gorini, 1967; Tanaka, 1975).

Glebomycin is produced by a new form of the strain Streptococcus hygroscopus (Okanishi et al., 1962; Miyaki et al., 1962). This sub-stance gives a positive Sakaguchi test. To our knowledge, its complete structure has not been reported. Glebomycin is active against acid-fast bacilli. Its antibiotic spectrum is similar to that of dihydro-streptomycin, but its specific activity is weaker. Glebomycin posses-ses comparatively low toxicity.

9.2. Bleomycin - Phleomycin group

Bleomycin B2 (89a) and bleomycin B4 (89b) are two guanidino glycopeptide antibiotics isolated by Umezawa et al. (1966) from the bleomycin complex produced by Streptomyces verticillus (reviews by Umezawa, 1975; Korzybski et al., 1978). They are closely related

89

a   R =   NH — (CH₂)₄ — NH — C — NH₂
                                       ‖
                                       NH

b   R =   NH — (CH₂)₄ — NH — C — NH — (CH₂)₄ — NH — C — NH₂
                                      ‖   ‖
                                      NH   NH

to the other guanidino antibiotics **phleomycin D1 (90a)** and **phleomycin E (90b)** isolated by Ikekawa et al. (1964) from the phleomycin complex which is produced by S. verticillus strain 831-1 (review by Korzybski et al., 1978).

The complete structure of the basic aminoglycosidic skeleton of the bleomycins and the phleomycins has been determined by hydrolytic studies, NMR spectroscopy and partial synthesis (Takita et al., 1978); it was confirmed by field desorption mass spectroscopy (Dell et al., 1980). The phleomycins differ from the bleomycins in the absence of a double bond in one of the bithiazole rings.

The bleomycins (and the phleomycins) differ from each other in the structure of their terminal amine (Fujii et al., 1973). These amines are guanidino amines in two bleomycins and in two phleomycins. Agmatine is the terminal amine for bleomycin B2 and phleomycin D1, and 1-(4-aminobutyl)-3-(4-guanidinobutyl)guanidine is the terminal amine for bleomycin B4 and phleomycin E. Addition of the terminal amine to the culture broth causes incorporation of this amine into bleomycin and suppresses production of other bleomycins (Umezawa,

90

a   R =   NH — (CH$_2$)$_4$ — NH — C — NH$_2$
                                   ||
                                   NH

b   R =   NH — (CH$_2$)$_4$ — NH — C — NH — (CH$_2$)$_4$ — NH — C — NH$_2$
                                  || ||
                                  NH   NH

1971). Thus, it is possible to produce artificial bleomycins.

Preliminary studies on the biosynthesis of the peptide part of bleomycins have been reviewed (Kurahashi, 1981).

pronounced antitumor activity (reviews by Umezawa, 1975, 1980; Korzybski et al., 1978). The antineoplasic properties of bleomycins are thought to arise from their interaction with cellular DNA and their ability to cause strand scissions of DNA. Bleomycins are used in the treatment of squamous cell carcinoma and of Hodgkin's disease. Phleomycins have not been developed for clinical utilization because of their renal toxicity (Dell et al., 1980, and cited references). utilization because of their renal toxicity (Dell et al., 1980, and cited references).

Victomycin (Takasawa et al., 1975a) and platomycins A and B (Takasawa et al., 1975b) produced by strains of Streptosporangium violaceochromagenes, belong to the bleomycin-phleomycin group. Their complete structure has not yet been established. Both compounds give a positive Sakaguchi test. Platomycins are chromatographically very similar to bleomycins. However, hydrolysis showed that platomycins differ from bleomycins in the bleomycic acid part of the molecule.

Victomycin is strongly active against gram-positive and gram-negative bacteria, including strains resistant to streptomycin and several other antibiotics. This substance has antitumor activity, but is inactive against leukemia L-1210. Platomycins have very similar properties.

9.3. Viomycin group

Viomycin (91) is produced by Streptomyces floridae (Bartz et al., 1951) and by S. puniceus (Finlay et al., 1951). Its complete structure

91

has been established (Bycroft et al., 1972a,b; Noda et al., 1972). Viomycin is a strongly basic polypeptide. The peptide chain contains a L-β-lysil residue. L-β-lysil residues have also been demonstrated in the antibiotics of the streptothricin group. The guanidine residue is incorporated into a cyclic guanidino carbinol system. This guanidinocarbinol structure also occurs in the pufferfish neurotoxin, tetrodotoxin. In viomycin, the cyclic guanidino carbinol structure exists in equilibrium with the acyclic guanidine aldehyde tautomer, which accounts for the positive Sakaguchi reaction shown. Viomycin is an antimicrobial and an antituberculous agent. Possible toxic-side effects are renal irritation and disturbance of the eighth cranial nerve.

Several antibiotics (e.g. **capreomycin, stendomycin, blasticidin S** and **tuberactinomycin**) comprise the same cyclic guanidine system as viomycin (Bycroft et al., 1972a and cited references).

Phthiomycin is produced by Streptomyces luteochromogenes (Maeda et al., 1953). Its structure has not been established. Its UV spectrum resembles that of viomycin in acidic medium but differs in alkaline solution. Phthiomycin has an actibacterial activity on acid-fast bacilli, but exhibits weaker activity on gram-negative microorganisms and is inactive on gram-positive microorganisms and fungi.

9.4. Streptothricin group

Streptothricin (92) (Waksman and Woodruff, 1942) and **streptolin** (93) (Rivett and Peterson, 1947) are two related antibiotics produced by different straïns of Streptomyces lavendulae. They have been reviewed by Korzybski et al., 1978). Their structure has been established by Van Tamelen et al.(1961). Both compounds have a

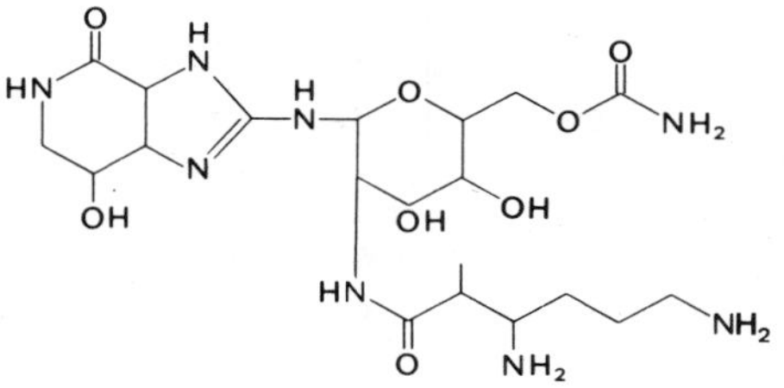

92

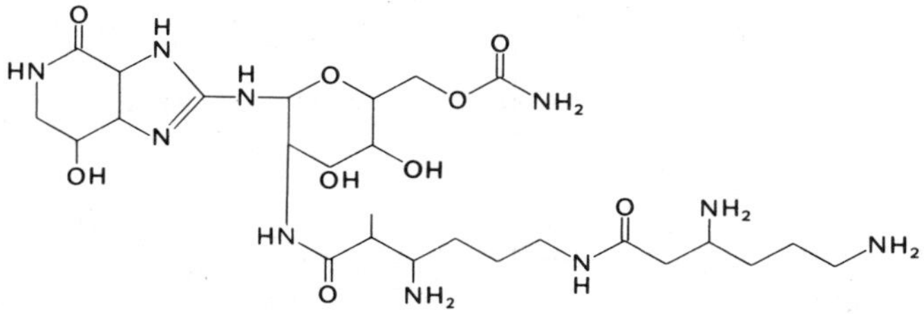

93

2-aminoimidazole ring linked with an aminosugar, D-2-gulosamine.
They differ in the residue bound to the amino group of the hexose;
L-β-lysil residue for streptothricine and L-β-lysil-L-β-lysil residue
for streptolin. As noted above, a β-lysil residue has also been found
in the structure of viomycin.

Streptothricin and streptolin have similar activities against
gram-positive and gram-negative microorganisms and against patho-
genic and saprophytic fungi. They have some toxic side-effect on
renal function.

Geomycin, produced by Streptomyces xanthophaeus nov. sp. (Brockmann and Musso, 1954) and roseothricin, isolated from cultures of
Streptomyces roseochromogenus (Hirata et al., 1955) also belong to
the streptothricin group of antibiotics. Like streptothricin and
streptolin, they contain L-β-lysil residues. The similarity between
the structures of streptolin, streptothricin, geomycin and roseothricin, suggests that the same basic structure may apply to all antibiotics of this group (Van Tamelen et al., 1961).

Geomycin is active against a number of microorganisms. The
substance is toxic and affects the renal function in the same way
as streptothricin.

ACKNOWLEDGEMENTS

    We would like to thank Dr. P. Desgrez for many helpful discus-
sions, Dr. N. Grant for pertinent comments in improving the English
text, Mr. R. Van Hove for his capable help with the photography and
Mrs. L. Vanhove for typing the manuscript. We wish also to thank
Dr. A. Mori, who has invited us to prepare this review.

BIBLIOGRAPHY

Ackermann, D., 1935a, Asterubin, eine schwefelhaltige Guanidinverbindung der belebten
    Natur, Z. physiol. Chem., 232 : 206.
Ackermann, D., 1935b, Synthese des Asterubins, Z. physiol.Chem., 234 : 208.
Ackermann, D. and Heinsen, H.A., 1935, Uber die physiologische Wirkung des Asterubins
    und anderer, zum Teil neu dargestellter, schwefelhaltiger Guanidinderivate, Z. phy-
    siol. Chem., 235 : 115.
Ackermann, D. and List, P.H., 1957, Konstitutionsermittlung des Herbipolins, einer
    neuen tierischen Purinbase, Z. physiol. Chem., 309 : 286.
Ackermann, D. and List, P.H., 1960, Zur Konstitution des Zooanemonins und des Herbi-
    polins, Z. physiol. Chem., 318 : 281.
Ackermann, D., and Menssen, H.G., 1959, Low-molecular nitrogenous constituents of the
    leather coral Alcyonium digitatum, Z. physiol. Chem., 317 : 144.
Ackermann, D. and Mohr, M., 1937, The occurrence of octopine, agmatine and arginine in
    the octopod, Eledone moschata, Z. physiol. Chem., 250 : 244.
Ackermann, D. and Müller, E., 1935, Zweite Synthese des Asterubins, Z. physiol. Chem.,
    235 : 233.
Ackermann, D. and Pant, R., 1961, Constituents of the sponge Calyx nicacensis, Z. phy-
    siol. Chem., 326 : 197.
Adams, W.S., Davis, F.W. and Hansen, L.E., 1964, Determination of serum creatinine by
    ion exchange chromatography and ultraviolet spectrophotometry, Anal. Chem., 36 :
    2209.
Albrecht, G.P., 1921, Chemical study of several marine molluscs of the pacific coast,
    J. Biol. Chem., 45 : 395.
Allan, J.D., Cusworth, D.C., Dent, C.E. and Wilson, V.K., 1958, A disease, probably
    hereditary, characterized by severe mental deficiency and a constant gross abnor-
    mality of amino acid metabolism, Lancet, 1 : 182.
Allende, C.C. and Allende, J.E., 1964, Purification and substrate specificity of argi-
    nyl-ribonucleic acid synthetase from rat liver, J. Biol. Chem., 239 : 1102.
Alles, G., 1926, The comparative physiological action of some derivatives of guanidine.
    J. Pharmacol. a. Therap., 28 : 251.
Andes, J.E., Linegar, C.R. and Myers, V.C., 1937, Guanidine-like substances in blood.
    II. Blood guanidine in nitrogen retention and hypertension, J. Lab. and Clin. Med.
    22 : 1209.
Angelini, C., Micoglio, G.F. and Trevisan, C., 1980, Guanidine hydrochloride in infan-
    tile and juvenile spinal muscular atrophy. A double blind controlled study, Acta
    Neurol., (Napoli), 6 : 460.
Armstrong, M.D. and Robinow, M., 1967, A case of hyperlysinemia: biochemical and cli-
    nical observations, Pediatrics, 39 : 546.
Auclair, M.C., Adolphe, M., Guillou, Y. and Robin, Y., 1976, Effet de la phascoline et
    de la phascolosomine, nouveaux dérivés guanidiques naturels, sur les cellules car-
    diaques de rat en culture, Compt. Rend. Soc. Biol., 170 : 65.
Audit, C., Viala, B. and Robin, Y., 1967, Biogénèse des dérivés diguanidiques chez la
    sangsue, Hirudo medicinalis L. I. Origine des groupements guanidiques et de la
    chaine carbonée, Comp. Biochem. Physiol., 22 : 775.
Baker, J.T., 1976a, Physiologically active substances from marine organisms, Aust. J.
    Pharm. Sci., 5 : 89.
Baker, J.T., 1976b, Some metabolites from australian marine organisms, Pure Appl. Chem.
    48 : 35.
Baker, J.T. and Murphy, V., 1976, in: "Handbook of Marine Science; Compounds from
    Marine Organisms", Vol.I., CRC Press, Cleveland, Ohio.
Baldwin, E., 1947, in: "Dynamic Aspects of Biochemistry", University Press, Cambridge,
    1st ed.
Baldwin, J. and Opie, A.M., 1978, On the role of octopine dehydrogenase in the adductor
    muscles of bivalve molluscs, Comp. Biochem. Physiol., 61B : 85.
Balinski, J.B., 1970, Nitrogen Metabolism in Amphibians, in: " Comparative Biochemistry
    of Nitrogen Metabolism ". 2. The Vertebrates., J.W. Campbell,ed., Academic Press,
    New -York.
Bannister, B.and Argoudelis, A.D., 1963, The chemistry of bluensomycin. II. The struc-
    ture of bluensomycin, J. Am. Chem. Soc., 85 : 234.
Barber, J.T. and Boulter, D., 1963, Argininosuccinic acid in germinating seeds of
    Vicia faba, Nature, 197 : 1112.
Barger, G. and White, F.D., 1922, The constitution of galegin, Biochem. J., 17 : 827.

Barlow, C.B. and Anderson, L., 1972, A study of the structure of bluensomycin with the tetramincopper reagent, J. Antibiotics, 25 : 281.

Barnes, R.L., 1962, Formation of γ-guanidinobutyric acid in Pine tissues, Nature, 193 : 781.

Bartz, Q.R., Ehrlich, J., Mold, J.D., Penner, M.A. and Smith, M.R., 1951, Viomycin, a new tuberculostatic antibiotic, Am. Rev. Tuberc., 63 : 4.

Beatty, I.M. and Magrath, D.I., 1959, Synthesis of DL- and L- lombricine, Nature, 183 : 591.

Beatty, I.M., Magrath, D.I. and Ennor, A.H., 1959, Occurrence of D-serine in lombricine, Nature, 183 : 591.

Beis, I. and Newsholme, E.A., 1975, The contents of adenine nucleotides, phosphagens and some glycolytic intermediates in resting muscles from vertebrates and invertebrates, Biochem. J., 152 : 23.

Bell, E.A., 1958, Canavanine and related compounds in Leguminosae, Biochem. J., 70 : 617.

Bell, E.A., 1960, Canavanine in the Leguminosae, Biochem. J., 75 : 618.

Bell, E.A., 1961, Isolation of a new amino acid from Lathyrus tingitanus, Biochim.Biophys.Acta., 47 : 602.

Bell, E.A., 1962a, α,γ-diaminobutyric acid in seeds of twelve species of Lathyrus and identification of a new natural amino acid, L-homoarginine, in seeds of other species toxic to man and domestic animals, Nature, 193 : 1078.

Bell, E.A., 1962b, The isolation of L-homoarginine from seeds of Lathyrus cicera, Biochem. J., 85 : 91.

Bell, E.A., 1963, New amino acid, γ-hydroxyhomoarginine in Lathyrus, Nature, 199 : 70.

Bell, E.A., 1964, The isolation of γ-hydroxyhomoarginine, as its lactone, from seeds of Lathyrus tingitanus, its biosynthesis and distribution, Biochem. J., 91 : 358.

Bell, E.A., 1965, Homoarginine, γ-hydroxyarginine and related compounds, in: "Comparative Biochemistry of Arginine and Derivatives", G.E.W. Wolstenholme and M.P. Cameron, eds., J. and A. Churchill, Ltd., London.

Bell, E.A. and Foster, R.G., 1962, Structure of lathyrine, Nature, 194 : 91.

Bell, E.A. and Przybylska, J., 1965, The origin and site of synthesis of the pyrimidine ring in the amino acid, Lathyrin, Biochem. J., 94 : 35p.

Bell, E.A. and Tirimanna, A.S.L., 1963, Occurrence of γ-hydroxyarginine in plants, Nature, 197 : 901.

Bell, E.A. and Tirimanna, A.S.L., 1964, The isolation of γ-hydroxyarginine, as its lactone, from seeds of Vicia sativa, and the identification of γ-hydroxyornithine as a naturally occurring amino acid, Biochem. J., 91 : 356.

Bergquist, P.R. and Hartman, W.D., 1969, Free amino acid patterns and the classification of the Demospongiae, Mar. Biol., 3 : 247.

Blass, J.P., 1960, The simple monosubstituted guanidines of mammalian brain, Biochem. J., 77 : 484.

Bloch, K. and Schoenheimer, R., 1941, The biological precursors of creatine, J. Biol. Chem., 138 : 167.

Böhles, H., Cohen, B.D. and Michalk, D., 1982, Guanidinosuccinic acid excretion in argininosuccinic aciduria, in : "Urea Cycle Diseases", A. Lowenthal, A. Mori and B. Marescau, eds., Plenum Press, New York.

Bomhoff, G., Klapwijk, P.M., Kester, H.C., Schilperoort, R.A., Hernalsteens, J.P. and Schell, J., 1976, Octopine and nopaline synthesis and breakdown genetically controlled by a plasmid of Agrobacterium tumefaciens, Mol. Gen. Genet., 145 : 171.

Borsook, H. and Dubnoff, J.W., 1940, The formation of creatine from glycocyamine in the liver, J. Biol. Chem., 132 : 559.

Borsook, H. and Dubnoff, J.W., 1941, The formation of glycocyamine in animal tissues J. Biol. Chem., 138 : 389.

Boulanger, P. and Osteux, R., 1956, Action de la L-aminoacide-dehydrogenase de foie de dindon (Meleagris gallopavo L.) sur les acides aminés basiques, Biochim. Biophys. Acta , 21 : 552.

Boulanger, P. and Osteux, R., 1960, Arginine metabolism in the white rat, Colloq.Intern. Centre Natl. Recherche Sci. (Paris), 92 : 326.

Brander, G. and Virtanen, A.I., 1964, α-keto-δ-guanidovaleriansaüre and γ-hydroxy-α-ketoglutarsaüre in grünen Teilen und Samen von Phlox Pflanzen, Acta Chem. Scand., 18 : 574.

Brnjevic, K. and Phillis, J.W., 1963, Iontophoretic studies of neurones in the mammalian cerebral cortex, J. Physiol., 165 : 274.

Brockmann, H. and Musso, H., 1954, Geomycin, ein neues, gegen gramnegative Bakterien wirksames Antibioticum, Naturwissenschaften, 41 : 451.

Buchwald, H.D., Durham, L., Fisher, H.G., Harada, R., Mosher, H.S., Kao, C.Y. and Fuhrman, F.A., 1964, Identity of tarichatoxin and tetrodotoxin, Science, 143 : 474.

Buergi, W., Colombo, J.P. and Richterich, R., 1966, Thin-layer chromatography of the acid- and ether-sol. DNP-aminoacids in urine, Klin. Wochschr., 43 : 1202.

Burns, D. and Sharpe, J.S., 1917, The parathyroids: Tetania parathyreopriva: Its nature, cause and relations to idiopathic tetany. V. Guanidine and methylguanidine in the blood and urine in tetania parathyreopriva and in the urine of idiopathic tetany, Quart. J. Exp. Physiol., 10 : 345.

Bycroft, B.W., Croft, L.R., Johnson, A.W. and Webb, T., 1972a, Viomycin. Part I. The structure of the guanidine-containing unit, J. Chem. Soc., Perkin I., p. 820.

Bycroft, B.W., Cameron, D., Croft, L.R., Hassanali-Walji, A., Johnson, A.W. and Webb,T. 1972b, Viomycin. Part II. The structure of the chromophore, J. Chem. Soc.,Perkin I. p. 827.

Cammarata, P.S. and Cohen, P.P., 1950, Scope of the transamination reaction in animal tissues, J. Biol. Chem., 187 : 439.

Cantoni, G.L. and Vignos, P.J., 1954, Enzymic mechanisms in transmethylation. V. Enzymic mechanism in creatine synthesis, J. Biol. Chem., 209 : 647.

Cariello, L., Crescenzi, S. and Prota, G., 1973, Zoanthoxanthin, a heteroaromatic base from Parazoanthus cfr. axinellæ (Zoantharia) : Structure confirmation by X-ray cristallography, J. Chem. Soc. Chem. Commun., p. 99.

Cariello, L., Crescenzi, S., Prota, G., Capasso, S., Giordano, F. and Mazzarella, L., 1974a, Zoanthoxanthin, a natural 1,3,5,7- tetracyclopent- (f) - azulene from Parazoanthus axinellae, Tetrahydron, 30 : 3281.

Cariello, L., Crescenzi, S., Prota, G. and Zanetti, L., 1974b, Zoanthoxanthins of a new structural type from Epizoanthus arenaceus, Tetrahydron, 30 : 4191.

Cariello, L., Crescenzi, S., Prota, G. and Zanetti, L., 1979, A survey on the distribution of zoanthoxanthins in some marine invertebrates, Comp. Biochem. Physiol., 63B : 77.

Carr, M.H. and Schloerb, P.R., 1960, Analysis for guanidine and methylguanidine in uremic plasma, Anal. Biochem., 1 : 221.

Carter, G.T. and Rinehart, K.L., 1978, Acarnidines, novel antiviral and antimicrobial compounds, J. Am. Chem. Soc., 100 : 4302.

Cheng, M.T. and Rinehart, K.L., 1978, Polyandrocarpidines : antimicrobial and cytotoxic agents from a marine tunicate (Polyandrocarpa sp.) from the gulf of California, J. Am. Chem. Soc., 100 : 7409.

Chevolot, L., 1981, Guanidine Derivatives, in : "Marine Natural Products", P.J.Scheuer, ed., Acad. Press, London. Vol.4.

Chevreul, M., 1835, Sur la composition chimique du bouillon de viande, J. Pharm. Sci. acces., 21 : 231.

Cimino, G., De Stefano, S. and Minale, L., 1974, Occurrence of hydroxyhydroquinone and 2- aminoimidazole in sponges, Comp. Biochem. Physiol., 47B : 895.

Cimino, G., De Stefano, S., Minale, L. and Sodano, G., 1975, Metabolism in porifera. III. Chemical patterns and the classification of the Demospongiae, Comp. Biochem. Physiol., 50B :279.

Cohen, B.D., 1970, Guanidinosuccinic acid in uremia. Arch. Intern. Med., 126 : 846.

Cohen, B.D. and Patel, H., 1982, Guanidinosuccinic acid and the alternate urea cycle, in : "Urea Cycle Diseases", A. Lowenthal, A. Mori and B. Marescau,eds., Plenum Press, New York.

Cohen, B.D., Stein, I.M. and Bonas, J.E., 1968, Guanidinosuccinic aciduria in uremia. A possible alternate pathway for urea synthesis, Amer. J. Med., 45 : 63.

Cohen, S.S., 1963, Sponges, cancer chemotherapy and cellular aging, Perspect. Biol. Med., 6 : 215.

Cohn, E.J. and Edsall, J.T., 1943, in : "Proteins, Amino Acids and Peptides as Ions and Dipolar Ions", Reinhold Publishing Corporation, New York.

Cooper, A.J.L. and Meister, A., 1978, Cyclic forms of the α- keto acid analogs of arginine, citrulline, homoarginine and homocitrulline, J. Biol. Chem., 253 : 5407.

Corrall,R.A., Orazi,O.O.and de Petruccelli, M.F., 1970, Synthesis of pterogynine and isolation of its isomer pterogynidine, a new guanidine alkaloid, J. Chem. Soc. D, p.556.

Corrall,R.A., Orazi, O.O. and de Petrucelli, M.F., 1972, Guanidine alkaloids of Pterogyne nitans Tul., Rev. Latinoamer. Quim., 2 : 178.

Corrigan, J.J., 1970, Nitrogen Metabolism in Insects, in : "Comparative Biochemistry of Nitrogen Metabolism". 1. The Invertebrates, J.W. Campbell ed., Academic Press, New York.

Curtis, D.R. and Watkins, J.C., 1960, The excitation and depression of spinal neurones by structurally related amino acids, J. Neurochem., 6 : 117.

Cusworth, D.C. and Dent, C.E., 1960, Renal clearances of amino acids in normal adults and in patients with amino-aciduria, Biochem. J., 74 : 550.

Davison, D.C. and Elliot, W.H., 1952, Enzymatic reaction between arginine and fumarate in plant and animal tissues, 169 : 313.

Davuluri, S.P., Hird, F.J.R. and Mc Lean, R.M., 1981, A re-appraisal of the function and synthesis of phosphoarginine and phosphocreatine in muscle, Comp. Biochem. Physiol., 69B : 329.

Dell, A., Morris, H.R., Hecht, S.M. and Levin, M.D., 1980, Characterisation of guanidino containing antibiotics: field desorption mass spectrometry of bleomycin B$_2$ and phleomycins D$_1$ and E, Biochem. Biophys. Res. Commun., 97 : 987.

Di Jeso, F., 1967, Sur la présence d'un phosphagène (phosphoarginine) et de la transférase assurant sa synthèse dans une bactérie: Escherichia coli, Compt. Rend. Soc. Biol., 161 : 584.

Dubnoff, J.W. and Borsook, H., 1941, A micromethod for the determination of glycocyamine in biological fluids and tissue extracts, J. Biol. Chem., 138 : 381.

Durzan, D.J. and Richardson, R.G., 1966, The occurrence and role of α-keto-δ-guanidinovaleric acid in white spruce (Picea glauca (Moench) Voss), Can. J. Biochem., 44 : 141.

Edwards, C. and Kuffler, S.W., 1959, The blocking effect of γ- aminobutyric acid (GABA) and the action of related compounds on single nerve cells, J. Neurochem., 4 : 19.

Eggleton, P. and Eggleton, G.P., 1927, Inorganic phosphate and a labile form of organic phosphate in the gastrocnemius of the frog, <u>Biochem. J.</u> , 21 : 190.
Ellington, W.R., 1980, Partial purification and characterization of a broadly-specific
  octopine dehydrogenase from the tissues of the sea anemone, Bunodosoma cavernata
  (Bosc), <u>Comp. Biochem. Physiol.</u>, 67B : 625.
Engeland, R., 1908, Uber den Nachweis organischer Basen im Harn, <u>Z. Physiol. Chem.</u>,
  57 : 49.
Ennor, A.H. and Morrison, J.F., 1958, Biochemistry of the phosphagens and related gua-
  nidines, <u>Physiol. Rev.</u>, 38 : 631.
Farley, J.M., Glavinovic, M.I., Watanabe, S. and Narahashi, T., 1979, Stimulation of
  transmitter release by guanidine derivatives, <u>Neurosci.</u>,4 : 1511.
Fearon, W.R. and Bell, E.A., 1955, Canavanine : detection and occurrence in Colutea
  arborescens, <u>Biochem. J.</u>, 59 : 221.
Findlay, L. and Sharpe, J.S., 1920, Adult tetany and methylguanidin; a metabolic study,
  <u>Quart. J. Med.</u>, 13 : 433.
Finlay, A.C., Hobby, G.L., Hochstein, F., Lees, T.M., Lenert, T.F., Means, J.A., P'an,
  S.Y., Regna, P.P., Routien, J.B., Sobin, B.A., Tate, K.B. and Kane, J.H., 1951,
  Viomycin, a new antibiotic active against mycobacteria, <u>Am. Rev. Tuberc.</u>, 63 : 1.
Fisher, B., Keller-Schierlein, W., Kneifel, H., König, W.A., Loeffler, W., Müller, A.,
  Muntwyler, R. and Zähner, H., 1973, Stoffwechselprodukte von Mikroorganismen.118
  Mitteilung. δ- N- hydroxy- L -arginin, ein Aminosäure-Antagonist aus Nannizzia
  gypsea, <u>Arch.Mikrobiol.</u>, 91 : 203.
Fiske, C.H. and Subbarow, Y., 1929, Phosphocreatine, <u>J. Biol. Chem.</u>, 81 : 629.
Forenza, S., Minale, L., Riccio, R. and Fattorusso, E., 1971, New bromo-pyrrole deri-
  vatives from the sponge Agelas oroides, <u>J. Chem. Soc. Chem. Commun.</u>, 1129.
Formstecher, P., 1978, L'argininosuccinylurie, in : "Le Cycle de l'Urée et ses Anoma-
  lies", J.P. Farriaux, ed., Doin, Paris.
Fort, L., Dando, P.R., Rouzé, P., Monneuse, M.O. and Olomucki, A., 1982, Immunological
  comparative studies of octopine dehydrogenase and other "pyruvate reductases"
  from different species, <u>Comp. Biochem. Physiol.</u>, 73B : 865.
Fühner, H., 1923, "Die Guanidingruppe", <u>Heffters Handb. d. exp. Pharmakol.</u>, 1 : 684.
Fujii, A., Takita, T., Maeda, K. and Umezawa, H., 1973, Chemistry of bleomycin. XI.
  Structures of the terminal amines, <u>J. Antibiot.</u>, 26 : 398.
Fujita, Y., 1959, γ-hydroxyarginine, a new guanidino compound from a sea cucumber,
  <u>Bull. Chem. Soc. Japan.</u>, 32 : 439.
Fujita, Y., 1960, γ-hydroxyarginine, a new guanidino compound from a sea cucumber. II.
  Determination of the configuration, <u>Bull. Chem. Soc. Japan</u>, 33 : 1379.
Gäde, G., 1980, Biological role of octopine formation in marine molluscs, <u>Marine Bio-
  logy Letters</u>, 1 : 121.
Gäde, G. and Zebe, E., 1973, Uber den Anaerobiosestoffwechsel von Molluskenmuskeln,
  <u>J. Comp. Physiol.</u>, 85 : 291.
Gale, E.F., 1946, The bacterial amino acid decarboxylases, <u>Adv. in Enzymol.</u>, 6 : 1.
Garcia, E.E., Benjamin, L.E. and Fryer, I.R., 1973, Reinvestigation into the structure
  of oroidin, a bromopyrrole derivative from marine sponge, <u>J. Chem. Soc. Chem.
  Commun.</u>, p. 78.
Garcia, I., Roche, J. and Tixier, M., 1956, Sur le métabolisme de la L-arginine chez
  les insectes.I., <u>Bull. Soc. Chim. Biol.</u>, 38 : 1423.
Giovannetti, S., Balestri, P.L. and Barsotti, G., 1973, Methylguanidine in uremia,
  <u>Arch. Intern. Med.</u>, 131 : 709.
Glasby, J.S., 1979, Encyclopaedia of antibiotics, John Wiley and Sons Ltd., Chichester,
  New York.
Goldmann-Menagé, A., 1970, Recherches sur le métabolisme azoté des tissus de crown-
  gall cultivés in vitro, <u>Ann. Sci. Nat. Bot., Paris</u>, 12ème série, 11 : 223.
Goldmann, A., 1977, Octopine and nopaline dehydrogenases in crown-gall tumors, <u>Plant
  Sci. Lett.</u>, 10 : 49.
Goldmann, A., Tempé, J. and Morel, G., 1968, Quelques particularités de diverses sou-
  ches d'Agrobacterium tumefaciens, <u>Compt. Rend. Soc. Biol.</u>, 162 : 630.
Goldmann, A., Thomas, D.W. and Morel, G., 1969, Sur la structure de la nopaline, méta-
  bolite anormal de certaines tumeurs de crown gall, <u>C. R. Acad. Sci., Paris</u>, 268 :
  852.
Goto, T., 1980, Bioluminescence of marine organisms, in : "Marine Natural Products",
  P.J. Scheuer,ed., Acad. Press, New York.Vol.3.
Greenwald, I., 1946, The presence of creatine in the testes of various invertebrates.
  The preparation of creatine phosphoric acid from fish testes, <u>J. Biol. Chem.</u>,
  162 : 239.
Guggenheim, M., 1951," Die Biogenen Amine", S. Karger Verlag, Basel, Suisse.
Guillou, Y. and Robin, Y., 1973, Phascoline (N-(3-guanidinopropionyl)-2-hydroxy-<u>n</u>-
  heptylamine) and phascolosomine (N-(3-guanidinoisobutyryl)-2-methoxy-<u>n</u>-heptyl-
  amine), two new guanidino compounds from sipunculid worms. Isolation and struc-
  ture, <u>J. Biol. Chem.</u>, 248 : 5668.
Guillou, Y. and Robin, Y., 1979, Présence de α-N-acétyl·agmatine chez des Cnidaires,
  Actinia equina et Actinia fragacea, <u>Compt. Rend. Soc. Biol.</u>, 173 : 576.
Guyon, P., Chilton, M.D., Petit, A. and Tempé, J., 1980, Agropine in "null-type" crown
  gall tumors : evidence for the generality of the opine concept, <u>Proc. Natl. Acad.
  Sci.</u>, USA, 77 : 2693.
Haurowitz, F. and Waelsch, H., 1926, Uber die chemische Zusammensetzung der Qualle
  Velella spirans, <u>Z. physiol. Chem.</u>, 161 : 300.

Hedin, S.G., 1895, Uber die Bildung von Arginin aus Protein Körpern, Z. physiol. Chem., 21 : 155.

Hiramatsu, C., 1980a, Guanidino compounds in mouse brain. I. Brain guanidino compound levels in twelve strains of mice, Okayama Igakkai Zasshi, 92 : 419.

Hiramatsu, C., 1980b, Guanidino compounds in mouse brain. II. Guanidino compound levels in brain in relation to convulsions, Okayama Igakkai Zasshi, 92 : 427.

Hirata, Y., Goto, T. and Hosoya, S., 1955, Structure of roseothricin, 3ème Congrès International de Biochimie, Résumé des communications, p. 95.

Hollenbeak, K.H. and Schmitz, F.J., 1977, Aplysinopsin : antineoplastic tryptophane derivative from the marine sponge Verongia spengelii, Lloydia, 40 : 479.

Hopkins, F.G., 1889, Note on a yellow pigment in butterflies, Nature, 40 : 335.

Horner, W.H., 1967, Streptomycin, in : "Antibiotics". Vol. II. D. Gottlieb and P.D. Shaw, eds, Springer Verlag, Berlin.

Horowitz, N.H. and Srb, A.M., 1948, Growth inhibition of Neurospora by canavanine and its reversal, J. Biol. Chem., 174 : 371.

Horowitz, H.I., Stein, I.M., Cohen, B.D. and White, J.G., 1970, Further studies on the platelet inhibitory effect of guanidinosuccinic acid and its role in uremic bleeding, Amer. J. Med., 49 : 336.

Hosotani, M., 1973, Guanidino compounds in brain by automatic liquid chromatography, Okayama Igakkai Zasshi, 85 : 373.

Hunter, A., 1928, "Creatine and Creatinine". Longmans Green, London.

Hurley, K.E. and Williams, R.J., 1955, Urinary amino acids, creatinine and phosphate in muscular dystrophy, Arch. Biochem. and Biophys., 54 : 384.

Ikekawa, T., Iwani, F., Hiranaka, H. and Umezawa, H., 1964, Separation of phleomycin components and their properties, J. Antibiotics., 17 : 194.

Irreverre, F. and Evans, R.L., 1959, Isolation of γ-guanidinobutyric acid from calf brain, J. Biol. Chem., 234 : 1438.

Irreverre, F., Evans, R.L., Hayden, A.R. and Silber, R., 1957, Occurrence of γ-guanidinobutyric acid, Nature, 180 : 704.

Irvin, J.L. and Wilson, D.W., 1939, Studies on octopine. II. The nitrogenous extractives of squid and octopus muscle, J. Biol. Chem., 127 : 565.

Iseki, T., 1931, Uber die basischen Extraktivstoffe des Octopodenmuskels, Z. physiol. Chem., 203 : 259.

Ito, K. and Hashimoto, Y., 1965, Occurrence of γ- (guanylureido) butyric acid in a red alga, Gymnogongrus flabelliformis, Agr. Biol. Chem., 29 : 832.

Ito, K. and Hashimoto, Y., 1966a, Gigartinine : a new amino acid in red algae, Nature, 211 : 417.

Ito, K. and Hashimoto, Y., 1966b, Distribution of gongrine and gigartinine in marine algae, Nippon Suisan Gakkaishi, 32 : 727.

Ito, K. and Hashimoto, Y., 1969, Syntheses of DL-gigartinine and gongrine, Agr. Biol. Chem. (Tokyo), 33 : 237.

Ito, K., Miyazawa, K. and Hashimoto, Y., 1967, Occurrence of γ-guanidinobutyric acid and concentration of gongrine and gigartinine in a red alga, Gymnogongrus flabelliformis, Nippon Suisan Gakkaishi, 33 : 572.

Jacoby, G.A. and Gorini, L., 1967, The effect of streptomycin and other aminoglycoside antibiotics on protein synthesis, in : "Antibiotics", Vol. I , D. Gottlieb and P.D. Shaw, eds., Springer Verlag, Berlin.

Jinnai, D., Sawai, A. and Mori, A., 1966, γ-Guanidinobutyric acid as a convulsive substance, Nature, 212 : 617.

Jinnai, D., Mori, A., Mukawa, J., Ohkusu, H., Hosotani, M., Mizuno, A. and Tye, L.C., 1969, Biological and physiological studies on guanidino compounds induced convulsions, Jpn. J. Brain Physiol., 106 : 3668.

Kakimoto, Y. and Akazawa, S., 1970, Isolation and identification of N^G,N^G- and N^G,N'^G-dimethylarginine, N^ϵ-mono-, di-, and trimethyllysine, and glucosegalactosyl- and galactosyl-δ-hydroxylysine from human urine, J. Biol. Chem., 245 : 5751.

Kalckar, H.M., 1941, Nature of energetic coupling in biological synthesis, Chem. Revs, 28 : 71.

Kalyankar, G.D., Ikawa, M. and Snell, E.E., 1958, The enzymatic cleavage of canavanine to homoserine and hydroxyguanidine, J. Biol. Chem., 233 : 1175.

Kao, C.Y., 1972, Pharmacology of tetrodotoxin and saxitoxin, Fed. Proc., 31 : 1117.

Karrer, P., Manunta, C. and Schwyzer, R., 1948, Uber ein Vorkommen von Purinen und eines Pterins in einer Ascidienart (Microcosmus polymorphus), Helv. Chim. Acta, 31 : 1214.

Kazlaukas, R., Murphy, P.T., Quinn, R.J. and Wells, R.J., 1977, Aplysinopsin, a new tryptophan derivative from a sponge, Tetrahydron Lett., p.61.

Kishi, Y., Goto, T., Hirata, Y., Shimomura, O. and Johnson F.H., 1966a, Cipridina bioluminescence I. Structure of Cipridina luciferin, Tetrehydron Lett., 3427.

Kishi, Y., Goto, T., Eguchi, S., Hirata, Y., Watanabe, E. and Aoyama, T., 1966b, Cipridina bioluminescence II. Structural studies of Cipridina luciferin by means of high resolution mass spectrometer and an amino acid analyzer, Tetrahydron Lett., 3437.

Kishi, Y., Goto, T., Inoue, S., Sugiura, S. and Kishimoto, H., 1966c, Cipridina bioluminescence III. Total synthesis of Cipridina luciferin, Tetrahydron Lett., 3445.

Kishi, Y., Fukuyama, T., Aratani, M., Nakatsubo, F., Goto, T., Inoue, S., Tanino, H. Sugiura, S. and Kakoi, H., 1972, Synthetic studies on tetrodotoxin and related compounds. IV. Stereospecific total synthesis of DL-tetrodotoxin, J. Am. Chem. Soc., 94 : 9219.

Kitagawa, M. and Tomiyana, T., 1929, A new amino-compound in the jack-bean and a cor-
    responding new enzyme, J. Biochem., 11 : 265.
Klinger, R., 1921, Beiträge zur pharmakologischen Wirkung der Guanidins, Arch. exp.
    Pathol. u. Pharmakol., 90 : 129.
Koch, W., 1913, Toxic bases in the urine of parathyreodectomized dogs, J. Biol. Chem.
    15 : 43.
Komoda, Y., Kanedo, S., Yamamoto, M., Ishikawa, M., Itai, A. and Itaka, Y., 1975,
    Structure of paragracine, a biologically active marine base from Parazoanthus
    gracilis, Chem. Pharm. Bull., 23 : 2464.
Korzybski, T., Kowszyk-Gindifer, Z. and Kurlowicz, W., 1978, "Antibiotics : Origin,
    Nature and Properties" (translated by E. Paryski), American Soc. for Microbiol.,
    Washington.
Kossel, A., 1910, Uber das Agmatin, Z. physiol. Chem., 66 : 257.
Kostir, J.V. and Pristoupil, T.I., 1953, Paper chromatography of urinary glycocyami-
    dine, Capsopis lekaru ceskych, 92 : 188. (Chem. Abstr., 51 : 18234c).
Krebs, H.A. and Henseleit, K., 1932, Untersuchungen über die Harnstoffbildung im Tier-
    körper, Z. physiol. Chem., 210 : 33.
Krnjevic, K. and Phillis, J.W., 1963, Iontophoretic studies of neurones in the mamma-
    lian cerebral cortex, J. Physiol., 165 : 274.
Kumagai, T., Kaway, S. and Shikinami, Y., 1928, Uber die Guanidinderivate, welche auf
    den Blutzucker senkend wirken, Proc. Imp. Acad. (Japan), 4 : 23.
Kurahashi, K., 1981, Biosynthesis of Peptide Antibiotics, in : "Antibiotics", Vol. IV.
    J.W. Corcoran, ed., Springer Verlag, Berlin.
Kutscher, F., 1907, Der Nachweis toxischer Basen im Harn, Z. physiol. Chem., 51 : 457.
Kutscher, F. and Ackermann, D., 1931, Uber das Arcain, Z. physiol. Chem., 203 : 132.
Lacombe, G., Thiem, N.V., Thoai, N.V. and Roche, J., 1964, Biogénèse de l'acide δ-gua-
    nidinovalerianique, Compt. Rend. Soc. Biol., 158 : 43.
Linneweh, F., 1931a, Uber die Spaltung des Arcains durch Mikroorganismen, Z. Physiol.
    Chem., 202 : 1.
Linneweh, W., 1931b, Uber das pharmakologische Verhalten des Arcains, Z. Biol., 92 : 163.
Lissitzky, S., Garcia, I. and Roche, J., 1954, Sur les dérivés guanidiques du muscle
    de scorpion, Androctonus australis, Compt. Rend. Soc. Biol., 148 : 436.
Lowenthal, A. and Marescau, B., 1981, Urinary excretion of monosubstituted guanidines
    in patients affected with urea cycle diseases, in : "Neurogenetics and Neuro-
    ophtalmology", A. Huber and D. Klein, eds., Elsevier/North-Holland Biomedical Press.
McLennan, H., 1959, The identification of one active component from brain extracts
    containing factor I., J. Physiol. (London), 146 : 358.
Maeda, K., Okami, Y., Utahara, R. and Umezawa, H., 1953, An antibiotic, phthiomycin,
    J. Antibiotics , 6 : 183.
Major, R.H. and Weber, C.J., 1927, Probable presence of increased amounts of guanidine
    in blood of patients with arterial hypertension, Bull. Johns Hopkins Hosp., 40 :
    85.
Makisumi, S., 1961, Guanidino compounds from a Sea-anemone, Anthropleura japonica
    Verril, J. Biochem., 49 : 284.
Marescáu, B., 1981, Analytische Studie van Guanidine-Derivaten in Urine van Patienten
    met Hyperargininemia, Thesis, Universiteit Antwerpen.
Marescau, B., Pintens, J., Lowenthal, A., Esmans, E., Luyten, Y., Lemière, G.,Dommise,
    R., Alderweireldt, F. and Terheggen, H.G., 1981, Isolation and identification of
    2-oxo-5-guanidinovaleric acid in urine of patients with hyperargininemia by chro-
    matography and gas chromatography/mass spectrometry, J. Clin. Chem. Clin. Biochem.
    19 : 61.
Marescau, B., Lowenthal, A., Terheggen, H.G., Esmans, E. and Alderweireldt, F., 1982,
    Guanidino compounds in hyperargininemia, in : "Urea Cycle Diseases", A. Lowenthal,
    A. Mori and B. Marescau, eds., Plenum Press, New York.
Matsumoto, M. and Mori, A., 1976, Effect of guanidino compounds on rabbit brain micro-
    somal Na$^+$-K$^+$ ATPase activity, J. Neurochem., 27 : 635.
Matsumoto, M., Kishikawa, H. and Mori, A., 1976a, Guanidino compounds in the sera of
    uremic patients and in the sera and brain of experimental uremic rabbits, Biochem.
    Medicine., 16 : 1.
Matsumoto, M., Kobayashi, K., Kishikawa, H. and Mori, A., 1976b, Convulsive activity
    of methylguanidine in cats and rabbits, IRCS Med. Sci., 4 : 65.
Matsumoto, M., Yokoi, I., Takeuchi, H. and Mori, A., 1976c, Effect of guanidino com-
    pounds on the electrical activity of giant neurons identified in subesophageal
    ganglia of the african giant snail, Achatina fulica Férussac, Comp. Biochem.
    Physiol., 54C : 123.
Matsumoto, M., Fujiwara, M., Mori, A. and Robin, Y., 1977, Effet des dérivés guani-
    diques sur la cholinestérase et sur l'acétylcholinestérase du cerveau de lapin,
    Compt. Rend. Soc. Biol., 171 : 1226.
Matsumoto, M., Kobayashi, K. and Mori, A., 1979, Distribution of guanidino compounds
    in bovine brain, J. Neurochem., 32 : 645.
Mayeda, H., 1936, Uber die Extraktivstoffe aus den Schliessmuskeln von Pecten (Patino-
    pecten) yessoensis Jay, Acta Schol. Med. Univ. Kyoto., 18 : 218.
Meister, A.,1965, "Biochemistry of the Amino Acids", A. Meister,ed., Acad. Press, New
    York.
Meister, A. and Wellner, D., 1963, Flavoprotein Amino Acid Oxidases, in : "The Enzymes"
    P.D. Boyer, H. Lardy and K. Myrbäk,eds., Academic Press, New York., Vol.7.,2d.ed.

Menagé, A. and Morel, G., 1964, Sur la présence d'octopine dans les tissus de crown-Gall, C. R. Acad. Sci. Paris, 259 : 4795.

Menne, F. and Beckmann, R., 1955, Creatine metabolism in children with dystrophia musculorum progressiva Erb., Klin. Wochschr., 33 : 556.

Meyer, K., 1949, Uber herzaktive Krötengifte (Bufogenine). Konstitution des Bufotalins, Helv. Chim. Acta, 32 : 1993.

Meyerhof, O., 1928, Über die Verbreitung der Argininphosphorsäure in der Muskulatur der Wirbellosen, Arch. Sci. biol. Napoli., 12 : 536.

Meyerhof, O. and Lohmann, K., 1928, Über die naturlichen Guanidophosphorsäuren (Phosphagene) in der quergestreiften Muskulatur, Biochem. Z., 196 : 22 and 49.

Miersch, J. and Reinbothe, H., 1966, Chromatographic separation of amino acids and guanidino compounds from fruit-bodies of higher fungi, Flora, 156 : 443.

Miyaki, T., Tsukiura, H., Wakae, M. and Kawaguchi, H., 1962, Glebomycin, a new member of the streptomycin class. II. Isolation and physicochemical properties, J. Antibiotics, 15 : 15.

Mizuno, A., Mukawa, J., Kobayashi, K. and Mori, A., 1975, Convulsive activity of taurocyamine in cats and rabbits, IRCS Med. Sci., 3 : 385.

Monneuse-Doublet, M.O., Lefebure, F. and Olomucki, A., 1980, Isolation and characterization of two molecular forms of octopine dehydrogenase from Pecten maximus L. Eur. J. Biochem., 108 : 261.

Moore, E. and Wilson, D.W., 1937, Nitrogenous extractives of scallop muscle. I. The isolation and study of the structure of octopine, J. Biol. Chem., 119 : 573.

Moore, R.E., 1977, Toxin from blue-green Algae, Bioscience, 27 : 797.

Morel, G. and Duranton, H., 1958, Le métabolisme de l'arginine par les tissus végétaux, Bull. Soc. Chim. Biol., 40 : 2155.

Mori, A., 1980, Natural occurrence and Analyses of Guanidino Compounds, Jpn. J. Clin. Chem., 9 : 232. Review (in Japanese).

Mori, A., 1983, Guanidino compounds and neurological disorders, Neurosciences (Kobe) 9 : 149. Review (in English).

Mori, A. and Ohkusu, H., 1971, Isolation and identification of alpha-N-acetyl-L-arginine and its effect on convulsive seizure, Adv. Neurol. Sci. (Tokyo), 15 : 303.

Mori, A., Tanaka, K., Tomita, T., Nakamura, K. and Hayashi, T., 1969, α-N-acetyl-γ-hydroxy-L-arginine in the human placenta, Biochim. Biophys. Acta, 192 : 255.

Mori, A., Hosotani, M. and Tye, L.C., 1974, Studies on brain guanidino compounds by automatic liquid chromatography, Biochem. Med., 10 : 8.

Mori, A., Hiramatsu, M., Takahashi, K. and Kohsaka, M., 1975, Guanidino compounds in rat organs, Comp. Biochem. Physiol., 51B : 143.

Mori, A., Ichimura, T. and Matsumoto, H., 1978, Gas chromatography-mass spectrometry of guanidino compounds in brain, Anal. Biochem., 89 : 393.

Mori, A., Ohkusu, H., Katayama, Y. and Watanabe, Y., 1979, Identification of guanidinosuccinic acid, guanidinoglutaric acid and homoarginine in the brain, Neuroscience Letters, suppl. 2, 14.

Mori, A., Akagi, M., Katayama, Y. and Watanabe, Y., 1980, α-guanidinoglutaric acid in cobalt-induced epileptogenic cerebral cortex of cats, J. Neurochem., 35 : 603.

Mori, A., Katayama, Y., Yokoi, I. and Matsumoto, M., 1981a, Inhibition of taurocyamine (guanidinotaurine)-induced seizures by taurine, in : "The Action of Taurine on Excitable Tissues", S.W. Schaffer, S.I. Baskin and J.J. Kocsis, eds., Spectrum Publications, New York.

Mori, A., Watanabe, Y., and Fujimoto, N., 1981b, Fluorometrical analysis of guanidino compounds in human cerebrospinal fluid, J. Neurochem., 38 : 448.

Mori, A., Watanabe, Y. and Akagi, M., 1982a, Guanidino compound anomalies in epilepsy, in : "Advances in Epileptology" : XIIIth Epilepsy Symposium, H. Akimoto, H. Kasamatsuri, M. Seino and A. Ward, eds., Raven Press, New York.

Mori, A., Watanabe, Y., Shindo, S., Akagi, M. and Hiramatsu, M., 1982b, α-guanidinoglutaric acid and epilepsy, in : "Urea Cycle Diseases", A. Lowenthal, A. Mori and B. Marescau, eds., Plenum Press, New York.

Morizawa, K., 1927, The extractive substances in Octopus octopodia, Acta Schol. Med. Univ. Imp. Kyoto, 9 : 285.

Mourgue, M. and Dokhan R., 1954, Les dérivés d'oxydation de l'arginine chez les végétaux, Compt. Rend. Soc. Biol., 148 : 1434.

Mourgue, M., Baret, R. and Dokhan, R., 1953, Sur la présence de dérivés guanidiques dans les graines de ricin (Ricinus communis), Compt. Rend. Soc. Biol. 147 : 1449.

Müller, H., 1925, Physiologische und chemische Studien über die Tanretsche Guanidinbase Galegin, Z. Biol., 83 : 320.

Murray, M. and Hoffmann, A.B., 1940, The occurrence of guanidine-like substances in the blood in essential epilepsy, J. Lab. Clin. Med., 25 : 1072.

Narahashi, T., 1972, Mechanism of action of tetrodotoxin and saxitoxin on excitable membranes, Fed. Proc., 31 : 1124.

Natelson, S. and Sherwin, J.E., 1979, Proposed mechanism for urea nitrogen re-utilization: Relationship between urea and proposed guanidine cycles, Clin. Chem., 25 : 1343.

Natelson, S., Stein, I. and Bonas, J.E., 1964, Improvements in the method of separation of guanidino amino acids by column chromatography. Isolation and identification of guanidinosuccinic acid from human urine, Microchem. J., 8 : 371.

Needham, A.E., 1970, Nitrogen Metabolism in Annelida, in : "Comparative Biochemistry of Nitrogen Metabolism", J.W. Campbell, ed., Academic Press, London. Vol.1.

Nichols, J.M., Adams, D.G. and Carr, N.G., 1980, Effect of canavanine and other amino-acid analogues on akinete formation in the cyanobacterium Anabaena cylindrica, Arch. Microbiol., 127 : 67.

Noda, T., Take, T., Nagata, A., Wakamiya, T. and Shiba, T., 1972, Chemical studies on tuberactinomycin. III. The chemical structure of viomycine (tuberactinomycine B), J. Antibiotics, 25 : 427.

Noguchi, T. and Hashimoto, Y., 1973, Marine toxins. XXXVIII. Isolation of tetrodotoxin from a goby Gobius criniger, Toxicon, 11 : 305.

Noguchi, T., Konosu , S. and Hashimoto, Y., 1969, Identity of the crab toxin with saxi-toxin, Toxicon, 7 : 325.

Obata, Y. and Iimori, M., 1952, Octopine biosynthesis by Saccharomyces cerevisiae, J. Chem. Soc. Jpn., 73 : 832.

Oesper, P., 1950, Sources of the high energy content in energy-rich phosphates, Arch. Biochem., 27 : 255.

Ohkusu, H. and Mori, A., 1969, Isolation of α-N-acetyl-L-arginine from cattle brain, J. Neurochem., 16 : 1485.

Ohtaka, Y. and Uchida, K., 1959, Studies on the constituents of sake. Part VII. Iden-tification of acetylagmatine (a new substance) and ethanolamine, J. Agr. Chem. Soc. Japan, 33 : 679.

Okanishi, M., Koshiyama, H., Ohmori, T. and Kawaguchi, H., 1962, Glebomycin, a new member of the streptomycin class. I. Biological studies, J. Antibiotics, 15 : 7.

Oriol-Audit, C., 1978, Polyamine-induced actin polymerization, Eur. J. Biochem., 87 : 371.

Otten, L.A.B.M., 1979, Lysopine dehydrogenase and nopaline dehydrogenase from crown gall tumor tissues, Thesis dissert. University Leiden.

Paik, W.K. and Kim, S., 1975, Protein methylation: chemical, enzymological and biolo-gical significance, Adv.Enzymol., 42 : 227.

Palmer, H.D., McNairScott, D.B. and Elliott, K.A.C., 1943, A note on the blood guanidine level in migraine subjects, J. Lab. Clin. Med., 28 : 735.

Patthy, A., Bajusz, S. and Patthy, L., 1977, Preparation and Characterization of N^G-Mono-, Di- and Trimethylated Arginines, Acta Biochim.and Biophys.Acad. Sci. Hung., 12 : 191.

Pauling, L., 1960,   "The Nature of the Chemical Bound", Cornell University Press, Ithaca.

Pavelka, L.A., Kim, Y.H. and Mosher, H.S., 1977, Tetrodotoxin and Tetrodotoxin-like compounds from the eggs of the Costa Rican frog, Atelopus chiriquiensis, Toxicon, 15 : 135.

Pearce, C.L. and Rinehart, K.L. Jr., 1981, Biosynthesis of Aminocyclitol Antibiotics, in : "Antibiotics", Vol. IV , J.W. Corcoran ed., Springer Verlag, Berlin.

Perez, G., Rey, A. and Schiff, L., 1976, The biosynthesis of guanidinosuccinic acid by perfused rat liver, J. Clin. Invest., 57 : 807.

Petit, A. and Morel, G., 1966, Le métabolisme de l'homoarginine par les tissus de crown-gall, Compt. Rend. Soc. Biol., 160 : 1806.

Petit, A. and Tempé, J., 1978, Isolation of Agrobacterium Ti-plasmid regulatory mutants, Mol. gen. Genet., 167 : 147.

Petit, A., Tempé, J. and Morel, G., 1968, Sur la présence d'un produit de transforma-tion de la canavanine dans les tissus tumoraux de Canavalia ensiformis, Compt. Rend. Soc. Biol., 162 : 632.

Petit, A.,Delhaye, S., Tempé, J. and Morel G., 1970, Recherches sur les guanidines des tissus de crown-gall. Mise en évidence d'une relation biochimique spécifique entre les souches d'Agrobacterium tumefaciens et les tumeurs qu'elles induisent, Physiol. Vég., 8 : 205.

Pettenkofer, M., 1844, Vorläufige Notiz über einen neuen stickstoffhaltingen Körper im Harne, Liebigs Ann. d. Chem., 52 : 97.

Pisano, J.J., Abraham, D. and Udenfriend, S., 1963, Biosynthesis and disposition of γ.guanidinobutyric acid in mammalian tissues, Arch. Biochem. Biophys., 100 : 323.

Prota, G., 1980, Nitrogenous Pigments in Marine Invertebrates, in : "Marine Natural Products", P.J. Scheuer ed., Academic Press, New York.Vol.3.

Pufahl, K. and Schreiber, K., 1961, Isolation of a new guanidine derivative from goat's rue, Galega officinalis, Experientia, 17 : 302.

Quadrifoglio, F., Crescenzi, V., Prota, G., Cariello, L., Di Marco, A. and Zunino, F., 1975, Interaction of natural tetra-azacyclopentazulene dyes with DNA and their effects on the DNA and RNA polymerase reactions, Chem. Biol. Interact., 11 : 91.

Quastel, J.H. and Witty, R., 1951, Ornithine transaminase, Nature . 167 : 556.

Rao, S.L.N., Ramachandran, L.K. and Adiga, P.R., 1963, The isolation and characteriza-tion of L-homoarginine from seeds of Lathyrus sativus, Biochemistry, 2 : 298.

Ratner, S., Petrak, B. and Rochovanski, O., 1953, Biosynthesis of urea. V. Isolation and properties of argininosuccinic acid, J. Biol. Chem., 204 : 95.

Reed, J.K. and Trzos, W., 1979, Interaction of substituted guanidines with the tetrodo-toxin-binding component in Electrophorus electricus, Arch. Biochem. Biophys., 195 : 414.

Regnouf, F. and Thoai, N.V., 1970, Octopine and lactate dehydrogenases in mollusc mus-cles, Comp. Biochem. Physiol., 32 : 411.

Reinbothe, H., 1963, Urea metabolism in Basidiomycetes. II. Formation of γ-guanidobuty-ric acid in fruit bodies of Lycoperdon, Phytochemistry, 13 : 327.

Reinbothe, H. and Mothes, K., 1962, Urea, ureides and guanidines in plants, Ann. Rev. Plant Physiol., 13 : 129.

Reiter, A.J. and Horner, W.H., 1979, Studies on the metabolism of guanidino compounds in mammals. Formation of guanidine and hydroxyguanidine in the rat, Arch. Biochim. Biophys., 197 : 126.

Reuter, G. Von, 1964, Zur Biochemie und Physiologie von Galegin in Galega officinalis L. Flora, 154 : 136.

Ritchie, J.M., 1980, Tetrodotoxin and saxitoxin and the sodium channels of excitable tissues, Trends Pharmacol. Sci., 1 : 275.

Rivett, R.W. and Peterson, W.H., 1947, Streptolin, a new antibiotic from a species of Streptomyces, J. Am. Chem. Soc., 69 : 3006.

Robin, Y., 1954, Répartition et métabolisme des guanidines monosubstituées d'origine animale, Thèse de Doctorat ès Sciences Naturelles, Paris.

Robin, Y., 1964a, Biological distribution of guanidines and phosphagens in marine anne- lida and related phyla from California, with a note on pluriphosphagens, Comp. Biochem. Physiol., 12 : 347.

Robin, Y., 1964b, Présence de l'acide β-guanidoisobutyrique libre et combiné chez des vers marins, Biochim. Biophys. Acta, 93 : 206.

Robin, Y., 1974, Phosphagens and molecular evolution in worms, Biosystems, 6 : 49.

Robin, Y., 1980, Les phosphagènes des animaux marins, in " Actualités de Biochimie Marine", Vol. 2 , Y. Le Gal, ed., Centre Natl. Rech. Sci., Paris.

Robin, Y., 1982, Metabolism of arginine in invertebrates : relation to urea cycle and to other guanidine derivatives, in : "Urea Cycle Diseases", A. Lowenthal, A. Mori and B. Marescau, eds., Plenum Press., New York.

Robin, Y. and Audit, C., 1966, Biogénèse des dérivés guanidiques chez Audouinia tenta- culata Mtg, Compt. Rend. Soc. Biol., 160 : 1410.

Robin, Y. and Guillou, Y., 1980, Contribution à l'étude des dérivés guanidiques de deux cnidaires, Actinia equina et A. fragacea, et de la bonellidine de l'échiurien Bonellia viridis, "Oceanis", 5 (Fasc. Hors Série): 575.

Robin, Y.and Roche, J., 1954, Sur la présence de taurocyamine (guanidotaurine) chez des Coelentérés et des Spongiaires, Compt. Rend. Soc. Biol., 148 : 1783.

Robin, Y. and Roche, J., 1965, Répartition biologique des guanidines substituées chez des vers terrestres et d'eau douce (oligochètes, Hirudinées, Turbellariées) récol- tés en Hongrie, Comp. Biochem. Physiol., 14 : 453.

Robin, Y. and Thoai, N.V., 1957, Métabolisme oxydatif de la L-arginine chez la limnée, Limnaea stagnalis L. I. Oxydation par la L-aminoacideoxydase, Compt. Rend. Soc. Biol., 151 : 2093.

Robin, Y. and Thoai, N.V., 1961a, Structure et synthèse de l'hirudonine (diamidino- spermidine ou N- (3-guanidopropyl)-4-aminobutylguanidine), C. R. Acad. Sci. Paris, 252 : 1224.

Robin, Y. and Thoai, N.V., 1961b, Métabolisme des dérivés guanidylés. X. Métabolisme de l'octopine: son rôle biologique, Biochim. Biophys. Acta, 52 : 233.

Robin, Y. and Thoai, N.V., 1962, Sur une nouvelle guanidine monosubstituée biologique, l'hypotaurocyamine (acide 2-guanidoéthane sulfinique) et le phosphagène correspon- dant, Biochim. Biophys. Acta, 63 : 481.

Robin, Y. and Viala, B., 1966, Sur la présence d'ATP : arginine phosphotransferase chez Tetrahymena pyriformis W. Cambridge, Comp. Biochem. Physiol., 18 : 405.

Robin, Y., Thoai, N.V. and Roche, J., 1957a, Sur la présence d'arcaïne chez la Sangsue, Hirudo medicinalis L., Compt. Rend. Soc. Biol., 151 : 2015.

Robin, Y., Thoai, N.V. and Pradel, L.A., 1957b, Métabolisme des dérivés guanidylés. VII. Sur une nouvelle guanidine monosubstituée biologique: l'hirudonine, Biochim. Biophys. Acta, 24 : 381.

Robin, Y., Audit, C. and Landon, M., 1967, Biogénèse des dérivés diguanidiques chez la sangsue, Hirudo medicinalis L. II. Mécanisme de la double transamidination, Compar. Biochem. Physiol., 22 : 287.

Roche, J., Thoai, N.V., Robin, Y., Garcia, I. and Hatt, J.L., 1952a, Sur la nature et la répartition des guanidines monosubstituées dans les tissus des Invertébrés. I. Présence de dérivés métaboliques de l'arginine chez des Mollusques, des Crus- tacés et des Echinodermes, Compt. Rend. Soc. Biol., 146 : 1899.

Roche, J., Thoai, N.V. and Glahn, P.E., 1952b, Sur la L-aminoacideoxydase de nombreux Invertébrés marins, Experientia, 8 : 428.

Roche, J., Thoai, N.V. and Robin, Y., 1957, Sur la présence de créatine chez les Inver- tébrés et sa signification biologique, Biochim. Biophys. Acta, 24 : 514.

Roche, J., Audit, C. and Robin, Y., 1965, Isolement et identification d'un nouveau dérivé diguanidique biologique, l'audouine (1,5-diamidinocadavérine) et de l'ar- caïned 1,4-diamidinoputrescine), chez une Annélide Polychète marine, Audouinia ten- taculata Montagu, Compt. Rend. Acad. Sci., Paris, 260 : 7023.

Rosenberg, H., 1959, Occurrence of guanidinoacetic acid and other substituted guani- dines in mammalian liver, Biochem. J., 72 : 582.

Rossiter, R.J., Gaffney, T.J., Rosenberg, H. and Ennor, A.H., 1960, The formation in vivo of lombricine in the earthworm (Megascolides cameroni), Biochem. J.,76:603.

Russel, F.E., 1967, Comparative pharmacology of some animal toxins, Fed. Proc., 26 : 1206.

Ryan, W.L. and Wells, I.C., 1964, Homocitrulline and homoarginine synthesis from lysine, Science, 144 : 1122.

Sakiyama, T., Suzuki, T., Owada, M. and Kitagawa, T., 1982, First case of argininosuccinic aciduria in Japan : clinical observations and treatment, in : "Urea Cycle Diseases", A. Lowenthal, A. Mori and B. Marescau, eds., Plenum Press, New York.

Schäfer, G., 1980, Guanidines and biguanides, Pharmac. Ther., 8 : 275.

Schantz, E.J., Mold, J.D., Stanger, D.W., Shavel, J., Riel, F.J., Bowden, J.P., Lynch, J.M., Wyler, R.S., Riegel, B. and Sommer, H., 1957, Paralytic shellfish poison. VI. Isolation and purification of the poison from toxic clam and mussel tissue, J. Am. Chem. Soc., 79 : 5230.

Schantz, E.J., Lynch, J.M., Vayvada, G., Matsumoto, K. and Rapoport, H., 1966, The purification and characterization of the poison produced by Gonyaulax catanella in axenic culture, Biochemistry, 5 : 1191.

Schatz, A., Bugie, E. and Waksman, S., 1944, Streptomycin, a substance exhibiting antibiotic activity against gram-positive and gram-negative bacteria, Proc. Soc. Exptl. Biol. Med., 55 : 66.

Scheuer, P.J., 1973, "Chemistry of Marine Natural Products", Acad. Press, New York.

Scheuer, P.J., 1975, Recent Developments in Chemistry of the Marine Toxins, Lloydia, 38 : 1.

Scheuer, P.J., 1977, Marine Toxins, Acc. Chem. Res., 10 : 33.

Scheumack, D., Howden, M.E.H., Spence, I. and Quinn, R.J., 1978, Maculotoxin : a neurotoxin from the venom glands of the Octopus Hapalochlaena maculosa identified as tetrodotoxin, Science, 199 : 188.

Schlessinger, D. and Medoff, G., 1975, Streptomycin, Dihydrostreptomycin and the Gentamycins, in : "Antibiotics", Vol. III , J.W. Corcoran and F.E. Hahn, eds., Springer Verlag, Berlin.

Schöpf, C., 1964, Die Anfänge der Pteridine-Chemie, in : "Pteridine Chemistry", W. Pfleiderer and E.C. Taylor, eds., The McMillan Co, New York.

Schotte, H. and Prieve, H., 1926, Synthese des N-Methyl-N-(ß-guanidino-äthyl)guanidins (Kutscher's Vitiatine), Z. physiol. Chem, 153 : 67.

Schulze, E., 1891, Über basische Stickstoffverbindungen aus der Samen von Vicia sativa und Pisum sativum, Z. physiol. Chem., 15 : 140.

Schulze, E. and Steiger, E., 1887, Über das Arginin, Z. physiol. Chem., 11 : 43.

Schwartz, J.H. and Maas, W.K., 1960, Analysis of the inhibition of growth produced by canavanine in Escherichia coli, J. Bacteriol., 79 : 794.

Scott-Emuakpor, A., Higgins, J.V. and Kohrman, A.F., 1972, Citrullinemia : a new case, with implications concerning adaptation to defective urea synthesis, Pediatric Research, 6 : 626.

Seki, Y., Nakamura, T. and Okami, Y., 1970, Accumulation of 2-aminoimidazole by Streptomyces eurocidicus, J. Biochem., 67 : 389.

Shaikin, R., Giatt, Y. and Berlyne, G., 1975, The presence and toxicity of guanidinopropionic acid in uremia, Kidney Internat., 7 : 302.

Sharma, G.M. and Burkholder, P.R., 1971, Structure of dibromophakellin, a new brominecontaining alkaloid from the marine sponge Phakellia flabellata, J. Chem. Soc. Chem. Commun., p. 151.

Sharma, G.M. and Magdoff-Fairchild, B., 1977, Natural products of marine sponges. 7. The constitution of weakly basic guanidine compounds, dibromophakellin and monobromophakellin, J. Org. Chem., 42 : 4118.

Sharma, G.M., Vig, B. and Burkholder, P.R., 1970, Antimicrobial substances of sponges. IV. Structure of a bromine-containing compound of a marine sponge, Proc. Conf. Food - Drugs from the Sea, p. 307. Marine Technology Society, Washington, D.C.

Shimizu, Y., 1978, Dinoflagellate Toxins, in : "Marine Natural Products", P.J. Scheuer, ed., Acad. Press, New York, Vol. 1.

Shimomura, O., Goto, T. and Hirata, Y., 1957, Crystalline Cypridina luciferin, Bull. Chem. Soc. Jpn., 30 : 929.

Shindo, S. and Mori, A., 1980, Biosynthesis of taurocyamine by mouse kidney transanidinase, IRCS Med. Sci., 8 : 91.

Smith, T.A. and Richards, F.J., 1962, The biosynthesis of putrescine in higher plants and its relation to potassium nutrition, Biochem. J., 84 : 292.

Späth, E. and Prokopp, S., 1924, Über das Galegin, Ber., 57 : 474.

Stafford, J.R. and Fann, W.E., 1977, Drug interactions with guanidinium hypertensives, Drugs, 13 : 57.

Stein, I.M., Cohen, B.D. and Kornhauser, R.S., 1969, Guanidinosuccinic acid in renal failure, experimental azotemia and inborn errors of the urea cycle, New Engl. J. Med., 280 : 926.

Stein, I.M. and Micklus, M.J., 1973, Concentrations in serum and urinary excretion of guanidine, 1-methylguanidine, and 1,1-dimethylguanidine in chronic renal failure, Clin. Chem., 19 : 583.

Stempien, M.F. (Jr.), Nigrelli, R.F. and Chib, J.S., 1972, Isolation and synthesis of physiologically active substances from sponges of the genus Agelas, 164 th ACS Meeting, Abstracts, MEDI 21.

Storey, K.B., 1976, Purification and properties of adductor muscle phosphofructokinase from the oyster, Crassostrea virginica. The aerobic/anaerobic transition: role of arginine phosphate in enzyme control, Eur. J. Biochem., 70 : 331.

Strecker, A., 1861, Etude sur la guanine, Compt. Rend. Acad. Sci., Paris, 52 : 1210.

Stumpf, P.K. and Green, D.E., 1944, L-amino acid oxidase of Proteus vulgaris, J. Biol. Chem., 153 : 387.

Subrahmanyan, P., Bhaskaran, K. and Satyanand, D., 1962, Phosphate and creatinine excretion in schizophrenics, Indian J. Psychiatr., 4 : 17.

Sullivan, M.X., 1911, The origin of creatinine in soils, J. Am. Chem. Soc., 33 : 2035.

Suwaki, S., 1978, Experimental model of hyperargininemia. II. Identification of guanidinosuccinic acid in urine of the arginine loaded rabbit and a possible pathway of its formation, Okayama Igakkai Zasshi , 90 : 1393.

Suzuki, T. and Muraoka, S., 1954, New guanidyl derivatives and amino acids in the extracts of Shellfish Cristaria plicata Leach, J. Pharm. Soc. Japan., 74 : 171.

Takasawa, S., Kawamoto, I., Okachi, R., Kohakura, M., Yahashi, R. and Nara, T., 1975a, A new antibiotic victomycin (XK 49-1-B-2). II. Isolation, purification and physicochemical and biological properties, J. Antibiotics, 28 : 366.

Takasawa, S., Kawamoto, I., Takahashi, I., Kohakura, M., Okachi, R., Sato, S., Yamamoto, M., Sato, T. and Nara, T., 1975b, Platomycins A and B. I. Taxonomy of the producting strain and production, isolation and biological properties of platomycins, J. Antibiotics, 28 : 656.

Takeuchi, A. and Takeuchi, N., 1975a, The structure -activity relationship for GABA and related compounds in the crayfish muscle, Neuropharmacology, 14 : 627.

Takeuchi, A. and Takeuchi, N., 1975b, Permeability changes of the crayfish muscle produced by beta-guanidinopropionic acid and related substances, Neuropharmacology, 14 : 635.

Takita, T., Muraoka, Y., Fujii, A., Itoh, H., Maeda, K. and Umezawa, H., 1972, The structure of the sulfur-containing chromophore of phleomycin and chemical transformation of phleomycin to bleomycin, J. Antibiotics, 25 : 197.

Takita, T., Muraoka, Y., Nakatani, T., Fujii, A., Umezawa, Y., Naganawa, H. and Umezawa, H., 1978, Chemistry of bleomycin. XIX. Revised structures of bleomycin and phleomycin, J. Antibiotics, 31 : 801.

Tanaka, N., 1975, Aminoglycoside Antibiotics. in : "Antibiotics", Vol. III, J.W. Corcoran and F.E. Hahn, eds., Springer Verlag, Berlin.

Tanino, H., Inoue, S., Aratani, M. and Kishi, Y., 1974, Synthetic studies on tetrodotoxin and related compounds. V. The protecting group of the C_9-hydroxy group, Tetrahydron Lett., p. 335.

Tanino, H., Nakata, T., Kaneko, T. and Kishi, Y., 1977, A stereospecific total synthesis of d,l-saxitoxin, J. Am. Chem. Soc., 99 : 2818.

Tanret, G., 1914, Sur un alcaloïde retiré de Galega officinalis, Bull. Soc. Chim., 15 : 613.

Tawara, Y., 1909, Study on a toxic compound of Tetrodontidae, Yakugaku Zasshi, 29 : 587.

Taylor, K.M., Baird-Lambert, J.A., Davis, P.A. and Spence, I., 1981, Methylaplysinopsin and other marine natural products affecting neurotransmission, Fed. Proc., Fed. Am. Soc. Exp. Biol., 40 : 15.

Tempé, J., 1982, Chemistry and Biochemistry of Open Chain Imino Acids, in : "Chemistry of Amino Acids, Peptides and Proteins", B. Weinstein,ed., M. Dekker, New York.Vol.7.

Tempé, J. and Goldmann, A., 1982, Occurrence and biosynthesis of opines in : "Molecular Biology of Plant Tumors", G. Kahl and J.S. Shell, eds., Academic Press.,New York.

Tepfer, D.A. and Tempé, J., 1981, Production d'agropine par des racines formées sous l'action d'Agrobacterium rhizogenes, souche A4, Compt. Rend. Acad. Sci., Paris, 292 :153.

Terheggen, H.G., Lavinha,F., Colombo, J.P., Van Sande, M. and Lowenthal, A., 1972, Familial hyperargininaemia, J. Génét. Hum., 20 : 69.

Terheggen, H.G., Lowenthal, A., Lavinha, F. and Colombo, J.P., 1975, Familial hyperargininaemia, Arch. Disease in Childhood, 50 : 57.

Thoai, N.V., 1965, Nitrogenous bases. in : "Comprehensive Biochemistry", M. Florkin and E.H. Stotz,eds., Elsevier Publishing Co, Amsterdam, London, New York. Vol. 6.

Thoai, N.V. and Desvages, G., 1963, Sur la nouvelle guanidine biologique végétale, la 4-hydroxy-galegine, Bull. Soc. Chim. Biol., 45 : 413.

Thoai, N.V. and Lacombe, G., 1958, Sur la présence de l'acide δ-guanido-n-valérianique dans les urines humaines, Biochim. Biophys. Acta, 29 : 437.

Thoai, N.V. and Olomucki, A., 1962, Arginine décarboxy-oxydase I. Caractères et nature de l'enzyme, Biochim. Biophys. Acta, 59 : 533.

Thoai, N.V. and Robin, Y., 1954a, Métabolisme des dérivés guanidylés.II. Isolement de la guanidotaurine (taurocyamine) et de l'acide guanidoacétique (glycocyamine) des vers marins, Biochim. Biophys. Acta., 13 : 533.

Thoai, N.V. and Robin, Y., 1954b, Métabolisme des dérivés guanidylés.IV. Sur une nouvelle guanidine monosubstituée biologique: l'ester guanidoéthylsérylphosphorique (lombricine) et le phosphagène correspondant, Biochim. Biophys. Acta, 14 : 76.

Thoai, N.V., and Robin, Y., 1959a, Métabolisme des dérivés guanidylés. VIII. Biosynthèse de l'octopine et répartition de l'enzyme l'opérant chez les Invertébrés, Biochim. Biophys. Acta, 35 : 446.

Thoai, N.V. and Robin, Y., 1959b, Sur la biogénèse de l'octopine dans différents tissus de Pecten maximus L., Bull. Soc. Chim. Biol., 41 : 735.

Thoai, N.V. and Robin, Y., 1961, Métabolisme des dérivés guanidylés. IX. Biosynthèse de l'octopine: étude du mécanisme de la réaction et de quelques propriétés de l'octopine synthetase, Biochim. Biophys. Acta, 52 : 221.

Thoai, N.V. and Robin, Y., 1969, Guanidine compounds and phosphagens, in : "Chemical Zoology" IV. Annelida, Echiura and Sipuncula., M. Florkin and B.T. Scheer eds., Acad. Press., New York.

Thoai, N.V. and Roche, J., 1960, Dérivés guanidiques biologiques, <u>Fortschr. Chem. org.</u> <u>Naturstoffe</u>, 18 : 83.

Thoai, N.V. and Roche, J., 1964, Diversity of phosphagens., in : "Taxonomic Biochemistry and Serology", Ch. A. Leone, ed., The Ronald Press, New York.

Thoai, N.V., Roche, J. and Robin, Y., 1953a, Métabolisme des dérivés guanidylés. I. Dégradation de l'arginine chez les Invertébrés marins, <u>Biochim. Biophys. Acta</u>, 11 : 403.

Thoai, N.V., Roche, J., Robin, Y. and Thiem, N.V., 1953b, Sur la présence de la glycocyamine (acide guanidylacétique), de la taurocyamine (guanidyltaurine) et des phosphagènes correspondants dans les muscles de vers marins, <u>Biochim. Biophys.</u> <u>Acta</u>, 11 : 593.

Thoai, N.V., Roche, J. and Olomucki, A., 1954, Sur la présence de la taurocyamine (guanidotaurine) dans l'urine de rat et sa signification biochimique dans l'excrétion azotée, <u>Biochim. Biophys. Acta</u>, 14 : 448.

Thoai, N.V., Hatt, J.L. and An, T.T., 1956a, Métabolisme des dérivés guanidylés. V. Oxydation enzymatique de l'arginine en guanidobutyramide, <u>Biochim. Biophys. Acta</u>, 22 : 116.

Thoai, N.V., Hatt, J.L., An, T.T. and Roche, J., 1956b, Metabolism of guanidyl derivatives. VI. Degradation of derivatives of guanidine in Streptomyces griseus, <u>Biochim. Biophys. Acta</u>, 22 : 337.

Thoai, N.V., Olomucki, A., Robin, Y., Pradel, L.A. and Roche, J., 1956c, Sur la présence de nombreux dérivés carbamylés et guanidiques dans les urines et sur leur signification biologique, <u>Compt. Rend. Soc. Biol.</u>, 150 : 2160.

Thoai, N.V., Robin, Y. and Pradel, L.A., 1957, Métabolisme oxydatif de la L-arginine chez la Limnée, Limnaea stagnalis L. II. Oxydation en guanidobutyramide, <u>Compt. Rend. Soc. Biol.</u>, 151 : 2097.

Thoai, N.V., Di Jeso, F. and Robin, Y., 1963a, Sur l'isolement et la synthèse d'une nouvelle guanidine monosubstituée biologique, l'acide guanidoéthyl-méthyl-phosphorique et sur le phosphagène correspondant, l'acide N'-phosphoryl-guanidoéthyl-méthyl-phosphorique, <u>Compt. Rend. Acad. Sci., Paris</u>, 256 : 4525.

Thoai, N.V., Zappacosta, S. and Robin, Y., 1963b, Biogénèse de deux guanidines soufrées: la taurocyamine et l'hypotaurocyamine, <u>Comp. Biochem. Physiol.</u>, 10 : 209.

Thoai, N.V., Di Jeso, F., Robin, Y. and Der Terrossian, E., 1966, Sur la nouvelle acide adénosine 5'-triphosphorique : guanidine phosphotransférase, l'ophéline kinase, <u>Biochim. Biophys. Acta</u>, 113 : 542.

Thoai, N.V., Regnouf, F. and Olomucki, A., 1967, Isolement d'un peptide phosphorylé et guanidique, l'aspartyllombricine, des muscles de Bonellia viridis, <u>Bull. Soc. Chim. Biol.</u>, 49 : 805.

Thoai, N.V., Robin, Y. and Guillou, Y., 1972, A new phosphagen, N'-phosphorylguanidino-ethylphospho- O - (α-N,N-dimethyl)serine (phosphothalassemine), <u>Biochemistry</u>, 11 : 3890.

Touitou, Y. and Perlemuter, L., 1976, " Dictionnaire Pratique de Pharmacologie Clinique", Masson, Paris.

Tsuda, K. and Kawamura, M., 1952, The constituents of the ovaries of globfish. VII. Purification of tetrodotoxin by chromatography, <u>J. Pharm. Soc. Japan,</u> 72 : 711.

Umezawa, H., 1971, Natural and artificial bleomycins. Chemistry and antitumor activity, Pure Appl. Chem., 28 : 665.

Umezawa, H., 1975, Bleomycin, in "Antibiotics", Vol. III, J.W. Corcoran and F.E. Hahn, eds., Springer Verlag, Berlin.

Umezawa, H., 1980, Recent progress in bleomycin studies, <u>Med. Chem. (Academic)</u>,16 : 147.

Umezawa, H., Suhara, Y., Takita, T. and Maeda, K., 1966, Purification of bleomycins, <u>J. Antibiotics</u>, 19A : 210.

Van Pilsum, J.F., Martin, R.P., Kito, E. and Hess, J., 1956, Determination of creatine, creatinine, arginine, guanidinoacetic acid, guanidine and methylguanidine in biological fluids, <u>J. Biol. Chem.</u>, 222 : 225.

Van Pilsum, J.F., Stephens, G.C. and Taylor, D., 1972, Distribution of creatine, guanidinoacetate and the enzymes for their biosynthesis in the animal Kingdom. Implications for phylogeny, <u>Biochem. J.</u>, 126 : 325.

Van Tamelen, E.E., Dyer, J.R., Whaley, H.A., Carter, H.E. and Whitfield, G.B. Jr., 1961, Constitution of the streptolin-streptothricin group of Streptomyces antibiotics, <u>J. Am. Chem. Soc.</u>, 83 : 4295.

Volcani, B.E. and Snell, E.E., 1948, The effects of canavanine, arginine and related compounds on the growth of bacteria, <u>J. Biol. Chem.</u>, 174 : 893.

Waksman, S.A. and Woodruff, J.B., 1942, Streptothricin, a new selective bacteriostatic and bactericidal agent particularly against gram-negative bacteria, <u>Proc. Soc. Exptl. Biol. Med.</u>, 49 : 207.

Walker, J.B., 1952, Argininosuccinic acid from Chlorella, <u>Proc. Natl. Acad. Sci. U.S.</u>, 38 : 561.

Walker, J.B., 1953, An enzymatic reaction between canavanine and fumarate, <u>J. Biol. Chem.</u>, 204 : 139.

Walker, J.B., 1955, Canavanine and homoarginine as antimetabolites of arginine and lysine in yeast and algae, <u>J. Biol. Chem.</u>, 212 : 207.

Walker, J.B. and Myers, J., 1953, The formation of argininosuccinic acid from arginine and fumarate, <u>J. Biol. Chem.</u>, 203 : 143.

Watanabe, C.K., 1918, Studies on the metabolic changes induced by administration of guanidine bases. I. The influence of injected guanidine hydrochloride upon blood sugar content, <u>J. Biol. Chem.</u>, 33 : 253.

Watanabe, Y., Ohara, S., Shindo, S. and Mori, A., 1983, Effects of α-keto-δ-guanidinovaleric acid on cAMP and cGMP contents in mouse brain, Neurosciences (Kobe), 9 : 42.

Watts, R.L. and Watts, D.C., 1968, Gene duplication and the evolution of enzymes, Nature, 217 : 1125.

Weber, C.J., 1935, The presence of glycocyamine in urine, J. Biol. Chem., 109, xcvl Proc.

Wesselow, de, O.L.V.S. and Griffiths W.J., 1932, Blood guanidine in hypertension, Brit. J. Exper. Path., 13 : 345.

Wiechert, P., Mortelmans, J., Lavinha, F., Clara, R., Terheggen, H.G. and Lowenthal, A., 1976, Excretion of guanidino derivatives in urine of hyperargininemic patients, J. Génét. hum., 24 : 61.

Wieland, H. and Alles, R., 1922, Uber der Giftstoff der Kröte, Ber., 55 : 1789.

Wieland, H., Hesse, G. and Hüttel, R., 1936, Toad poisons. IX. Further consideration of constitutional problems, Ann., 524 : 203.

Zammit, V.A. and Newsholme, E.A., 1976, The maximum activities of enzymes of carbohydrate utilization in muscles from marine invertebrates, Biochem. J., 160 : 447.

PARTICIPANTS

ABE, H.
The 1st Dept. of Med.
Osaka Univ. Hospital
Fukushima-ku, Osaka 553
Japan

ABE, S.
Dept. of Anesthesiology
Okayama Univ. Med. Sch.
Okayama 700
Japan

ANDO, A.
The 1st Dept. of Med.
Osaka Univ. Hospital
Fukushima-ku, Osaka 553
Japan

AOYAGI, K.
Inst. of Clinical Med.
Univ. of Tsukuba
Sakura-mura, Ibaraki-ken 305
Japan

ARAKAWA, M.
The 2nd Dept. of Med.
Niigata Univ. Sch. of Med.
Niigata 951
Japan

ASANO, H.
Sendai Shakai-Hoken Hospital
Tsutsumi-machi, Sendai 980
Japan

BABA, S.
Dept. of Med. Tech.
College of Biochemical Tech.
Niigata Univ., Niigata 951
Japan

BONHAUS, D.
Dept. of Pharmacology
Univ. of Arizona
Tucson, AZ 85724
U.S.A.

CHANG, J. S.
Inst. for Neurobiology
Okayama Univ. Med. Sch.
Okayama 700
Japan

CHIKU, K.
Dept. of Nutritional Chemistry
Tokushima Univ. Med. Sch.
Tokushima 770
Japan

CHITRA, K.C.
Dept. of Zoology
Sri Venkateswara Univ.
Tirupati-517 502
India

COHEN, B.D.
Bronx-Lebanon Hospital
Bronx, N.Y.
U.S.A.

439

AZUSHIMA, C.
Div. of Nephrology
Juntendo Univ. Sch. of Med.
Bunkyo-ku, Tokyo 113
Japan

EDAMATSU, R.
Inst. for Neurobiology
Okayama Univ. Med. Sch.
Okayama 700
Japan

ENDO, Y
Dept. of Nutritional Chemistry
Tokushima Univ. Med. Sch.
Tokushima 770
Japan

FUJII, M.
The 1st Dept. of Med.
Osaka Univ. Hospital
Fukushima-ku, Osaka 553
Japan

FUJIWARA, S.
The 2nd Dept. of Internal Med.
Jikei Univ. Sch. of Med.
Minato-ku, Tokyo 108
Japan

FUKUDA, T.
Dept. of Pharmacology
Faculty of Med., Kagoshima Univ.
Kagoshima 890
Japan

FUKUI, M.
Inst. for Neurobiology
Okayama Univ. Med. Sch.
Okayama 700
Japan

GEJO, F.
The 2nd Dept. of Med.
Niigata Univ. Sch. of Med.
Niigata 951
Japan

HAYAKAWA, C.
Dept. of Pediatrics
Nagoya Univ. Sch. of Med.
Showa-ku, Nagoya 466
Japan

DE DEYN, P.
Laboratory of Neurochemistry
Born-Bunge Foundation
U.I.A., 2610 Wilrijk
Belgium

HIGASHIDATE, S.
JASCO
Japan Spectroscopic Co., Ltd.
Ishikawa-Cho, Hachioji 192
Japan

HIRAMATSU, M.
Inst. for Neurobiology
Okayama Univ. Med. Sch.
Okayama 700
Japan

HOSHINO, T.
Pharmaceutical Inst.
Sch. of Med., Keio Univ.
Shinjuku-ku, Tokyo 160
Japan

HUANG, Y.-L.
Faculty of Pharmaceutical Sci.
Kyushu Univ.
Higashi-ku, Fukuoka 812
Japan

HUXTABLE, R. J.
Dept. of Pharmacology
Univ. of Arizona
Tucson, AZ 85724
U.S.A.

IIDA, N.
Kideney Disease Center
Osaka Prefectural Hospital
Sumiyoshi-ku, Osaka 558
Japan

IIDA, S.
Inst. of Clinical Med.
Univ. of Tsukuba
Sakura-mura, Ibaraki-Ken 305
Japan

IMAI, E.
The 1st Dept. of Med.
Osaka Univ. Hospital
Fukushima-ku, Osaka 553
Japan

INDIRA, K.
Dept. of Zoology
Sir Venkateswara Univ.
Tirupati-517 502
India

INOUCHI, M.
The 1st Dept. of Internal Med.
St. Marianna Univ. Sch. of Med.
Miyamae-ku, Kawasaki 213
Japan

ISHIDA, M.
The 1st Dept. of Internal Med.
St. Marianna Univ. Sch. of Med.
Miyamae-ku, Kawasaki 213
Japan

ISHIZAKI, M.
Sendai Shakai-Hoken Hospital
Tsutsumi-machi, Sendai 980
Japan

ISODA, K.
The 2nd Dept. of Internal Med.
Jikei Univ. Sch. of Med.
Minato-ku, Tokyo 105
Japan

ITABASHI, H.
Division of Nephrology
Juntendo Univ. Sch. of Med.
Bunkyou-ku, Tokyo 113
Japan

ITANO, Y.
Dept. of Anesthesiology
Okayama Univ. Med. Sci.
Okayama 700
Japan

IZUMI, K.
Dept. of Pharmacology
Faculty of Med., Kagoshima Univ.
Kagoshima 890
Japan

JAGANNATHA RAO, K. S.
Dept. of Zoology
Sri Venkateswara Univ.
Tirupati-517 502
India

KABUTO, H.
Inst. for Neurobiology
Okayama Univ. Med. Sch.
Okayama 700
Japan

KADOYA, T.
Inst. for Neurobiology
Okayama Univ. Med. Sch.
Okayama 700
Japan

KAI, M.
Faculty of Pharmaceutical Sci.
Kyushu Univ.
Higahsi-ku, Fukuoka 812
Japan

KATO, T.
Dept. of Pediatrics
Nagoya Univ. Sch. of Med.
Shouwa-ku, Nagoya 466
Japan

KAWASHIMA, T.
Kidney Disease Center
Osaka Prefectural Hospital
Sumiyoshi-ku, Osaka 558
Japan

KIKUCHI, T.
Morishita Pharm. Co., Ltd.
Dosho-machi, Higashi-ku
Osaka 541
Japan

KIMURA, Y.
The 1st Dept. of Internal Med.
St. Marianna Univ. Sch. of Med.
Miyamae-ku, Kawasaki 213
Japan

KISHITA, C.
Dept. of Pharmacology
Faculty of Med., Kagoshima Univ.
Kagoshima 890
Japan

KISHORE, B. K.
The 2nd Dept. of Med.
Niigata Univ. Sch. of Med.
Niigata 951
Japan

KITAMURA, H.
Sendai Shakai-Hoken Hospital
Tsutsumi-machi, Sendai 980
Japan

KOIDE, H.
Div. of Nephrology
Juntendo Univ. Sch. of Med.
Bunkyou-ku, Tokyo 113
Japan

KOJA, T.
Dept. of Pharmacology
Faculty of Med., Kagoshima Univ.
Kagoshima 890
Japan

KOKUBA, Y.
The 1st Dept. of Med.
Osaka Univ. Hospital
Fukushima-ku, Osaka 553
Japan

KOSAKA, F.
Dept. of Anesthesiology
Okayama Univ. Med. Sci.
Okayama 700
Japan

KŌSOGABE, Y.
Div. of I.C.U.
Okayama Univ. Hospital
Okayama 700
Japan

KURODA, M.
The 2nd Dept. of Internal Med.
Sch. of Med., Kanazawa Univ.
Kanazawa 920
Japan

LAIRD, H. E.
Dept. of Pharmacology
Univ. of Arizona
Tucson, Arizona 85724
U.S.A.

LETARTE, J.
Centre de Recherche pediatrique
Hopital Ste. Justine
Montreal, Que. H3T 1C5
Canada

LOWENTHAL, A.
Laboratory of Neurochemistry
Born-Bunge Foundation
U.I.A., 2610 Wilrijk
Belgium

MAEHARA, M.
Dept. of Pediatrics
Nagoya Univ. Sch. of Med.
Shouwa-ku, Nagoya 466
Japan

MAEKUBO, T.
JASCO
Japan Spectroscopic Co., Ltd.
Ishikawa-Cho, Hachioji 192
Japan

MARESCAU, B.
Laboratory of Neurochemistry
Born-Bunge Foundation
U.I.A., 2610 Wilrijk
Belgium

MATSUDA, R.
Div. of I.C.U.
Okayama Univ. Hospital
Okayama 700
Japan

MIKAMI, H.
The 1st Dept. of Med.
Osaka Univ. Hospital
Fukushima-ku, Osaka 553
Japan

MIURA, K.
Sendai Shakai-Hoken Hospital
Tsutsumi-machi, Sendai 980
Japan

MIYAHARA, T.
The 2nd Dept. of Internal Med.
Jikei Univ. Sch. of Med.
Minato-ku, Tokyo 105
Japan

MIYAZAKI, M.
Inst. of Clinical Med.
Univ. of Tsukuba
Sakura-mura, Ibaraki-Ken 305
Japan

MIZUTANI, N.
Dept. of Pediatrics
Nagoya Univ. Sch. of Med.
Shouwa-ku, Nagoya 466
Japan

MOHANACHARI, V.
Dept. of Zoology
Sri Venkateswara Univ.
Tirupati-517 502
India

MORI, A.
Inst. for Neurobiology
Okayama Univ. Med. Sch.
Okayama 700
Japan

MURAMOTO, H.
The 2nd Dept. of Internal Med.
Sch. of Med., Kanazawa Univ.
Kanazawa 920
Japan

NAGASE, S.
Inst. of Clinical Med.
Univ. of Tsukuba
Sakura-mura, Ibaraki-Ken 305
Japan

NAKAGAWA, K.
Kyoto Women's College
Higashiyama-ku, Kyoto 605
Japan

NAKAJIMA, M.
Research & Development Div.
Kikkoman Corporation
Noda 278
Japan

NAKAMURA, K.
Research & Development Div.
Kikkoman Corporation
Noda 278
Japan

NAKANISHI, I.
Kidney Disease Center
Osaka Prefectural Hospital
Sumiyoshi-ku, Osaka 558
Japan

NAKANO, K.
Dept. of Biochemistry
Sch. of Med., Kagoshima Univ.
Kagoshima 890
Japan

NAKANO, T.
The 2nd Dept. of Internal Med.
Jikei Univ. Sch. of Med.
Minato-ku, Tokyo 105
Japan

NAKAO, T.
Saiseikai Central Hospital
Minato-ku, Tokyo 108
Japan

NARITA, M.
Inst. of Clinical Med.
Univ. of Tsukuba
Sakura-mura, Ibaraki-ken 305
Japan

NATORI, Y.
Dept. of Nutritional Chemistry
Tokushima Univ. Med. Sch.
Tokushima 770
Japan

NEERAJA, P.
Dept. of Zoology
Sri Venkateswara Univ.
Tirupati-517 502
India

NISHITANI, K.
Dept. of Anesthesiology
Okayama Univ. Med. Sci.
Okayama 700
Japan

OCHIAI, Y.
Div. of I.C.U.
Okayama Univ. Hospital
Okayama 700
Japan

OHBA, S.
Inst. of Clinical Med.
Univ. of Tsukuba
Sakura-mura, Ibaraki-Ken 305
Japan

OHKURA, Y.
Faculty of Pharmaceutical Sci.
Kyushu Univ.
Higashi-ku, Fukuoka 812
Japan

OKADA, A.
The 1st Dept. of Med.
Osaka Univ. Hospital
Fukushima-ku, Osaka 553
Japan

OKAMOTO, H.
Morishita Pharm. Co., Ltd.
Dosho-machi, Higashi-ku
Osaka 541
Japan

OKAZAKI, H.
Sendai Shakai-Hoken Hospital
Tsutsumi-machi, Sendai 980
Japan

ORITA, Y.
The 1st Dept. of Med.
Osaka Univ. Hospital
Fukushima-ku, Osaka 553
Japan

OWADA, S.
The 1st Dept. of Internal Med.
St. Marianna Univ. Sch. of Med.
Miyamae-ku, Kawasaki 213
Japan

OZAWA, S.
The 1st Dept. of Internal Med.
St. Marianna Univ. Sch. of Med.
Miyamae-ku, Kawasaki 213
Japan

PASANTES-MORALES, H.
Centro de Investigaciones
Fishiologia Celular
Univ. of Mexico, Mexico City
Mexico

PATEL, H.
Bronx-Lebanon Hospital
Bronx, N.Y.
U.S.A.

PLUM, C. M.
Aldersrogade 43F
DK 2200
Kobenhavn N
Denmark

QUERSHI, I. A.
Centre de Recherche pediatrique
Hopital Ste. Justine
Montreal, Que. H3T 1C5
Canada

RINNO, H.
Div. of Nephrology
Juntendo Univ. Sch. of Med.
Bunkyou-ku, Tokyo 113
Japan

ROBIN, Y.
E.P.H.E., Dpt. de Physiologie
Faculte de Pharmacie
92290, Chatenay-Malabry
France

SAHEKI, T.
Dept. of Biochemistry
Sch. of Med., Kagoshima Univ.
Kagoshima 890
Japan

SAITO, M.
JASCO
Japan Spectroscopic Co.,Ltd.
Ishikawa-Cho, Hachioji 192
Japan

SASE, M.
Dept. of Biochemistry
Sch. of Med., Kagoshima Univ.
Kagoshima 890
Japan

SATYAVELU REDDY, K.
Dept. of Zoology
Sir Venkateswara Univ.
Tirupati-517 502
India

SEKI, S.
Dept. of Physiology
Kanagawa Dental Coll.
Yokosuka 238
Japan

SENDA, M.
JASCO
Japan Spectroscopic Co.,Ltd.
Ishikawa-Cho, Hachioji 192
Japan

SHIMIZU, T.
Dept. of Pharmacology
Sch. of Med., Kagoshima Univ.
Kagoshima 890
Japan

SHINDO, S.
Inst. for Neurobiology
Okayama Univ. Med. Sch.
Okayama 700
Japan

SHIROKANE, Y.
Research & Development Div.
Kikkoman Corporation
Noda-Shi 278
Japan

SUZUKI, S.
Dept. of Pediatrics
Nagoya Univ. Sch. of Med.
Shouwa-ku, Nagoya 466
Japan

SUZUKI, Y.
The 2nd Dept. of Med.
Niigata Univ. Med. Sch.
Niigata 951
Japan

SWAMI, K. S.
Dept. of Zoology
Sri Venkateswara Univ.
Tirupati-517 502
India

TAKAHASHI, H.
Sendai Shakai-Hoken Hospital
Tsutsumi-machi, Sendai 980
Japan

TAKEDA, R.
The 2nd Dept. of Internal Med.
Sch. of Med., Kanazawa Univ.
Kanazawa 920
Japan

TOFUKU, Y.
The 2nd Dept. of Internal Med.
Sch. of Med., Kanazawa Univ.
Kanazawa 920
Japan

TOJO, S.
Inst. of Clinical Med.
Univ. of Tsukuba
Sakura-mura, Ibaraki-Ken 305
Japan

TOMA, J.
Inst. for Neurobiology
Okayama Univ. Med. Sch.
Okayama 700
Japan

TOMOBUCHI, M.
Kidney Disease Center
Osaka Prefectural Hospital
Sumiyoshi-ku, Osaka 558
Japan

TSUBAKIHARA, Y.
Kidney Disease Center
Osaka Prefectural Hospital
Sumiyoshi-ku, Osaka 558
Japan

WATANABE, K.
Dept. of Pediatrics
Nagoya Univ. Sch. of Med.
Shouwa-ku, Nagoya 466
Japan

WATANABE, Y.
Inst. for Neurobiology
Okayama Univ. Med. Sch.
Okayama 700
Japan

WATANABE, Y.
Dept. of Med. Tech.
College of Biochemical Tech.
Niigata Univ., Niigata 951
Japan

YAGI, Y.
Dept. of Biochemistry
Sch. of Med., Kagoshima Univ.
Kagoshima 890
Japan

YAMADA, T.
Dept. of Anesthesiology
Okayama Univ. Med. Sci.
Okayama 700
Japan

YOKOGAWA, T.
Kidney Disease Center
Osaka Prefectural Hospital
Sumiyoshi-ku, Osaka 558
Japan

YOKOI, I.
Inst. for Neurobiology
Okayama Univ. Med. Sch.
Okayama 700
Japan

YOSHINO, M.
Dept. of Pediatrics
Kurume Univ. Sch. of Med.
Kurume 830
Japan

YUASA, S.
Kidney Disease Center
Osaka Prefectural Hospital
Sumiyoshi-ku, Osaka 558
Japan

Abbreviations used: ARA; arginic acid, ARF; acute renal failure, Arg; arginine, BUN; blood urea nitrogen, Ccr; creatinine clearance, CRF; chronic renal failure, CRN; creatinine, CSF; cerebrospinal fluid, CTN; creatine, GAA; guanidinoacetic acid (glycocyamine), GBA; γ-guanidinobutyric acid, GES; 2-guanidinoethane sulfonate (taurocyamine), GPA; β-guanidinopropionic acid, GSA; guanidinosuccinic scid, Gua; guanidine, HArg; homoarginine, MGua; methylguanidine, NAA; α-N-acetylarginine,